America's Water Crises

David E. McNabb · Carl R. Swenson

America's Water Crises

The Impact of Drought and Climate Change

palgrave
macmillan

David E. McNabb
Shelton, WA, USA

Carl R. Swenson
Oak Park, IL, USA

ISBN 978-3-031-27379-7 ISBN 978-3-031-27380-3 (eBook)
https://doi.org/10.1007/978-3-031-27380-3

Cover credit: Trigga/gettyimages

This Palgrave Macmillan imprint is published by the registered company Springer Nature Switzerland AG
The registered company address is: Gewerbestrasse 11, 6330 Cham, Switzerland

Books by David E. McNabb and Carl R. Swenson

Collaboration in Government

Disaster Management Policies and Practices

Water Books by David McNabb

Global Pathways to Water Sustainability

Water Resource Management

Acknowledgments and Authors Focus

In preparing this book, we have depended upon the previous work of many federal, state, and academic researchers and water system administers. These knowledgeable professionals have published their investigations into the state of water and the water crisis now accelerating in the United States. This book could not have been written without their analyses and findings. In addition, we owe special thanks to at Palgrave Macmillan Publishers and our supporter from day one, Rachael Ballard. Throughout our preparation of this, the third in our study of the water resources, we have sought to make available to a wider audience some of the current research findings and data on the perilous state of the nation's surface and groundwater resources. Clearly, this is a time of climate change. The extreme drought that is occurring in many sections of the nation is having a big impact on the Nation's water resources. We have depended upon the research conducted by and sponsored by federal, state, and local government agencies for their research on these changes and how they are affecting freshwater in the coterminous United States.

Our descriptions of the water resource systems are based upon on research carried out and the findings published by the water scientists in the United States Geological Survey, the Environmental Protection Agency, the U.S. Corps of Engineers, and water-related agencies of the Departments of Agriculture and of the Interior and state and local water management agencies. With others, these organizations publish research findings about the many different natural and human caused hazards that threaten the Nation's critical water resource, as well as the energy, mineral, and other natural resources that are impacted by climate and land-use change.

For this and for all our previous publications, we add our gratitude to the School of Business and the library personnel at my academic home, Pacific Lutheran University, for making available my access to the global contributions to the research on the crises status now underway in the Earth's freshwater supply.

The Authors' Focus

The primary focus of the authors of this book is the way it blends their academic and practical public management experience together with examples of current climate change pressures now affecting the state of a critical public water management focus—America's public water resource. The authors have gained decades of experience in public administration from both the practical and academic points of view. In addition to his academic career, retired management professor Dr. David McNabb has served as the manager of an economic development program for a California city government and as an elected public utility commissioner. He has been the sole author and the joint author of more than 30 scholarly texts on such topics as research methods and management and marketing topics. He taught in American universities and European business schools and universities and in graduate degree management training programs for the U.S. Department of Defense.

Carl Swenson, BA and MPA, brings examples and practical illustrations from his 35 years in city government that illuminate this academic foundation. He has served as the Public Administration manager for communities in Washington, Arizona, and Illinois. His experience in these communities in three different environments granted him exceptional regional administration management perspectives. In this book, authors McNabb and Swenson have combined administrative and program management perspectives together to create a descriptive narrative of the current problems facing public organization managers and the general public need to know about the state that highlights how the concepts of governmental and public sector collaboration work in practice. In addition to anyone concerned over the effects of changing climate and its effects on the Nation's already over-stressed public water resource, this book will be of value to academics, students, and practitioners looking for a text that bridges the academic and practical realms of public sector resource governance.

Contents

Part I The Water World

1	Water: America's Most Valuable Natural Resource	3
2	From Water Stress to a Water Crisis	29
3	Water Problems and Quality Controls	55

Part II Shape of the Water Resource

4	Drought's Role in a Water Crisis	75
5	Surface Water: The Fading Everyday Resource	91
6	Groundwater: The Disappearing Resource	109
7	Stored Water: Failing Dams and Other Storage Crises	129
8	Repairing Infrastructure and Reusing Water	157

Part III Regional Problems and Solutions

9	Water Crises in the West	179
10	Water Crisis in the Southwest	203
11	Water Crises in the High Plains	225
12	Water Use in the Great Lakes Region	241

13 Water Crises in the South and Southcentral Regions 257
14 Water Crises in the Appalachian Mountain Region 279
15 Water Stress in Mid-Atlantic Region 297
16 Water Stress in New England 315

Bibliography 331
Index 361

Acronyms

AGS	Arkansas Geological Survey
AGU	American Geophysical Union
ASCI	American Society of Civil Engineers
ASDSO	Association of State Dam Safety Officials
CAST	Council for Agricultural Science and Technology
CDC	Center for Disease Control
CDWR	California Department of Water Resources
CGS	Colorado Geological Survey
CIA	Central Intelligence Agency
CRT	Climate Resilience Toolkit
EGLE	Michigan Department of Environment, Great Lakes and Energy
EIA	U.S. Energy Information Administration
EPA	Environmental Protection Administration
ESRI	Environmental Systems Research Institute
FEMA	Federal Emergency Management Administration
GAO	U.S. Government Accountability Office
IDNR	Indiana Department of Natural Resources
KDOA	Kansas Department of Agriculture
KGS	Kansas Geological Survey
MARISA	Mid-Atlantic Regional Integrated Sciences and Assessments
MCA	Maine Conservation Alliance
NASA	National Aeronautics and Space Administration
NCA	National Climate Assessment
NCCOS	National Center for Coastal Ocean Science
NCEI	National Centers for Environmental Information
NDEA	Nebraska Department of Environment and Energy

NDMC	National Drought Mitigation Center
NGWA	National Ground Water Association
NIDIS	National Integrated Drought Information System
NOAA	National Oceanic and Atmospheric Administration
NPS	National Park Service
NRCS	Natural Resources Conservation Service
NWS	National Weather Service
SETP	Society of Experimental Test Pilots
SPE	Society of Petroleum Engineers
TWDB	Texas Water Development Board
UCS	Union of Concerned Scientists
UDWR	Utah Division of Water Resources
UGS	Utah Geological Survey
UNESCO	United Nations Educational, Scientific and Cultural Organization
UNW	United Nations-Water
USACE	U.S. Army Corps of Engineers
USBR	U.S. Bureau of Reclamation
USCRT	U.S. Climate Reliance Toolkit
USDA	United States Department of Agriculture
USDA/NASS	U.S. Agriculture Dept/National Agricultural Statistics Service
USDM	United States Drought Monitor
USFWS	United States Fish and Wildlife Service
USGCRP	U.S. Global Change Research Program
USGS	United States Geological Survey
WAECY	Washington State Department of Ecology
WNA	World Nuclear Association
WRDS	Water Resources Data System
WSGS	Wyoming State Geological Survey
WWC	World Water Council

List of Figures

Fig. 7.1	Map of the Snake and Columbia rivers with dam sites and elevation identified (*Source* U.S. Wheat Association, June 14, 2022)	145
Fig. 9.1	The northern five states of the West Region (*Source* www.freeworldmaps.net)	181
Fig. 9.2	The Northwestern Volcanic Aquifer study area with sub-regions (Key: Dots: research well site. Sub-regions: MC Middle Cascades; SC Southern Cascades; SN Sierra Nevada; WP Walla Walla Plateau; HB Harney Basin; BR Basin and Range; BM Blue Mountain; SR Snake River) (*Source* USGS; Curtis et al. [2020])	184
Fig. 9.3	California Central Valley Aquifer system with major surface water basins (*Sources* USGS 2021: Public Domain)	194
Fig. 10.1	Map of the Snake River in Idaho and other states (*Source* Wikimedia Commons; Map of the Snake River watershed, USA. Intended to replace older File:SnakeRiverNicerMap.jpg. Created using public domain USGS National Map data. *Source* = Own work Author = [User:Shannon1|Sh])	215
Fig. 10.2	The two basins of the Colorado River (*Source* USGS Public Domain)	217
Fig. 10.3	USGS map of the Colorado River Southwest to Mexico (USGS: Colorado Water Science Center 2012)	219
Fig. 11.1	Extent of the High Plains aquifer (*Source* USGS Nebraska Water Science Center [2014])	226

Fig. 12.1	The location of the Great Lakes states (*Source* ESRI Base Map from ESR)	242
Fig. 12.2	Extent of the Missouri River Basin (*Source* NIDIS [2022])	245
Fig. 13.1	Range of the Mississippi Embayment aquifer system (*Source* USGS)	258
Fig. 14.1	Extent of the Appalachian Plateau aquifer system	280
Fig. 14.2	Range of the North Atlantic Coastal Plain aquifer system (*Source* Public Domain map, Area of the Northern Atlantic Coastal Plain Aquifer)	288
Fig. 15.1	Sections of the Mid-Atlantic and the Northeast regions (*Source* No source. Available at https://mapsof.net/uploads/static-maps/USA_Mid_Atlantic_map.png)	298
Fig. 15.2	USGS Public domain map of the Delaware River Basin (*Source* USGS 2010)	302
Fig. 15.3	NRCS map of the full scope of the Chesapeake Bay Watershed	310
Fig. 16.1	NCDC map of New England, showing the 15 NOAA weather units (*Source* NCEI; Fisk [2019])	316

List of Tables

Table 1.1	Adjusted regions of the U.S. with generally similar climates	4
Table 1.2	Ten of the most populous nations in 2021 and 2050 (pre-COVID-19 estimates)	14
Table 1.3	Water withdrawals in 2015 by use category in billion gallons per day	15
Table 1.4	Water use data in the U.S, 2000–2015	17
Table 2.1	Standard climate change temperature reference points	40
Table 2.2	Long-term storm and precipitation event changes	41
Table 3.1	U.S. participation in global change research programs	64
Table 4.1	USDM drought index and PDSI range of drought severity scores	78
Table 4.2	Percent of area in drought for CONUS in 2020	78
Table 5.1	Example stormwater management practices	99
Table 5.2	Agricultural water needs in gallons and acre-feet	104
Table 6.1	List of important U.S. groundwater aquifers and basins	110
Table 6.2	States with estimates of the most irrigated agriculture fields	114
Table 7.1	The 8 continuous mountain states 2018–2019 winter snowfall (inches)	138
Table 7.2	Winter snowfall (inches) of the 3 west coast states, 2018–2019	138
Table 7.3	Sample of recent failures of U.S. dams	146
Table 9.1	Shares of state area included in each drought categories July 2022	181

Table 9.2	Water use in Montana	182
Table 10.1	Drought in the Southwest states first week of September 2022	204
Table 10.2	New Mexico water withdrawals in 2015 (billion gallons/day) and source	212
Table 10.3	Lake Powell chance or reaching critically low reservoir elevations in 2022	218
Table 10.4	Lake Mead chance of critical reservoir elevations in 5 Future Water Years	220
Table 11.1	Northern High Plains drought by area under drought level, September 2022	229
Table 11.2	Central/Southern High Plains levels of drought in last week of summer 2022	231
Table 12.1	Great Lakes Region levels of drought by category, June 2022	243
Table 13.1	Drought conditions in the South, Spring 2022	260
Table 13.2	Water withdrawals in Alabama by water use category in million gallons/day	267
Table 13.3	Tennessee surface and groundwater use in MGD by use from 2015	270
Table 14.1	August 2022 drought conditions in the Appalachian Mountain Region	281
Table 14.2	Percentages of all groundwater withdrawals by sector in CCPCUA study area	293
Table 15.1	Mid-Atlantic region drought levels, August 2022	299
Table 16.1	Sixty-plus off-and-on cycles of droughts in New England with section affected	317
Table 16.2	State levels of drought by the category, August 2022	318
Table 16.3	Public drinking water systems use of groundwater and surface water, 2014 and 2022 estimates	319

List of Boxes

Box 1.1	Components of the Hydrological Cycle	20
Box 1.2	U.S. National Weather Service weather forecast 2022	22
Box 1.3	The Clean Water Act (CWA)	25
Box 2.1	Terms Used to Describe a Water Crisis	35
Box 3.1	Fourth report of the study of the impact of climate change in the U.S.	62
Box 4.1	Drought and the impact of climate change	80
Box 5.1	Prolonged drought and U.S.-Mexico water sharing agreements	93
Box 5.2	EPA's 2020 report on sources of drinking water pollution	100
Box 6.1	Washington groundwater level status, Spring 2016	124
Box 7.1	US Fish and Wildlife service description for dam removals	132
Box 7.2	USACE description of benefits from existing dams on the Snake River	142
Box 7.3	Heavy rains result in failures of two dams	150
Box 8.1	The state of the nation's drinking water infrastructure	159
Box 12.1	Failing infrastructure in the Great Lakes water region	247
Box 12.2	Factors influencing groundwater in the states of the Great Lakes region	248
Box 13.1	Saltwater intrusion process and impact	262
Box 13.2	2020 Notices of noncompliance to the Safe Drinking Water Act	265
Box 13.3	Charges and court decisions on the Alabama-Georgia-Florida water cases	273

Box 14.1	The effects of mountain-top mines and valley fills on aquatic ecosystems	281
Box 14.2	Why the Atlantic coast particularly vulnerable to sea level water crises	287
Box 15.1	Turn of the century drought in the Delaware River Basin	303
Box 15.2	Water shortage by infrastructure failures	304

PART I

The Water World

CHAPTER 1

Water: America's Most Valuable Natural Resource

Freshwater is fast becoming the most valuable ecosystem of the United States and, even more important, the chief asset of the world. There are substitutes for oil and gold, but none for water. That value is based on the availability, the quality of the product of the resource, and the state of its demand. The availability of freshwater is influenced by weather and the water cycle and the extent to which the increasingly stressed supply is mismanaged. The quality of freshwater is influenced by three forces: (1) the degree to which natural and human actions have influenced pollution of the resource, (2) the geological and hydrographic structure of the region, and (3) the conditions of the climate in which the water resource is located. The demand for freshwater is shaped by population growth and increasingly, upon the impact of climate change.

The quality of the water is influenced by the degree of pollutants it contains. These pollutants can be from wastewater discharges, stormwater, roads, highways and parking lots, runoff from agricultural lands, heat from soil and air, removal of shade trees and brush along a watercourse, industrial wastes, and many other sources.

Both surface water and groundwater resources are exploited for human consumption, irrigation, power generation, and industry at rates far more rapidly than then can be naturally replaced. In the following pages, we identify what may be the most critical crisis conditions of surface and groundwater in selected nine state regions. The list here is similar to the

D. E. McNabb and C. R. Swenson, *America's Water Crises*,
https://doi.org/10.1007/978-3-031-27380-3_1

Table 1.1 Adjusted regions of the U.S. with generally similar climates

Chapter region states included		
10	The Northwest	Washington, Oregon, Idaho, Montana, Wyoming, California, Nevada
11	The Southwest	Arizona, New Mexico, Utah, Colorado
12	The Great Plains	North Dakota, South Dakota, Nebraska, Kansas, Oklahoma, Texas
12	Great Lakes	Wisconsin, Illinois, Missouri, Minnesota, Michigan, Iowa, Indiana, Ohio
13	The South	Georgia, North Carolina, South Carolina, Kentucky, Florida, Tennessee,
14	Atlantic Coastal and Appalachian Plateau	Alabama, Georgia, Tennessee, Kentucky, Maryland, Ohio, Pennsylvania, Virginia, West Virginia, and New York
15	Mid-Atlantic	Delaware, Maryland, New York, New Jersey, Pennsylvania, Virginia, West Virginia
16	Northeast	Connecticut, Maine, Massachusetts, New Hampshire, Rhode Island, Vermont,

climate regions provided by NOAA as shown in Table 1.1, somewhat modified to focus more on water resource factors.

The number of climate regions in the coterminous United States and the states in each region varies from time to time and the nature of the climates characteristic of the region (or zone). They range from seven to eleven different zones. Each climate zone includes the states that are part of the specific area because they share common weather patterns during each season. Eight of the nine-region model of NOAA's National Climate Data Center are shown in Table 1.1.

Water and Climate Change in the U.S.

Every 10 years, NOAA releases an analysis of U.S. weather over the past three decades. The three decade norms provide the average values for temperature, rainfall, and other conditions. Measurements include the

30-year combinations of coterminal 48 states, plus Alaska, the Hawaiian Islands, and possessions. The 30-year analyzes the *U.S. Climate Normals*. The current 30-year average represents the new "normal" that can be compared with previous normals to identify the changes occurring in the climate. The most recent Normal covers climate measurement that has occurred from 1991 to 2021. Overall, the United States is becoming warmer and wetter, although not everywhere. The West and middle regions of the contiguous 48 states continue to be drier while the East and portions of the South regions are becoming wetter. Collectively, while the drier states are suffering from prolonged drought, wetter regions must cope with severe storms and highly destructive floods. The following chapter focuses on drought conditions and the third chapter looks at the impact of more and more severe storms. Sources for the drought discussion are the annual weather and climate series released by the National Centers for Environmental Information (NCEI), an office of NOAA. Global temperature records start around 1880 because observations did not sufficiently cover enough of the planet prior to that time. NOAA now used data collected from more than 20,000 locations on land and sea to identify temperatures and evaluate changes.

Climate changes in the most recent series of U.S. Climate Normals since 1961 are compared with earlier normals. The comparisons reveal that all North America is continuing to get warmer compared to the overall twentieth-century average. As the climate continued to grow warmer, most of the eastern states in the coterminal U.S. (the CONUS) are becoming wetter while states in the West continued to be both warmer and drier. The wetness is linked to overall climate warming and extreme storm crises.

While climate warming is adding to the droughts in the West and Southwestern states, rising temperatures add to the decline in water resources by allowing more water to evaporate from the ocean, land surfaces, and water storage reservoirs. The normal pattern of winds carries this collected western moisture to the East, releasing it over the East, often during rainstorms, resulting in more severe storms in the central and eastern states. Throughout the Southwest, average surface water resources have been decreasing and are expected to continue well into the next century, resulting in even longer mega droughts.

This book is arranged in three sections: the first consists of four chapters that introduce the current conditions of our existing water world and seeks to establish a basis for the evolution of water stress to water crisis and even to water disasters. The second section establishes the four legs of the water world: surface water, groundwater, stored water, and reused water, each in its own chapter. Each chapter describes the existing condition of each of these water resources. The third section includes eight chapters, each of which focuses on some aspect of water stress or crisis that may be in place or looming in eight regions of the coterminous United States. Throughout, chapters focus on individual states with what may be considered to exemplify the problems in place at the first quarter of the twenty-first century. A continuing topic in all regions is individual state's drought status.

Throughout attention is also given to the inequities found in the allocation of this scarce resource, as well as the uneven funding for maintenance and operation of the nation's water distribution system across the country. Increasingly, this lack of attention to aging and deteriorating water infrastructure significantly exacerbates the problems associated with a lack of dependable clean and safe drinking water for public consumption.

Shaping Demand for Water

This book is about the various factors that shape the finite supply of freshwater available in North America for use in all the purposes for which water is necessary, with particular attention paid to the stresses to freshwater occurring in sections of the continental (coterminous) United States. The events and conditions that are amplifying the damage accruing from record-breaking heat, prolonged periods of drought, and the excessive dependence upon groundwater as a replacement for the stresses placed on supplies of surface water. Groundwater has quickly become one of if not the most important and highly stressed natural resources. As the nature and scope of surface water changes with global warming, many parts of the world are increasingly subject to a limited supply of freshwater. These water stress conditions were described in research that addressed the issue from a supply and demand framework. The problem of water is considered a consequence of climate change that has resulted in changes in the hydrological cycle, precipitation patterns, intensity, amounts, and the timing and form of precipitation. These changes have caused water to become scarce in many regions of the world. Increasing demand from rapid population and urbanization growth, with economic growth, have resulted in increasing demand for freshwater…and the contamination of both surface water and groundwater (Ding and Ghosh 2017, 91). As demand has grown, the supply of both surface water and groundwater in many parts of the U.S. has declined and continues to decline despite increasing efficiency in water uses. Climate change is one of the contributors to these declines (Averyt et al. 2013).

Some of the more critical stresses under which the water supply is changing and how the water crises associated with these changes keep growing and outpacing the ability to mitigate the resulting damages are discovered in the following chapters. These are topics everyone should be aware of and doing what we can to alleviate the damage that they bring about. Two forces are amplifying the necessity for achieving and maintaining a sustainable supply of the surface water and groundwater resources. In this book, we describe what is happening to the nation's freshwater resources in areas of water scarcity stress, risk, and crisis. While also addressing problems in surface water supplies and the degradation of these resources by pollution, we give special emphasis on the over-withdrawals of groundwater.

Throughout the remainder of the book, we will approach the discussion of the water resource according to three fundamentals of water resource management: the availability of the resource, maintaining the quality of the supply, and providing for the accessibility of clean, safe, and affordable supplies of the freshwater resource. Together, these three management tasks constitute the three pillars of water resource governance. We begin with this brief introduction to water resource management pillars as they appeared to us at the time of our writing.

Access to the information on the nation's water supply sources is made available by scientists associated with the United States Geological Survey, the U.S. Army Corps of Engineers, the United Nations, and many other nation's agencies with some responsibility for protecting the world's water resources. The information that U.S. agencies provide on the global water crises is part of the public domain and accessible to everyone. Credit for the other sources cited in our writing of the book is included to the best of our ability, with access locations included in the bibliography. Any misconceptions, disagreements, omissions, or errors in the interpretations and descriptions of the USGS studies and other sources included in this study are exclusively ours and we take full responsibility for the errors. Our interpretation of these sources is shaped by our experiences as participants in the administration of municipal organizations that include domestic water supply and wastewater treatment systems.

Water Crisis Concepts

Ours is, indeed, a wet, warm world, one of the unknown numbers of "water rich planets," enriched "over billions of years by comets and asteroids that have collided with Earth, enriching our planet with water" (NASA, n.d.; Noack et al. 2016). However, one of the problems with being a warm world is that many parts of the earth are fast getting warmer. The year 2021 was the warmest it has been for more than 1,200 years. And, as well as getting warmer, it is also getting drier in much of the United States. Warmer climate means greater demands for freshwater.

When people talk about the weather, they say either good things or bad things. They often do the same when the conversation is about water. In that conversation, it is there is too much water or not enough water when or where it is needed. The sun is too warm or it is another cold day, or there is too much rain or not enough and so on. When we talk about water, they are seldom concerned about whether the water they drink is from a river or stream, a lake or a hole in the ground. They want the water to come out of their faucet clean and safe to drink. Most don't care if it came from a surface water source or it was groundwater pumped from a well. It's all the same. Hardly, anyone stops to ask whether his or her water had been used once before, if it was safe to drink, of where the water in their glass came from.

In less than 30 years from our writing this book in 2022, 40 of the 50 U.S. states expect to have to contend with some degree of water shortages. Although the shortages will occur in most of the nation, several areas of the United States will be particularly affected, including parts of the Great Plains, the South, the Southwest, the Midwest, the Rocky Mountain States, and California.

One of us spent a working lifetime coping with public utility water agencies as part of municipal water systems. The other, who grew up with all the untreated water used by his family coming from the well in the backyard, many years later becomes an elected commissioner for a water and wastewater utility. Both authors have been exposed to and to cope with many of the issues we write about here. These experiences enable us to focus on a wide variety of issues and concerns, and through these different experiences, we collectively give tremendous credit to the men and women who have devoted their careers to what it takes to acquire, treat, and distribute water in small and large communities.

Water World's Common Problems

This study of the water world in the contiguous United States suggests the similarities in the changes taking place to the water world due to the hydrologic structure of the earth, and the climate and weather above justify dividing the nation into four sections with common characteristics and somewhat common water problems: the West, Center, Northeast, and the Southeast.

The West ranges from generally mild weather shaped by the Pacific Coast, passes over mountain ranges that deposit the snow and ice that provides surface water for spring and summer, to continue eastward to where they pass the weather to the Great Plains. The mountain ranges run from Canada south to Mexico can block coastal weather. East of the mountain ranges, the winter snowmelt moves eastward, feeding the streams and rivers that prove surface water for Central area farms, and then passes across the dry, semiarid regions beyond.

The Center includes the Great Plains and some of what is generally referred to as the Midwest. The High Plains section can receive little rain, resulting in dependence upon groundwater for municipal and agricultural use. Summers can often be very hot, and the winters can be extremely cold. The Northeast reflects the weather patterns that are evolving together as recipients of a wetter future with increasing occurrence of severe water events. The weather in the Northeast provides warm and fairly moderate temperatures in the summer, with a few very hot and humid days. Winters in the Northeast are moderately rainy and can see heavy snow and freezing rain. Surface water is generally sufficient to meet most needs throughout this region, but some drier periods can occur.

Much of the South has hot humid summers with most winter precipitation falling as rain. The occurrence of severe rain storms and hurricanes and tornadoes is possible. Summers in the more western states of this region are hot and humid, with temperatures ranging from 85 to 95 degrees; temperatures can often exceed 100 degrees.

Increasing demand for access to the limited accessible freshwater resource is resulting in increasing maltreatment of many water resources, resulting in more water scarcity. This is leading to what many water scientists see is a water crisis forming in many parts of the United States. Many water resources are in what are close to or may already have reached a crisis state. From the point of view of those who treat and deliver freshwater for the nation, their water systems can be found functioning in one or more of the positions on this brief management list:

- Water resource treats: Threats to freshwater systems include channelization (i.e., straightening or redirecting natural streams in an artificially modified or constructed streambed), groundwater pumping, course diversion, dam building, industrial and human-caused pollution, human-induced climate change, and overexploitation of natural resources (Malmqvist and Rundle 2002).
- Water scarcity: Scarcity of readily available freshwater is major threat to many water resources. Scarcity occurs when the volume of a freshwater resource does not meet the long-term needs of the activities for which water is a critical resource. As scarcity continues, it generates water stress.
- Water stress: Water that is stressed is what happens to the freshwater resource when the supply of freshwater is curtailed for any reason for a critical time period. Among the major forces causing stress are the increase in the human population, increases in urban development, industry, agricultural activities and water abstraction, diversion and damming of streams and rivers for reservoirs. If the problematic conditions forcing the stress are not resolved, the stressed water supply is at risk of becoming a crisis.
- Water crisis: When demand for water is greater than supply of usable water or when the populations that depend on the resource no longer have access to an existing or a reliable alternative source, the water crisis can become a disaster.
- Water disaster: When the crisis continues and the water needs of its human and environmental populations have not been met, a disaster ensues. This can occur from prolonged drought, changes in precipitation patterns, pest infestation, and other sources.

Water Resources Under Threat

The story of the nation's miss-treatment of its freshwater resource that is taking place in many locations begins with the need to cope with the scarcity of the resource. The Clean Water Act (1972, revisions in 1977, 2006 and 2020) of 2006 listed a variety of human activities that are potential threats to the quality of water and two that are potential threats to the quantity of water. The adverse conditions are brought about by a short-term combination of events or the results of a change to or returning to a climate pattern not understood by users of the resource. An example is the climate conditions that led to the dust storms from 1930 to 1940 on the High Plains region of the United States, a region with a history of a dry climate. The dust storms would not cease until farmers changed the way they farmed the land. They needed help and that meant they needed water to survive. The Federal Government produced the help beginning in 1933 with passage of the Farm Relief Act and the Farm Mortgage Act (McElvaine 1982). The rest of the question was water. The answer was groundwater and large-scale irrigation agriculture (Green 1992).

The condition of the exceptionally warm weather, lower natural precipitation amounts and earlier melting of snow levels, severe windstorms, and flooding since the 1950s have seen lower streamflows in many rivers and reservoirs, resulted in increasing reliance upon groundwater for irrigated agriculture freshwater resource. It is increasingly apparent that the scarcity of freshwater exists in the United States in finite and, in many locations, is becoming even more scarce. Water scarcity exists in critical conditions in the American West and Southwest and it has also become a problem in many regions. According to U.S. weather records, every part of the Southwest experienced higher average temperatures between 2000 and 2020 than the average from 1895 to 2020. Drought conditions have become common in this six state area since weekly drought monitor records began in 2000. For the periods from 2002 to 2005 and from 2012 to 2020, nearly the entire region was abnormally dry (NOAA data).

Drought and other climate events are leading to water scarcity in what had long been favorably blessed with an abundance of readily accessible freshwater in many sections of the United States. The water has long been scarce in the Southwest, where people are particularly dependent on surface water supplies from a few surface water streams and lake reservoirs which are vulnerable to evaporation. The availability of large amounts of freshwater that could be imported from other resources alleviated the extent of scarcity. However, those outside sources can no longer be depended upon to meet the demand exacerbated by climate change and population increase. The decrease in precipitation in this already arid region can seriously threaten natural systems and society. Droughts also contribute to increased pest outbreaks and wildfires, both of which damage local economies, and they reduce the amount of water available for generating electricity. Drought conditions in the Southwest have varied since 1895. Beginning in the 1900s, the Southwest has experienced wetter than average conditions during three periods: the 1900s, 1940s, and 1980s. Drier conditions occurred through the 1920s, the 1930s, again in the 1950s, and since 1990, when the Southwest has seen some of the most persistent droughts on record.

While more than 70% of earth's surface is covered by water, only 2.5% is usable freshwater. The remaining 97.5% is seawater or brackish water. Surface water stored and transmitted in rivers, lakes, canals, and reservoirs continues to be the main usable source of freshwater. When held in permanent ice, such as mountain and continental glaciers, freshwater has long been the earth's largest natural freshwater reservoir. However, global warming is resulting in the melting of large amounts of what has long been permanent ice. The remaining freshwater is groundwater, which is stored underground in aquifers and the moist soil. Much of the existing groundwater is being withdrawn from the earth far faster than it can be replenished.

Water Scarcity

The United States is one of three countries responsible for nearly 38% of the world's freshwater withdrawals: China and India are the others. The three major purposes for which water for these three nations is used are agriculture and industry, which includes power generation, and municipalities, which includes domestic use. In the following chapters, we describe examples of the groundwater crisis fast approaching and already occurring in many locations of the United States.

Groundwater constitutes anywhere from approximately 22 to 30% of the world's freshwater, with ice accounting for most of the remaining 78 to 70%. Freshwater stored in aquifers is increasingly the chief source for water needed for agriculture, municipal services, and industry. In becoming critical for its many uses, the groundwater that was once inexpensive to acquire and readily available has become overused. Many areas are now closely monitoring groundwater depletion and are actively recharging the aquafers, but too many users give little or no thought to refilling the aquifers from which it was withdrawn. This over-taxing of major aquifers has put much of the nation on the fast road to a major water crisis, with particular danger to the nation's food supply.

This world of ours is increasingly becoming a crowded world, one that is fast becoming a warmer planet. Estimates of the population of the ten most populous nations in 2021 and 2050 are shown in Table 1.1. In less than 80 years, the world's population is estimated by some forecasters to double to 16 billion. However, not everyone agrees that the world's population will continue to increase. Other forecasters predict that rather than increasing the lack of resources and natural disasters such as the COVID-19 pandemic, it is more likely to result in a decrease in population to somewhere between 6 billion and 10.9 billion by the end of this century. The global fertility rate is expected to be 1.9 births per woman by 2100, down from 2.5 in 2019. The replacement fertility rate of 2.1 births per woman will be reached by 2070 to the stresses now associated with accessible freshwater in many parts of the world (Cilluffo and Ruiz 2019).

Although the world is not about to run out of water, the freshwater that was once commonly accessible in is either on the brink of disappearing or getting difficult to come by and more expensive when it can be withdrawn. The surface sources upon which have relied for freshwater are simply overused, over stressed and harder to keep clean. Turning to groundwater to avoid these problems has resulted in many a water crisis that cannot easily be resolved. Although there are several possible ways of augmenting the supply of surface water, there are no easy or inexpensive substitutes. Desalination of seawater and brackish lakes and ponds or re-using water that has been used more times is costly to collect, clean, and disinfect.

Less than 3.0% of the world's water is the freshwater that must serve the world's 7.9 billion people—and that is getting harder to do. In many parts of the world, we seem to be using most of the accessible freshwater and destroying much of the mechanisms nature has developed for replenishing the supply. Along with global warming, changing traditional weather patterns have produced more droughts in some areas and wasteful flooding many other regions. In addition to these weather changes, the world continues to be crowded. By 2050, the world population is predicted to have increase to between 9.7 and 9.8 billion. Table 1.1 lists the 2021 and 2050 population estimates for ten of the world's most populous nations, all of whom will need freshwater to survive.

How Water Is Used

Water use in the United States in 2015 was about 322 billion gallons per day (BGD), a 9% decline in water uses from 2010 to 2015. The 2015 estimates saw total withdrawals at the lowest level since 1970, following an overall trend of decreasing total withdrawals seen from 2005 to 2010. Freshwater withdrawals were 281 BGD, or 87% of total withdrawals, and saline-water withdrawals were 13% of total withdrawals.

Fresh surface-water withdrawals were 14% less than in 2010, but groundwater withdrawals were about 8% greater than in 2010. Saline surface-water withdrawals in 2015 were 38.6 BGD or 14% less than in 2010. Total saline groundwater withdrawals in 2015 were 2.34 BGD. The shares of freshwater withdrawals for most uses declined between 2010 and 2015, with withdrawals in 2015 for thermoelectric power leading the decline with 18% less than occurred in 2010. Other withdrawal reductions over these five years were: public-supply withdrawals by 7%, self-supplied domestic uses by 8%, self-supplied industrial uses by 9%, aquaculture by 16%, irrigation by 2%, and mining, withdrawal declined by 1%. Livestock withdrawals remained essentially the same in 2015 compared to 2010. Thermoelectric power, irrigation, and public-supply withdrawals accounted for 90% of total withdrawals in 2015. Estimated annual 2015 withdrawals of freshwater for major uses by major use categories are presented in Table 1.2.

Together, the surface water stored in glaciers, lakes, reservoirs, ponds, rivers, streams, wetlands, and underground make up less than 1% of the

Table 1.2 Ten of the most populous nations in 2021 and 2050 (pre-COVID-19 estimates)

Country	*2021 population*	*2050 population est.*
China	1,443,870,568	1,402,405,167
India	1,392,450,780	1,639,176,036
United States	**332,776,413**	**379,419,097**
Indonesia	276,158,589	330,904,672
Pakistan	224,893,735	338,013,193
Brazil	213,891,631	228,950,400
Nigeria	211,023,908	401,314,996
Bangladesh	166,189,015	192,867,778
Russia	145,913,489	135,824,486
Mexico	130,167,167	155,150,813

Source Estimates from worldmeters.info, populationpyramid.net and the U.S. Census Bureau

world's total surface area. Despite their importance to life, sustaining crops through irrigation, growing food, and powering homes and industry and stored behind dams and other reservoirs many sources of natural freshwater supply resources are fading at an alarming rate. As a result, groundwater is increasing being used to augment if not replace many former uses of surface water. Recycled water and desalinated seawater are also increasingly important for uses where access to traditional surface sources of freshwater is restricted or altogether eliminated. The amount of freshwater for drinking and agriculture is particularly important. Freshwater exists in lakes, rivers, groundwater, and frozen as snow and ice. Estimates of groundwater are particularly difficult to make, and they vary widely. The USGS assesses total withdrawals by use category every five years; 2020 data were still not available in 2022 (Table 1.3).

According to the World Atlas, 68.7% of all the world's freshwater is stored in glaciers and ice caps, 30.1% is at various depths as groundwater, and 1.2% is fresh surface water, stored precariously in lakes, rivers, and reservoirs. Of all the earth's surface water, a third is in the soil and frozen in permafrost; nearly 21% is in lakes; 3.8% is in the soil; 2.6% is in swamps and marshes. Less than one-half of one percent is in rivers, 0.26% is contained in living plants and animals, and 3.0% is in the air and on the surface as water vapor. Global warming is taking big bites out of almost all this stored surface water.

Table 1.3 Water withdrawals in 2015 by use category in billion gallons per day

Rank	*Use category*	*Withdrawals in BGD*
1	Thermoelectric generation	133.00
2	Agriculture Irrigation	118.00
3	Public supply	39.00
4	Self-supplied industrial uses	14.80
5	Aquaculture	7.55
6	Mining	4.00
7	Self-supplied domestic use	3.26
8	Livestock	2.00

Source USGS (2015)

Freshwater is withdrawn from three primary sources: rivers, lakes, and constructed reservoirs. Replenishment of these sources is natural precipitation, with some natural precipitation. Surface water is withdrawn from lakes, rivers, and reservoirs. Rain and snow are also important sources in many regions. Natural underground aquifers are the source for groundwater. Groundwater is pumped from where it is stored and often then applied where it leaves the pump. Groundwater is absorbed through the soil and eventually stored in the cracks and gaps underground.

In some locations, once-used water from any source is returned to the earth's aquifer storage through a recharge process. This recycled water, also known as reclaimed water, must be cleaned and purified prior to reuse. The reclaimed water is frequently used for irrigation and municipal landscaping or for similar non-human consumption with minimal treatment. However, in many dry regions in the Southwest, recycled water is used for aquifer recharging where it is naturally mixed with or to augment naturally existing groundwater. The mixed water is then withdrawn, again treated to a higher standard and again used for various purposes. These approaches are used for recharging aquafers in the highly populated dry Southwestern states and have been common for many years in Arizona and Southern California. Globally, irrigation is currently responsible for about 70% of freshwater withdrawals worldwide. Although as much as 90% of the groundwater withdrawn for residential and industrial uses is eventually returns to the aquifer, only about one-half of the water used for irrigation is reusable. In arid and semiarid climate regions as the High Plains and the Southwest, evaporation, evapotranspiration from crops and water delivery losses such as from leaky pipes, as much as half of the groundwater withdrawn is forever removed from the water cycle (Kelly 2020).

Coming Water Stress

Withdrawals of all types of water in the U.S. in 2015 were estimated to be about 322 billion gallons per day. Despite increasing population and more agricultural irrigation, this was 9% less than what occurred in 2010. This estimate was the lowest level since before 1970, following the same overall trend of decreasing total withdrawals observed from 2005 to 2010. Freshwater withdrawals were 281 BGD or 87% of total withdrawals, and saline-water withdrawals were 13% of the total. Fresh surface-water withdrawals of 198 BGD were 14% less than in 2010, and

Table 1.4 Water use data in the U.S, 2000–2015

Estimated water withdrawals billion gal/day 5-year estimates						
Category	*2000*	*2005*	*2010*	*2015*	*% of 2015 total*	*2010–15$ change*
All water total	408	410	355	332	100%	n/c
Freshwater	333.1	349	306	281	87%	85%
Saline Water	74.9	270	48.3	41.0	13%	15%
Fresh surface water	n/a	270	230	198	n/a	14% less
Fresh groundwater	100%	79.6	76.0	82.3	n/a	8% more
Saline surface water	n/a	n/a	45.0	38.6	n/a	14% less
Saline groundwater	n/a	n/a	1.51	2.34	n/a	0.0

Source USGS 5-Year Estimates

fresh groundwater withdrawals (82.3 BGD) were about 8% greater than in 2010. Table 1.4 compares USGS water use totals for three 5-year surveys of water use in the United States with selected breakdowns of surface water and groundwater billion gallon per day estimated uses.

These estimates are evidence that surface water supplies in most of the US are no longer continuing to decline as an entirely reliable source for meeting the freshwater needs for water. While some areas are experiencing a growing number of extreme weather-related water crises, droughts create a different water crisis. The longer the drought, the less that surface water is available for agriculture and greater is the demand for pumped groundwater. Weather researchers agree that these drought and the resulting environmental disasters that result occur by the effects of climate change.

Climate-Related Events

The U.S. National Centers for Environmental Information (NCEI) time series data show there were 22 separate billion-dollar weather and climate disaster events in the U.S. in 2020, breaking the previous annual record of the storm events that occurred in 2017 and in 2018. The 2020 costs were $95.0 billion, with Hurricane Laura, the August derecho, and the historic Western wildfires as the most costly events that occurred during the decade. A water problem is a situation where the supply of available unpolluted freshwater from any available resource within a specified region is less than that region's demand for water. A crisis occurs when the region's governing bodies are unable to rectify the water supply problem.

The groundwater stored at different locations in the soil and layers underground in most of the world is highly stressed. Although often stored deeply underground, the water continues to move, although usually very slowly. It is still part of the water cycle. Most of the water in the ground is from rain and melted snow and ice that filters down from the land surface. The portion of the groundwater in the upper layer of the soil is considered the *unsaturated zone*. Water here seldom saturates the soil present, remaining in varying amounts that change with the seasons. Below this layer is the *saturated zone*, where the pores, cracks, and spaces between rock particles are saturated with water. Where the two meet is known as the *water table*. The term groundwater is used to describe water from this area. In addition, the region of aquifers is storehouses of earth's water and people all over the world depend on groundwater in their daily lives. Where the surface unsaturated zone and where groundwater occurs is the location of the water table and where the saturated water supply is measured. In 2015, more than 50% of the total withdrawals from groundwater in the U.S. occurred in 12 States: California, Texas, Idaho, Florida, Arkansas, New York, Illinois, Colorado, North Carolina, Michigan, Montana, and Nebraska (Dieter et al. 2018).

In 2015, California accounted for almost 9% of the total groundwater withdrawals, including a similar share of freshwater withdrawals in the United States. Most of the state's groundwater withdraws are used for irrigation in the state's Central Valley. Texas accounted for almost 7% of the total annual groundwater withdrawals, using that water in a mix of thermoelectric power generation, agriculture irrigation, and public water supply.

Florida accounted for 23% of the total saline-water withdrawals in the United States, mostly from surface-water sources for thermoelectric power. Together, Texas and California accounted for 59% of the total saline groundwater withdrawals in the United States, mostly for mining. All state's withdrawals of saline water for any purpose are not included in Table 1.4, but for groundwater, they are largest for Texas with 1,030 million gallons per day (MGD) and Florida for surface water with 9,400 MGD.

Hydrocycle Cycle Changes

Changes in the supply of freshwater are influenced by variations in the hydrocycle (Box 1.1). The hydrological cycle is the cycle of changes under which water comes and goes. First, liquid water evaporates into water vapor, condenses to form clouds, and precipitates back to earth in the form of rain and snow. Also, the hydrological cycle involves factors such as water storage through snow, ground, and vegetation. Ocean currents and large-scale dynamics transport energy across the planet and regulate global climate patterns. Hydroclimate phenomena such as drought, flooding, and precipitation can have societal impacts such as crop damage and loss of life.

The sustainability of groundwater resources, for example, is influenced by such factors as depletion of stored groundwater, reductions in surface streamflow, loss of wetland and riparian ecosystems, land subsidence, saltwater intrusion, and changes in groundwater quality. Groundwater aquifer systems and development situations are unique, and each requires an analysis adjusted to the nature of the existing water issues (Alley et al. 1999).

The advent of a water crisis is heightened whenever condition upsets the cycle. The impacts of climate change and population growth greatly intensify changes, intensifying the effects of climate change and exacerbating ongoing water resource degradation. This has set up a dynamic that requires decisive and thoughtful action if we are to assure there is ample water for basic human needs going forward. In many areas, this dynamic has already stressed social, economic, and political systems, and to date, most attempts to protect and preserve the water resources have literally been in a "drop in the bucket" to what is needed.

These modest conservation efforts have focused mostly on domestic conservation, while industrial and agricultural water consumption has increased, far exceeding the gains made from this conservation. Moreover, deforestation and pollution have further degraded and depleted the surface and groundwater available for human consumption. These conservation measures are a good beginning, but a great deal more attention to water supply management is needed to overcome the dual challenge of climate change and population growth.

Although long-term weather trends are often more or less regular over long periods, predicting changes in the weather varies from day to day and from region. Weekly predictions of weather for longer periods are, of course more variable, and vary from region to region. Sections of the National Weather Service's estimates of the last nine months of 2022 illustrate the generalization effects of period weather predictions. The key points of the March 2022 predictions are the forecasts for the onset and continuation of stress in regional areas (Box 1.2).

In the following chapters, we focus on the current conditions depleting and degrading surface and groundwater resources and on measures that can be taken to restore and protect those resources for future generations. Many of these measures will require changes in behavior and water use trade-offs which will be difficult to make. Established and entrenched economic forces will push for a continuation of water use practices that are not sustainable. Many of these practices are based on out-of-date assumptions regarding the availability of water that are no longer valid due to the rapid alteration of precipitation and temperature patterns climate change has brought. Others are based on short-term financial reward structures that hold deep vested interests that are steadfastly resistant to change.

Box 1.1 Components of the Hydrological Cycle

The five principal components of the hydrological cycle—evaporation, transpiration, condensation, precipitation, and runoff—each play a major role in sustaining global life. Components of the hydrological cycle take place at different timescales. Processes such as evaporation, transpiration, and condensation are constantly occurring at the surface and atmosphere, and only require several days for evaporated water to fall back to the earth as precipitation. In contrast, oceanic currents, snowpack and groundwater change on timescales of months to years to centuries. Evaporation is the change of state of water from a liquid to a gas, which requires energy through the sun, the atmosphere, earth, or objects created by humans. Similarly, transpiration is the evaporation of water from plants through stomata, which are small openings found on the underside of leaves that are connected to vascular plant tissues. Both of these steps are key to the creation of clouds, as they transfer water from the ground to the atmosphere. Precipitation occurs after condensation, when atmospheric water vapor created through evaporation and transpiration undergoes a phase change. Precipitation is the primary way the earth's surface receives freshwater. Falling in the form of rain, hail, snow, or sleet precipitation replenishes groundwater and reservoirs to allow crops to grow. Rivers and lakes can be formed as a result of runoff, which is a process that occurs when the ground is saturated and cannot absorb more water. In mountainous areas, snowmelt can be the primary source of runoff during the spring and summer months. Many reservoirs rely on heavy winter snows to produce spring runoff to replenish their water supplies. The evaporation of this runoff water begins the hydrological cycle again.

Source NOAA (2020)

Stopping wasteful environmental and ecological practices, curtailing environmental degradation, and altering consumption patterns will not come easily or cheaply, but will nonetheless be needed to confront this looming crisis. It is our hope that, in this book, others will find a point of departure and join us in finding and applying solutions to the water resource challenges we will face no and in the near future.

The Resource Under Stress

Nature does not discriminate when it comes to where and when a water resource comes under stress, let alone when a water crisis will occur. It can happen anywhere and is likely to happen any time. Water crises may take place more often than others; they are not felt in equal measure throughout the nation. There are significant geographic and climatological differences determining how or when an area experiences a water crisis. Much as emergency response system reactions to natural disasters are geared to react to the vulnerability differences from one area to another, reactions to a water crisis will depend of resources and planning and where the water crisis occurs. The differences result from variation in population growth, climate, the source of the water, the age of the infrastructure used to acquire and deliver the treated water and the willingness of leaders to invest in water acquisition. The response will depend on three management factors: planning, nature of the resource and sustainability of the water resource. Our purpose in writing this book was to help water system managers be prepared to meet the water crisis when it occurs.

Box 1.2 U.S. National Weather Service weather forecast 2022

In March of 2022, nearly 55 percent of the contiguous U.S. continues to experience drought conditions since late November 2021. Long-term drought is entrenched across much of the western half of the lower 48 states, with additional areas across the Upper Midwest, parts of the Carolinas and Maine, experiencing moderate to severe drought conditions. Seasonal snowpack is currently a mixed bag for much of the West, with above-normal snowpack over parts of the Northwest, but near to below-normal snowpack across the Great Basin and Sierra Nevada. However, farther east in the High Plains, snowpack is completely absent from eastern-Central Montana to the South-Central High Plains. Ongoing La Niña favors drought improvement for eastern Washington and northern Idaho. Improvements are likely to continue across parts of Minnesota through the end of May.

Ongoing La Niña conditions in the tropical Pacific Ocean are tilting odds toward warmer and drier conditions as the March–April–May (MAM) season progresses, particularly across the southern Great Basin and Four Corners region. Parts of the Central and Southern Plains saw record warmth this winter, accompanied by much below-normal precipitation, leading to drought expansion over these regions. Drought is likely to persist and expand in coverage, as warmer and drier conditions are typical during La Niña seasons across these areas. Persistence is also favored across the western Gulf Coast and coastal portions of the Carolinas, with development likely along the western Gulf Coast, Atlantic Coast, and Florida for the same reasoning, and further supported by these areas entering into a climatological time of year for soil moisture discharge.

From the interior Carolinas northeastward to New England, drought persistence and small areas of development are likely by the end of May, due to lack of precipitation signal and above-normal temperatures across these areas favored for much of the MAM season, also further supported by these areas entering into a climatological time of year for soil moisture discharge. Alaska is likely to remain drought-free for the coming season. Drought improvement/removal (one class improvement) is favored for Hawaii. Drought removal is likely for Puerto Rico."

Source US NWS. Forecaster: Yun Fan. March 17,m 2022\

Regardless of its final usage, water can be captured as storm runoff, as the flow of a stream or stored in a reservoir. It can also be pumped from an underground aquifer. The means of acquiring the water from the source differ in the technology, planning and management techniques needed to acquire, treat and distribute the water. Surface water is collected in watershed areas or drawn from existing bodies of water such as lakes and rivers. Water supply systems that rely on surface water are challenged by the need to assure that there is ample precipitation and other sources to supply the needed runoff. Also necessary is the need to prevent the resource from pollution to the surface water resource. The use of surface water runoff also holds inherent challenges related to the need of assuring that the water resource is not depleted so as to negatively impact alternate priority water uses to sustain wildlife, allow for recreation, or support economic interests for industrial cooling and consumption.

The acquisition of groundwater pumped from wells holds difficulties of a different kind. Here, the primary challenges include the need to understand the character and sustainability of the aquifer being accessed, the rate at which it can be recharged, the source of that recharge, and the cost of drilling and maintaining the wells, as well as building and operating the treatment facilities. Geotechnical information needed to answer these questions is, in many cases, difficult and expensive to acquire which can limit the availability of critically information. The less information available to the user, the more vulnerable the source becomes as pressure to draw down the aquifer occurs. This is particularly difficult when the aquifer underlies a metropolitan region or areas with extensive population growth.

Maintaining Source Sustainability

Obviously, these two primary water sources hold a myriad of different localized circumstances that impact the availability and cost of acquiring the water and the sustainability of the source. For example, surface water sources can be depleted as climate change decreases precipitation just as growth pressure requires an increase in water. Excessive groundwater withdrawals can deplete aquifers rapidly as development increases demand while simultaneously hindering recharge by paving land surfaces needed for water absorption and recharge. Regardless of the method of acquiring the water, stressed water resources everywhere are in stress today or may be affected. The nature and severity of the stressors will vary greatly, as will be the ability of the resource management capabilities to protect the resource from further stress. Similarly, the area of the country and the age of the delivery systems also impact the availability of useable freshwater. Large rivers, lakes, and large aquifers were able to meet the demands for agriculture, industry, and growing cities.

Many older municipalities in the Eastern and Central section of the country rely on these large municipal water suppliers that operated older surface water acquisition and delivery systems. In the past, surface water availability in these parts of the country also tended to be more reliably consistent due to weather patterns that regularly brought necessary amounts of precipitation. Until the 1960s, this was sufficient to resupply the large watersheds that provided adequate surface water. By the 1990s, many of the users that relied on reliable and attainable water resources found that water could no longer be reliable. In many areas, supplies of both surface water and groundwater can no longer be relied upon.

Protecting Against Water Pollution

In addition to not having access to enough freshwater to survive, a host of other water problems affects this critical resource. Modern efforts at assuring the water we can expect to receive from public and private providers began in 1948 with passage of the Federal Water Pollution Control Act of 1948. This law was the first major U.S. law to address water pollution. Growing public awareness and concern for controlling water pollution led to amendments in 1972 that became commonly known as the Clean Water Act (CWA).

Box 1.3 The Clean Water Act (CWA)

The Clean Water Act was adopted to prevent and repair water pollution in the United States. The Environmental Protection Agency, authorized by the CWA, uses two methods to protect the quality of water: monitoring water quality standards and controlling discharge from point sources. Water quality standards are jointly regulated by the EPA and by the states. States are required to classify bodies of water by their designated use (swimming, fishing, water supply, navigation, industrial waste disposal) and must adopt a plan to ensure the water meets the standards for that given use. These plans must then receive the approval of the EPA. Achieving this approval requires that pollution limits are sufficient to ensure the water is of high enough quality for its designated use and that the designated uses will not result in further degradation of waterways.

Water quality plans are determined by both technology-based limitations (requiring polluters to use the best available technology—BAT) and quality-based standards (by setting Total Maximum Daily Limits—TMDL). These TMDLs approximate the maximum amount of daily pollution that can be discharged into waterways while still meeting EPA standards. If a state fails to create standards that satisfy EPA requirements, the EPA may develop a plan itself and require the state to comply.

In addition to monitoring water quality standards, the CWA requires the regulation of point sources or stationary locations that discharge pollution into water. Under CWA §301, discharge of a pollutant from a point source into waters of the United States is illegal without a National Pollution Discharge Elimination System (NPDES) Permit.

- Established the basic structure for regulating pollutant discharges into the waters of the United States.
- Gave EPA the authority to implement pollution control programs such as setting wastewater standards for industry.
- Maintained existing requirements to set water quality standards for all contaminants in surface waters.

- Made it unlawful for any person to discharge any pollutant from a point source into navigable waters, unless a permit was obtained under its provisions.
- Funded the construction of sewage treatment plants under the construction grants program.
- Recognized the need for planning to address the critical problems posed by non-point source pollution.

Conclusion

Climate change and population growth are two of the pressing social problems that underlie the critical state of the decline in access to the freshwater resources of the world. Achieving sustainability of the world's water supply is a goal high on the objectives list of the United Nations and other global peace organizations. Far-thinking government leaders recognize climate warming's contribution to the shrinking supplies of accessible freshwater not only exists but will assuredly increase in the near future. A global problem needs a global commitment if the water crisis is to be resolved. Until that occurs, however, it is up to individual nations to do what can be done to reverse this trend. Resolution of the local drivers of the declining water sustainability water problem may be high in the commitment of the UN and other global collaborative to limit global warming and reverse agricultural and industrial activities that result in excessive withdrawals of the limited supply of freshwater. In an effort to increase awareness of the perilous condition of the United States freshwater resources, this book includes stories and reports on water crises that exist or which are soon to contribute to the crises in the near future. Their selection reflects the perspectives of the authors.

References

Alley, William M., Thomas E. Reilly, and O. Lehn Franke. 1999 (and 2013 update). Sustainability of groundwater resources. USGS Survey Circular 1186. Available at https://pubs.usgs.gov/circ/circ1186/pdf/p1-5.pdf.

Averyt, Kristen, James Meldrum, Peter V. Caldwell, Ge Sun, Steven G. McNulty, and Annette Huber-Lee. 2013. Sectoral contributions to surface water stress in the coterminous United States. *Environmental Research Letters* 8(2-013). Available at https://doi.org/10.1088/1748-9326/8/3/035046/pdf.

Cilluffo, Anthony, and Neil G. Ruiz. 2019. World's population is projected to nearly stop growing by the end of the century. *Pew Research Center*. Available at www.pewresearch.org/fact-tank/2019/06/17/worlds-population-is-projected-to-nearly-stop-growing-by-the-end-of-the-century/.

Dieter, Cheryl A., Molly A. Maupin, Rodney R. Caldwell, Melissa A. Harris, Tamara I. Ivahnenko, John K. Lovelace, Nancy L. Barber, and Kristin S. Linsey. 2018. Estimated use of water in the United States in 2015. USGS Circular 1441. Available at https://pubs.er.usgs.gov/publication/cir1441.

Ding, Grace K.C., and Sumita Ghosh. 2017. Sustainable water management—A strategy for maintaining future water. *Encyclopedia of Sustainable Technologies* 4. Available at https://doi.org/10.1016/8975-0-12-409548-991201741-X.

EPA (Environmental Protection Administration). 2022a. History of the Clean Water Act. Available at www.epa.gov/laws-regulations/history-clean-water-act.

EPA (Environmental Protection Administration). 2022b. Basic information about water reuse. Available at www.epa.gov/waterreuse/basic-information-about-water-reuse.

EPA (Environmental Protection Administration). 2022c. What is green infrastructure? Available at www.epa.gov/green-infrastructure/what-green-infrastructure.

EPA (Environmental Protection Administration). 2022d. Harmful algal blooms. Available at www.epa.gov/nutrientpollution/harmful-algal-blooms.

EPA (Environmental Protection Administration). 2022e. What is radon? Available at /www.epa.gov/radon/what-radon.

Green, Donald E. 1992. A history of irrigation technology used to exploit the Ogallala aquifer. In *Groundwater exploitation in the high plains*, ed. David E. Kromm and Stephen E. White, Chapter 2: 28–43. Lawrence, Kansas: University Press of Kansas.

Kelly, Morgan. 2020. *Expansion, environmental impacts of irrigation y 23050 greatly underestimated*. Princeton Environmental Institute. Available at www.princeton.edu/news/2020/05/05/expansion-environmental-impacts-irrigation-2050-greatly-underestimated.

Malmqvist, Björn, and Simon Rundle. 2002. Threats to the running water ecosystems of the world. *Environmental Conservation* 29 (2): 134–153. Available at https://doi.org/10.1017/S0376892902000097.

NASA (National Aeronautics and Space Administration). n.d. Ocean worlds: Water in the solar system and beyond. Available at https://www.nasa.gov/specials/ocean-worlds/#banner.

NOAA (National Oceanic and Atmospheric Administration). 2020. The hydroclimatic cycle. Available at www.weather.gov/media/climateservices/Hydroclimate.pdf.

Noack, L., D. Höning, A. Rivoldini, C. Heistracher, N. Zimov, B. Journaux, H. Lammer, T. Van Hoolst, and J. H. Bredehöft. 2016. Water-rich planets: How habitable is a water layer deeper than on Earth? *Icarus* 277: 215–236. Available at https://doi.org/10.1016/j.icarus.2016.05.009.

McElvaine, Robert S. 1982. *The Great Depression*. NY: Times Books.

USGS (United States Geological Survey). 2015. Update for the South Carolina Atlantic Coastal Plain Groundwater Availability Model. Available at www.usgs.gov/centers/sa-water/science/update-south-carolina-atlantic-coastal-plain-groundwater-availability-0?qt-science_center_objects=0#qt-science_center_objects.

CHAPTER 2

From Water Stress to a Water Crisis

Water stress is what happens when there is not enough water to meet the needs of people who depend on it. It can happen to a watershed or any local water source. It can happen when the water source is overused, no longer accessible, or polluted beyond the ability of users of the water to meet their needs. We are seeing water stress continuing to affect more people as an artifact of changes in the climate and how those changes affect the water resources. As this continues, water stress has become a crisis for the people, a community, a state, or a region. A water crisis is running out of water. Water can be used more than once and still returned to the source from which it was first taken. On the other hand, water can be used and lost or even lost without any use.

The terms used to identify the state in which water is used or lost are non-consumptive, partially consumptive, and consumptive. Non-consumptive water is water that is used and returned without any damage. For example, water that is withdrawn from a surface water resources such as a stream or a lake is used in a hydropower system for cooling and then returned to the resource when cooled. Another is water used in cities. Increasing amount of used domestic water is recycled for reuse. Water that is partially consumptive is water that is withdrawn, used for a purpose, during which all is expected to be non-coverable; however but some in returned to the system. An example is water withdrawn for irrigation. Some of this water may be collected runoff and may be used

D. E. McNabb and C. R. Swenson, *America's Water Crises*,
https://doi.org/10.1007/978-3-031-27380-3_2

again. Consumptive water is water lost in use. Examples include a large percentage of the water used for agriculture. Much of this water is transpired by plants or evaporates and is thus transferred to the atmosphere. Another example is water lost in thermoelectric power generation, where much of the water also evaporates.

According to the World Water Council, freshwater resources are, indeed, becoming scarce, but to become a crisis, it doesn't have to be a water scarcity. The source can become polluted or simply overused. There is some kind of a water crisis occurring today in many parts of the United States, just as there is in many parts of the world. But the crisis is not one of having a lack of water, but increasingly it manifests due to a failure to manage the water available in a way that enables its human use. That is, it is a crisis of managing water so badly that the people who need the water find themselves unable to use available water resulting from pollution or mismanagement of the available supply. In Flint, Michigan, the crisis was the public water supply becoming harmful because those charged with assuring safe water for the City put cost reduction above public health and the water became too polluted for human use. When the system's management changed to an alternate water source to cut costs, the altered chemistry of the new water caused the distribution pipes to corrode, releasing highly toxic levels of lead leaching into the city's water supply.

In another case, a crisis occurred when farmers, who have suffered under prolonged extreme drought, had to withdraw groundwater from ancient aquifers that cannot be replaced. This scenario became a crisis for a small town in central California that had seen two wells go dry and water from their remaining well become too contaminated with agricultural pesticides to consume. All of this occurred after the state had undergone the most severe drought since records were kept for 125 years.

These scenes are being repeated across the country with increasing frequency. Whether it is a water system neglect in Jackson Mississippi, or well pollution and groundwater overuse in rural agricultural areas of California, the result is the same: not enough clean safe water for those who need and depend on it for life. This crisis is resulting in many more people and communities facing the devastating prospects of no safe water. Tragically, this crisis exacerbates deeper long-standing social in inequities, as frequently the crisis befalls communities where a history of neglect and prejudice has left them most vulnerable.

Approaching a Water Disaster

As population continues to grow and climate changes become more severe, community leaders across the nation find themselves forced to find new ways to capture and store freshwater. As they do, they will encounter more and more severe water problems and greater competition for access to the already limited resources. The country's leaders will have to understand that stress is not unidimensional. Rather, it is a multifaceted problem affecting everyone. Because water can be obtained in many ways and tainted by many different natural and human-caused pollutants, the problems will differ in nature and scope, with causes and cures that vary from community to community. For people in the arid Southwest and the Rocky Mountains, the crisis is based on a lack of a dependable supply that meets the growing demand. Rapid population growth is occurring simultaneously with changing climate that is altering traditional precipitation patterns, due to higher temperatures and prolonged droughts. These changes should challenge all previous assumptions about how much growth can be supported in these regions. Record high temperatures have exacerbated the drought by increasing evaporation from the artificially formed lakes, irrigation canals, and reservoirs established for agriculture and municipal water storage. Similarly, the warmer temperatures increase evaporation of the water stored for landscape and recreation use. In the upper Midwest, the crisis is not one of supply but one of inadequate investment in managing the water supply and maintenance of aging distribution system infrastructure.

The City of Chicago water supply system is a good example of this circumstance. By the 1920s, the City of Chicago had created a large system designed and built for future growth and expansion of the city and surrounding area using water drawn from Lake Michigan. Lake Michigan is hydrologically connected to Lake Huron creating an enormous body of water covering a surface area of over 45,410 square miles covering roughly seven billion acre-feet of water. The city constructed significant water extraction, treatment, pumping, and distribution systems for their anticipated need well into the future and not only draws down, treats, and distributes enough water for the city and dozens of suburban systems in the surrounding area. Precipitation pattern is the geographically huge watershed area that encompasses Lake Michigan has proven to be adequate even recognizing fluctuations in total lake volumes. In fact, in 2020, water levels in Lake Michigan have been some of the highest since records were kept in 1918.

Moreover, the population within the area served by the City of Chicago system is more stable than many other areas of the country where water scarcity is a way of life. As a result, demand for finished water seldom fluctuates. In the fast growing Southwest and Rocky Mountain West, water suppliers have had to seek ways to balance increasingly stressed water supply with demand from agriculture and population growth as they must simultaneously cope with years of extreme drought and higher temperatures. While water supply is adequate, the aging system infrastructure requires ever increasing investment to assure the available water is safe and distributed to those who rely on this supply.

Arizona is an example of how many water suppliers in the Southwest have had to invest large sums on creative means for maintaining freshwater supplies and on conservation technology for stretching the already limited supplies. With one of the nation's driest and hottest climates and fastest growing populations, leaders in Arizona have worked to assure that adequate water will be available for future needs by heavily investing in acquiring and transporting supplies will exist where water is needed. With these investments, Arizona has created a comprehensive allocation and prioritization system with redundancies and contingencies to assure water will be available for human consumption, agriculture, and industry in the future. However, the effects of rapid climate change have caused this system to become highly stressed resulting in an increasingly urgent need to revisit the growth assumptions underlying their allocation decisions.

A Resource at Risk

A water resource at risk is a signal that the resource is in a situation where it is open to a damaging if not fatal condition. Examples of this can occur are an infrastructure failure, a discharge of toxic chemicals to a water resource, surface pollution from an extreme flooding event, water well failure or pump break downs, depletion of a groundwater aquifer system, catastrophic tornado, a tsunami, a forest fire, and many others. To assure a level of preparedness, water systems serving the public are required to prepare and annual Risk and Resilience Assessment.

The Clean Water Act, 2006, identified 21 different activities that represent existing or potential threats to drinking water. These include livestock grazing or pasturing, the operation of waste disposal sites and septic systems, the handling and storage of fuel, pesticides and commercial fertilizers, and the use of road salt and aircraft de-icing chemicals. To prepare for coping with a disaster, every five years water system suppliers must conduct a Risk and Resilience Assessment and prepare an Emergency Response Plan for the EPA. This provides the system with a summary of the reliability, redundancy, risk assessment, and resilience of the water system along with the recommended actions. The actions are those needed in the unlikely event of a severe emergency with water supply, public health, and safety resulting from bioterrorism, vandalism, aquifer depletion, equipment or well failure, or any naturally occurring event. These assessments are required for all community water systems that serve a population of 3300 and more.

Measuring Water Stress

Water stress is measured as the ratio of total water withdrawals to available renewable supply in an area. The water resources vulnerability index (WRVI) is a commonly used indicator of the degree to which a surface water watershed is under stress. It is a ratio of total water withdrawals to the available renewable supply. The index calculates the ratio of annual water withdrawals to annual runoff (renewable freshwater supply). Watersheds where withdrawals are less than 20% of the annual supply indicate a ratio of less than 0.2 at low but not water stress. A ratio between 0.2 and 0.4 indicates a medium stress. Watersheds where withdrawals are greater than 0.4 have high stress. In extremely high-stress areas, the ratio is 0.8 or higher (Alcamo et al., 2007; Hanasaki et al., 2008; Maddocks and Reig 2014).

Stress to a water resource may be a product of greater demand caused by increases in population, of warmer climate that leads to a need for more water for agriculture irrigation, of greater need for water for use in new or increases in a thermoelectric facility, or any combination of these and other demands for water (Arata 2020). If the stressed water resource is not properly managed, it quickly become a disaster.

Water Crisis Examples

In addition to water crises caused by excessive withdrawals from a limited resource, natural phenomena such as extreme storms, flood and forest fires, and water stress can also occur due to human error, mismanagement of political decisions, the failure of aging infrastructure, and/or economic decisions. To solve eventual water stress problems, the conditions causing the stress problems in an area must be resolved before becomes a water crisis. An example of a human-caused crisis is the mismanagement action that occurred in the water supply of Flint Michigan. Here, financial pressure on the city budget collided with political discord. An appointee of the Governor with little knowledge of water system operations enabled a series of decisions, all made for financial expedience, that resulted in a crisis where residents were harmed by drinking dangerously polluted water. A different type of water crisis is occurring in the Southwestern states now as climate change, population control, and diminished supply and perceived over use of a water resources resource long period of drought result are coming together to form a water crisis of significant effect. Here too, political decision making plays a significant role in water management decisions. In the rapidly growing Phoenix area, for example, old assumptions about the availability of water continue to be used, which may well overstate the amount of water that will actually be available for all the new residents as the speed and impact of climate change are being realized.

The crisis can also be from pollution of the existing supply. When available supplies remain less than demand for extended time periods, conditions require the region find itself on the threshold of a water crisis. Arrival of a water crisis may begin with a prolonged drought or severe storm, a forest fire, or by failure of aged water infrastructure. It may be the result of excessive use of existing but limited and irreplaceable water resources. That resource may be a supply of surface water as in a lake or a river or a supply of water that is withdrawn from an underground aquifer.

A potential crisis is often preceded by the occurring of water stress, which occurs when the demand for water is greater than the amount of water available for a certain period in time, or when water is of poor quality that restricts its use. Water stress is deterioration in either or both of the quantity of available water or the quality of available water available is polluted by either natural or human pollutants affecting its source, making it no longer safe for use. This often results from what is known as eutrophication, which is when pond, or lake or the stream or river becomes enriched in dissolved nutrients such as phosphates. These nutrients may stimulate aquatic plant life growth results in the depletion of dissolved oxygen. Eutrophication is often caused by runoff from the land and causes excessive plant growth, or pollution from inorganic or organic matter, excessive animal waste, saline intrusion, and other fluids (safeopedia.com 2017). Similarly, groundwater can become unsafe to drink if certain naturally occurring chemicals reach high levels. Arsenic is one such substance that is commonly found in groundwater across the country. Regular texting assures it is identified and treated commonly. Over time, and in certain circumstances, it may increase in well water to a point where a well must be taken offline resulting from levels that are too high to reasonably treat.

Maintaining the resource for this and future generations means achieving sustainability. Groundwater sustainability entails the "development and use of groundwater in a manner that can be maintained for an indefinite time without causing unacceptable environmental, economic, or social consequences" (Alley et al. 1999 and 2013). Without sustainability of this critical resource, the crisis affects us all. If the crisis is not mitigated, it can become a disaster.

Changing weather patterns are contributing to the nation's water crisis. Changes in precipitation and the water supplies that shape to water supplies and the water cycle that results in the supply of water can also result in a disaster and a calamity. The differences between these events are described in Box 2.1.

Box 2.1 Terms Used to Describe a Water Crisis

Water emergencies are natural events that are signals that an expensive and physically harmful crisis is on the way and, if insufficiently addressed, threatens to become a crisis. Emergencies are precursors of crises in that they call for immediate action to prevent significant harm to people, property, or the environment. They cannot be put aside for another day or other individuals to resolve.

A **water crisis** is a critical situation that threatens the functioning or survival of a government agency, a community, organization, or state available supply of freshwater. It can be caused by almost anything and strike anyone, anytime, anywhere. Crises in water management left unresolved can quickly evolve into a disaster. When the price we have to pay for water is more than we can afford, it can become a financial disaster or a personal calamity. Water crises are unexpected events with some, but not much more than a little warning. A crisis has three key features: It is unexpected, unique, and largely uncontrollable without significant effort.

Water-related disasters have the greatest potential for damage or pain. A disaster is an event that has occurred and must be dealt with. Disasters can bring harm to many people and be more difficult and costly to mitigate than emergencies or crises. Because a disaster will often develop from a series of events or in a combination of jurisdictions, a disaster mitigation often requires collaborative support from private and regional and national level stakeholders. For example, a city forced to deal with a hurricane for flood disaster will often require the help of many different responders and resource providers. Communities hit with disasters must often seek help from private and faith-based civil organizations.

A **calamity** is an often used term to describe either a crisis or a disaster that affects a community or an industry. It identifies the effects of a serious accident or an event that caused extensive damage or physical injury. Synonyms often used to indicate a calamity include catastrophe, a tragedy, cataclysm, or a disaster.

Sources Boin (2017) and Trias (2020)

Another of the growing severity of the water crisis is population growth and the expanding urban construction that is amplifying the groundwater crisis is reduction in such aquifer water recharging areas as farms, parks, and forests. The disappearance of traditional groundwater recharging water bodies and the growing need for irrigated farmlands that often return less groundwater than is taken from the aquifers, are adding to the growing water crisis.

We are all stuck with the amount of freshwater now on earth and contained as water vapor in earth's atmosphere. What we have is all we are going to get. Sometimes we get too much and other times not enough. The results include floods, droughts, and declining aquifers. Although the hydrologic cycle cleans and regularly replenishes most of the planet's water, it is often unpredictable as well as counter protective.

Barriers to Acquiring Clean Water

Acquiring all the clean freshwater we need is getting harder to find and, if available and can be maintained, is becoming more costly to come by. Water needs for human consumption, to generate electricity for light and power, for growing and processing the food we eat, for keeping industry working, and all the other many uses for which water is necessary are getting harder and more costly, even while the sources are decreasing. For those living on an island surrounded by seawater or stored in river or a lake, or coping with heavy rainwater and subsequent floods, it is hard to believe that America has all the water it needs. The problem with that conclusion, however, is the salt must be removed before it can be used. In all the world's oceans, there is not a drop of freshwater to drink.

Climate change and overuse are regularly reducing the supply of available freshwater. Storm and flood water result in more damage than benefit. This abundance of water on the planet is threatened by a lack of clean drinking water, but not by a lack of saltwater. Converting saltwater to freshwater is possible but is still an extremely expensive process. The World Water Council's brief introduction to water crises began with this undated statement that coincides with our goal in writing this book: "There is a water crisis today. But the crisis is not about having too little water to satisfy our needs. It is a crisis of managing water so badly that billions of people - and the environment - suffer badly."

Water from a Hole in the Ground

For many families, farms, and industries, the water used is pumped from a relatively shallow aquifer. It is relatively easy to clean, treated to government health standards, and delivered through old and cheaply built piping infrastructure. We have to be careful in the way and how much we use our water because only 3% of the planet's water is freshwater and ours comes just the one percent that is accessible to humans and things (World Atlas 2022).

The beauty of public acknowledgment of the need for better care of our water resources is that most of us are becoming better stewards of the water we take out of the earth. Better care of the resource is occurring for most water managers but not all, as the experience in Flint, Michigan, proved. Flint's water management's unwillingness to acknowledge the veracity of and react to claims of lead in their drinking water serves as a reminder for all of us that the fragile stock of freshwater must be cared for religiously. Unless we do, we can depend on facing a crisis and a threat to human health, just as happened in Flint.

Gaining Water Knowledge

Freshwater is clearly the world's most critical resource. However, most people don't know enough about it, where it comes from other than a faucet over the sink, and that it is getting harder to come by. We all need to know more about the resource and what is taking place now and what is planned for the future, and never consider there is a need now to protect and preserve what water we have. Determining the extent of the water available in the nation, this was followed by assessing the state of the resource. The U.S. government and individual state government partners have a long history of both of these actions. The first task has been accomplished. Every five years the U.S. Geological Survey compiles county, state, and national water withdrawal and water use data for a number of water use categories. At the time of writing this book, data for the 2020 survey were being analyzed and prepared for distribution. Therefore, the data for 2015 served as the base rate for most of the descriptions of water used and consumed. For example, in 2015, the average water used in the U.S. for all major categories, in average billions of gallons per day (BGD), was reported in the following way:

- Public-supply withdrawals from all sources in 2015 were 39.0 BGD, or 7% less than in 2010, continuing the declines observed from 2005 to 2010.
- Self-supplied industrial withdrawals were 14.8 BGD in 2015, a 9% decline from 2010, continuing the downward trend since the peak of 47 BGD in 1970.
- Total aquaculture withdrawals were 7.55 BGD in 2015, or 16% less than in 2010, and surface water, with 79%, was the primary source.
- Total mining withdrawals in 2015 were 4.00 BGD, or about 1% of total withdrawals from all uses and 2% of total withdrawals from all uses, excluding thermoelectric.
- Livestock withdrawals in 2015 were 2.00 BGD, the same as in 2010.

The world' annual rainfall, its lakes, rivers, streams, and underground aquifers have long supplied people living and working in cities and towns and on farms with abundant safe freshwater. Modern cities could not exist without adequate supplies of freshwater. In a 2016 report to Congress, the U.S. Government Accountability Office began with an observation that, of all services provided by municipalities, delivering a safe and adequate supply of water is perhaps the most essential. However, due in part to climate change and population growth, this necessary resource is no longer always available when and where it is needed or in the amount or quality desired. The fading availability of clean, safe freshwater is what international organizations such as the United Nations have identified as the major environmental crisis of the twenty-first century. Well into the third decade of this century, large sections of the nation have experienced more and more severe droughts, while other regions continue to suffer through powerful storms and the floods that follow the storms.

From Stress to Crises

For much of the world's population, some type of water stress occurs from one year to another and crisis is often just a day away from it happening. The surface water they depend on disappears early in the year and for which they have no alternative but turn to groundwater where they can find it. Much of America's and the rest of the world's freshwater supplies are facing one or more water crises from any one or more of the three related adverse weather and hydrologic events that are becoming more a calamity than "just another water crisis." At the core of these crises are first the increasing number and length of droughts and the global warming that is affecting weather pattern. The changes to weather patterns associated with global warming are more and more severe floods in some areas along with the droughts in others. In other locations, the populations are suffering through heat waves with temperatures never before recorded in many locations.

In the past, if properly managed, the potential for a water crisis may have been addressed and resolved. It was deemed an emergency and remedial action was taken. Droughts could have been taken as a signal and action taken before could become a calamity. Today, however, the people who depend on a limited water source do not have the resources needed to make changes. The mediation could have reduced the extent of the human and structural damage brought on by the adverse weather event. The water crisis has become a calamity (Lall et al. 2008).

Effects of Weather Change

Weather change on the environment and the population in America's Southwest have been exacerbated by the prolonged drought that occurred

Table 2.1 Standard climate change temperature reference points

Climate factor	*Effects and conditions*
Average Temperature	Average temperatures have risen across the 48 states since 1901. Eight of the top 10 warmest years have occurred since 1998. All of the top 10 warmest years on record worldwide have occurred since 2005
Winter Temperature	Since 1896, average winter temperatures across the contiguous 48 states have increased by nearly 3 °F. Spring temperatures have increased by about 2 °F, while summer and fall temperatures have increased by 1.4 °F
Heat waves	Heat waves are occurring more than they used to in major cities across the United States, occurring three times more often than they did in the 1960s—about six per year compared with two per year. The average heat wave season is 47 days longer, lasting longer and becoming more intense
Droughts	Average drought conditions in the 1930s and 1950s were the most widespread droughts; weather since the1970s have generally been wetter than average. By region, the West has generally experienced more drought while the Midwest and Northeast have become wetter. Since 2000, roughly 20 to 70% of the U.S. land area experienced conditions that were at least abnormally dry at any given time

Source EPA (2021)

from 2015 to 2019. There simply was not surface water or enough precipitation to replenish the surface water supply withdrawn for regional agriculture (Table 2.1).

Temperature Changes

Average temperatures have risen across the contiguous 48 states since 1901, with an increased rate of warming over the past 30 years. Eight of the top 10 warmest years on record have occurred since 1998. Average global temperatures show a similar trend, and all of the top 10 warmest years on record worldwide have occurred since 2005. Within the United States, temperatures in parts of the North, the West, and Alaska have increased the most. Weather research indicates that extreme weather events such as heat waves and large storms are likely to become more frequent or more intense with human-induced climate change. The changes in climate now taking place are accompanied with changes in weather patterns. Table 9.1 describes differences in precipitation and the severity of rainstorm events. The main differences are the higher temperatures and droughts in the West and rainstorms and floods in the East (Table 2.2).

For hundreds of years, historical records have revealed that prolonged droughts resulting in wide-reaching forest fires and declines of available surface water for all uses were the norm in the West Coast states. Only in the twentieth century did these events result in water crises that resulted in environmental disasters. Until advances in powered pump technology made it possible to rely on withdrawals from deep aquifers, agriculture production relied on precipitation and existing surface rivers and lakes. Resupply of shallow aquifers occurred by natural precipitation that collected on farmland and in rivers and streams where it was either

Table 2.2 Long-term storm and precipitation event changes

Weather event	*Description and conditions*
Precipitation increase	Total annual precipitation has increased over land areas in the United States and worldwide. Since 1901, precipitation has increased at an average rate of 0.1 inches per decade over land areas worldwide. However, shifting weather patterns have caused certain areas, such as the Southwest, to experience less precipitation than usual
Intense one-day events	A higher percentage of precipitation in the United States has come in the form of intense single-day events. The prevalence of extreme single-day precipitation events remained fairly steady between 1910 and the 1980s but has risen substantially since then. Nationwide, nine of the top 10 years for extreme one-day precipitation events have occurred since 1996
Tropical storms	Tropical storm activity in the Atlantic Ocean, the Caribbean, and the Gulf of Mexico has increased during the past 20 years and sea surface temperature in the tropical Atlantic and has risen noticeably during that time. However, records since the late 1800s suggest that the actual number of hurricanes per year has not increased
Floods	Changes in the frequency and magnitude of river flood events vary by region. Floods have generally become larger across parts of the Northeast and Midwest and smaller in the West, southern Appalachia, and northern Michigan. Large floods have become more frequent across the Northeast, Pacific Northwest, and parts of the northern High Plains, and less frequent in the Southwest and the Rockies

Source EPA (2021)

absorbed into the soil or stored behind low dams to be used in the next season. Or, it simply evaporated in the natural water life cycle.

When the storms were exceptionally severe such as during a hurricane or tornado, little of the precipitation could be collected and stored. In normal dry periods, farmers and city water managers could always depend on available groundwater to cover the short surface water supply. In 2000, the freshwater pumped from underground in all U.S. states for all purposes was estimated to be approximately 83 billion gallons per day (Hutson et al. 2004), about 8% of the estimated 1 trillion gallons per day needed for the natural recharge of the nation's groundwater systems. Water withdrawn from all sources in the United States in 2015 was estimated to be about 322 billion gallons per day. Fresh surface-water withdrawals (198 BGD) were 14% less than in 2010. However fresh groundwater withdrawals (82.3 BGD) were about 8% greater than in 2010 (Dieter et al. 2018). In California alone, groundwater accounts for 41% of the state's total annual water supply on an average basis and up to 58% of the total annual water supply in drought years. About 83% of Californians depend on groundwater for some portion of their water supply and many communities are 100% reliant on groundwater for all their water needs, California Department of Water Resources (CDWR 2021).

The Disappearing Supply

The loss of a steady supply of water from any dependable source or sources is particularly devastating for irrigation and domestic uses in arid or remote areas. In parts of Texas, Arizona, New Mexico, and California, for example, surface water is periodically in short supply or far enough toward a crisis to be impossible to regularly depend on in these areas. Water managers and agriculturalists have turned to groundwater to make up the lost differences. However, the groundwater they used must be replenished before the well runs dry. That replacement is most often left to nature. Some although not all annual rainfall and water from melting snow and ice can filter back into the soil. However, when rainfall soaks into the ground, it can take hundreds to thousands of years before it replaces what has been extracted. In arid areas, high demand for groundwater has resulted in declining of the water table by hundreds of feet.

Freshwater supply is affected by the type and extent of the regional environmental conditions and events that occur within the reach of a region. To limit the damage to human life and property from crises and disasters, greater preparation and planning for coping with these events are occurring through the nation. The need for planning and preparing for mitigating and recovering from a crisis, an emergency, or a disaster at some time in their tenure faces all public sector managers. Efficient and effective governance of the response and recovery efforts for those incidents is likely to be critical. Much of that research and crisis planning was slowed due to the need for coping with the COVID-19 pandemic.

Weather's Effect on Supply

August 2019 was a mixed bag of weather-related crisis in America. Extreme drought and abnormal dryness expanded or intensified across much of the Southwest, southern Plains, and upper Midwest, much of southern coastal Alaska, and along the East Coast, in Hawaii and in Puerto Rico. Drought and abnormal dryness contracted or became less intense in east-Central Alaska and parts of the Plains, Pacific Northwest, Southeast, Ohio Valley, Puerto Rico, U.S. Virgin Islands, Hawaii, and Micronesia. Moderate-to-exceptional drought footprint rose from 3.5% of the continental U.S. at the end of July to 10.0% at the end of August (from 6.9 to 11.3% for the 50 States and Puerto Rico). According to the Palmer Drought Index, about 8.2% of the nation was in moderate to extreme drought at the end of August. Ten states were listed in the driest third of the historical record, including Arizona with driest summer on record, Utah the sixth driest, California the ninth driest, and New Mexico tenth driest. Although far fewer than in 2018 when 3,976 fires burned 621,784 acres, in 2019 3,409 forest fires continued to rage in California (NOAA 2019).

Extreme Weather Events

The United States is experiencing a growing number of extreme weather events, the advent of each can be a crisis or even a disaster for some people some place. These crises and environmental disasters that result are brought about or significantly increased by the effects of climate change and what people are doing to our supplies of freshwater. By 2020, for example, the U.S. had sustained 285 weather and climate disasters since 1980 where the overall costs of damages reached or exceeded $1 billion in

2020 dollars. The U.S. National Centers for Environmental Information (NCEI) time series data show that there were 22 separate billion-dollar weather and climate disaster events in the U.S. in 2020, breaking the previous annual record of 16 events that occurred in 2017 and 2011. The 2020 costs were $95.0 billion, with Hurricane Laura, the August derecho (a line of intense, widespread, and fast-moving windstorms and sometimes thunderstorms that moves across a great distance), and the record-breaking heats and wildfires as the most costly events that occurred during the decade.

The Looming Water Crisis

In another June 2021 indication of the water crisis resulting from the Western U.S. drought, federal water managers shut off irrigation water to farms from a reservoir on the Klamath River. The U.S. Bureau of Reclamation announcement said it would not release water in the 2021 season into the main canal that feeds the Klamath Reclamation Project, the first such event in the programs 114-year history. Agency spokespersons said they would not release water into the Klamath River. In addition to lack of water for irrigation, the freshwater is needed to cool the river, which is necessary for the survival of infant coho salmon, an already endangered species. The lack of freshwater has resulted in deaths of large numbers of juvenile salmon. This is a calamity to the Yurok Tribe that depends on the returning salmon.

Groundwater Losses

Groundwater is a valuable resource in the United States and in many parts of the world. It accounts for approximately 24% of all the nation's annual freshwater withdrawals, making it doubly important where surface water supplies are scarce or unavailable. In the United States, groundwater meets the drinking water needs of about half the total U.S. population and nearly 100% of the rural residents' water needs. It is particularly critical in times of drought. This dependence upon groundwater resources makes it almost too popular. Too many withdrawals with little or no replenishment is resulting in rapid depletion. In fact, in many locations, the nation is experiencing groundwater depletion with a water crisis already in hand (USGS 2018).

Withdrawing water from the ground faster than it can be replenished causes many problems, not the least of which is the end of supply. Aquifer depletion, the amount of water stored underground with little or no replenishment, is occurring with more frequency in many areas. Sustained groundwater withdrawals can result in the following potential crises:

- Lowering the water table and drying up of wells.
- Less water in streams, lakes, and surface ponds and swamps.
- Deterioration of water withdrawn for human and livestock use.
- Deeper wells and increased pumping costs.
- Land subsidence.

Groundwater withdrawals for irrigation in areas without adequate precipitation or a constant source of surface water are what's causing the need to sink deeper in the well to access the freshwater in the aquifer over which people live. To find the water freshwater head, it is necessary to drill new, deeper wells before additional resources are available, if any exist. Over-drafts of water in the High Plains (the Ogallala) aquifer and Colorado Plateau aquifers are removing water that took some 15,000 years to accumulate. Halting removals and recharging of these aquifers before another 15,000 years passes by is not possible. The Ogallala Aquifer, for example, is being both depleted and polluted, some by the chemicals used to break open oil deposits. Withdrawals for agriculture are the biggest user of the undergroundwater in these two major groundwater systems. Although agricultural water is discharged on top of the soil, little of it is replaced by recharge. Only about one-half of the water used in irrigation is reusable. The rest is lost by evaporation into the air, evapotranspiration from plants, or simply lost as it rolls off the soil and disappears in catch basins.

Since large-scale irrigation began in the 1940s, water levels have declined more than 100 feet in parts of Kansas, New Mexico, Oklahoma, and Texas. In the 1980s and 1990s, the rate of groundwater mining, or overdraft, lessened, but still averaged approximately 2.7 feet per year.

Thermopower's Water Use

Power generation and irrigation are the two largest uses of water. However, in the 2015 USGS analysis of water consumption, it is found that total withdrawals decreased for thermoelectric power, but withdrawals had increased for agriculture irrigation, the second largest amount of water use. Withdrawals for thermopower were 133 BGD in 2015 and represented the lowest levels since before 1970.

Thermal power generation is the process of obtaining thermal energy by burning fuels to produce very hot and high-pressure steam that produces electric energy in power generating facilities. Thermal power generation is thus steam power generation. Water is heated to generate steam and then more water is used to cool the steam and equipment. Depending on the fuel type and thermodynamic process employed, the overall efficiency of this conversion typically ranges from 30 to 40%, resulting in the significant loss of energy. The unused energy is the thermal energy in the hot exhaust gases from the combustion process. In a combined cycle application, electrical efficiencies can be raised to 60%. In the first and subsequent cycles, combined cycle operations employ a Heat Recovery Steam Generator (HRSG) that captures heat from the high temperature exhaust gases to produce more steam, which is then supplied to a steam turbine to generate additional electric power.

Combined cycle power generation is a high efficiency method of power generation combining a gas turbine and a steam turbine. (1) High thermal efficiency is achieved by combining two methods of power generation: gas turbine power generation from rotating a generator by generating combustion gas by burning fuels in compressed air and (2) steam power generation with a steam turbine using the residual heat of the exhaust gas. In changes in demand or an emergency, power generation can continue by can immediately switching to just one of the sources. A Nuclear Power Plant (NPP) employs thermal power generating station using a nuclear reactor as the heat source for generating the steam that drives a steam turbine connected to a generator that produces electricity.

Thermal Power's Water Use

Every year, the United States generates an average of 4,000 million MWh of electricity from utility-scale sources. A megawatt-hour (MWh) is the amount of electricity generated by a one-megawatt (MW) electric generator operating or producing electricity for one hour. A majority of the fuels used in generating electricity in 2020 were fossil fuels such as natural gas (32.1%) and coal (29%). Other minor heat sources used range from biomass to geothermal and nuclear. Almost all thermal power generating methods require huge amounts of water for cooling purposes.

Thermoelectric power plants account for 40% of the freshwater withdrawn every year in the U.S. and 43% in Europe. This is approximately the same as the agriculture sector. Although most of the water is not consumed and is returned to the water source, these huge volumes of water withdrawn by the power sector have an impact on the ecosystem and on the water resources of a region. The growing water scarcity from economic and population growth and climate change has become a potential barrier to power industry growth of power generation.

Water is used in two methods for generating electricity: thermal power and hydroelectric power generation. Thermal power is generated by using heat from nuclear fission or from the combustion of natural gas, coal, oil, or biomass to generate steam that then drives power generating turbines. Surface water withdrawals for this purpose accounted for more than 99% of total thermoelectric power withdrawals in 2015 with 72% of the surface-water withdrawals withdrawn from freshwater sources such as rivers and lakes. Saline surface-water withdrawals for thermoelectric power accounted for 97% of total saline surface-water withdrawals for all uses (Dieter et al. 2018).

Hydroelectric power generating plants convert the gravitational energy of stored or flowing water to generate electricity with hydraulic turbines. The water that passes through the turbines is not consumed in any way. Rather, water loss occurs through the evaporation of the water stored in reservoirs or flowing in the stream from which was it withdrawn.

Power plants require water for several generating processes (steam cycle, ash handling, flue gas desulfurization systems, among others); most of the water requirements (usually about 90% of the total) are for cooling purposes. Because the high-energy heated steam expands in driving the generator, to cool it down before discharging, it must be cool. More water is used to cool the steam. From a regional water consumption perspective, non-cooling plant processes are usually negligible contributors to water loss totals. However, when the used water is returned to a stream or other body of water, it is hotter than the receiving water. Also, the water to be returned to the source may contain a variety of pollutants. To reduce or eliminate environmental damage, it must be treated as well as cooled before being returned to the source. Thus, thermopower plants can incur very significant costs associated with treating wastewater for discharge under stringent EPA and local regulations (Global Water Forum 2015).

Uses of Water in Hydropower Generation

Water used in hydroelectric generation process does not necessarily result in the loss of freshwater. However, the storage of large amounts of surface water behind the nation's many dams can have significant negative impact on the accessible supply of water for all uses. The USFS defines hydropower, or hydroenergy, as a form of renewable energy that uses the water stored in dams, as well as flowing in rivers to create electricity in hydropower plants. The water rotates blades of a turbine, which then spins a generator that converts the mechanical energy into electrical energy. Although hydroelectric power produces only a small share of the total electricity generated in the U.S. with 7% of the total generated, it is a significant component of renewable electricity production in the United States, accounting for 52% of the nation's renewable electricity.

Hydropower generation does not pollute the water or the air. However, hydropower facilities can have large environmental impacts by the withdrawals of water, thereby changing the environment and affecting land use, homes, and natural habitats in the dam area. Operating a hydroelectric power plant may also change the water temperature and the river's flow. These changes may harm native plants and animals in the river and on land. Reservoirs may take over and/or destroy people's homes, important natural areas, agricultural land, and archaeological sites. Building dams can usually require relocating people. Methane, a strong greenhouse gas, may also form in some reservoirs and be emitted to the atmosphere (USGS, n.d.). Clearly, hydroelectric power is not perfect; it does have some disadvantages, including the following items identified by the USGS with explanations by such sources as American rivers.org, Hydro Review.org, the U.S. EIA, and other state and local reports:

- High investment costs. In a U.S. EIA 2018 report, hydropower facilities had the highest average construction cost per kilowatt (CPK) of any generating technology installed in 2016. At $5,312 CPK, it was more than double solar construction at $2,434 CPK and natural gas at $895 CPK. The total cost for all hydroelectric construction was $2.5 billion in 2016, compared with solar at nearly $20 billion and wind power construction at nearly $15 billion.
- Hydrology dependent. Climate change effect on precipitation patterns; drought reduces stream and storage, with resulting limits in power generation capacity. Drought conditions in 2020 and 2021

included below-normal precipitation and snowpack accumulation, very dry soil and higher-than-normal temperatures in most of the Western United States; in June 2021, 100% of the California was experiencing some degree of drought. About a third of the state was categorized as under exceptional drought, the most intense classification. The drought conditions severely affected California's water supply levels and its hydropower plants. The lower the water supply available in the summer months resulted in a 19% loss in power generation.

- Possible inundation of land and wildlife habitat. Dams change the way rivers function. They often trap sediment, burying rock riverbeds where fish spawn. Gravel, logs, and other important food and habitat features can also become trapped behind dams. This negatively affects the creation and maintenance of more complex habitat (such as riffles and pools) downstream. These effects modify generation capacity.
- In some cases, loss or modification of fish habitat. Slow-moving or still water in reservoirs can heat up, resulting in abnormal temperature fluctuations which can affect sensitive fish species, and can lead to algal blooms and decreased oxygen levels. Other dams decrease normal surface water temperatures by releasing cooled, oxygen-deprived water from the reservoir bottom, thereby impacting downstream environments.
- Fish entrainment or passage restriction. Fish entrainment occurs whenever streams, creeks, or rivers are diverted for irrigation and other uses; fish follow the diversion of water NS are transported along with the flow of water and out of their normal environments and become stranded and die.
- Often, changes in reservoir and stream flow can adversely affect water quality. Surface water withdrawn for power generation may result in downstream pollution to steams that contain returned surface water. One of the most destructive can be including temperature pollution.
- In some cases, displacement of local populations. As reservoirs fill, they inundate surrounding space, forcing movement of communities. In extreme weather and precipitation patterns, forced release of stored water to protect the dam can result in floods downstream, curtailing if not eliminating power production during the disaster period.

Where Are All the U.S. Hydroelectric Reservoirs?

The U.S. Army Corps of Engineers 2018 national inventory of dams lists of from 79,777 to 90,0.000 dams exist in the United States, including Puerto Rico and the U.S. Virgin Islands. The list includes all dams that are either (1) 50 feet or more in height, (2) with a normal storage capacity of 5,000 acre-feet or more, or (3) with a maximum storage capacity of 25,000 acre-feet of water or more. Of all the dams listed, at the time of the survey, just 2,198 active hydroelectric power generating plants were producing power in the United States. Dams built for hydropower alone constitute just 4% of the nearly 90,000 dams in the U.S. Of the rest, just 11% are multipurpose dams. Run-of-river dams do not include a reservoir, depending on the river's flow to power the generator (USACE 2018).

Although some hydropower generation facilities can be found in almost all U.S. states, about half of total U.S. utility-scale conventional hydroelectricity generation capacity is concentrated in Washington, California, and Oregon. Washington has the most conventional hydroelectric generating capacity of any state and is the site of the Grand Coulee Dam, the largest U.S. hydropower facility, and the largest U.S. power plant in generation capacity. New York has the largest conventional hydroelectricity generation capacity of all states east of the Mississippi River, followed by Alabama. The top five states and their percentage shares of total U.S. conventional hydroelectricity generation in 2020 were Washington 26%, Oregon 12%, New York 11%, California 7%, and Alabama 4% (EIA 2021).

Conclusion

Access to the limited supply of freshwater in our increasingly crowded environment that we are continuing to do the things that m0ake the world warmer. Together, these conditions are making a water-related disaster certain. All life on the planet must have water to survive. We must do what is necessary to ensure the nation's water supply is sustainable. A sustainable supply of water is necessary for all plants and animals and, of course, the growing numbers of people on earth.

Meanwhile, the environment that dictates our need for freshwater is being attacked by the very events and conditions that make freshwater necessary. Agriculture is the second largest user of freshwater. As the climate gets warmer and drier, farmers need more water to sustain the growth of edible crops. As population continues to increase, the demand for water will increase commensurately. Thus, the two drivers of this water crisis, population increase and the climate change, are making prolonged drought more widespread. Meanwhile, changing precipitation patters are associated with more and more severe climate-driven extreme weather events bring destruction to structures and urban-related infrastructure, including the infrastructure needed to obtain, treat, and distribute water. These events range from floods in some areas and severe freshwater shortages in others. Overpopulation and unsustainable growth in some areas drive a looming water crisis, particularly in the arid areas of the Southwest. Underfunding for system maintenance and upgrades of older water distribution systems, and short-sighted or ill-informed cost-cutting decisions by political leaders most commonly in the mid-west and deep south are increasingly placing vulnerable urban residents in danger of having polluted drinking water, inadequate water pressure, or simply no water at all.

References

Alcamo, Joseph, Martina Flörke, and Michael Maerker. 2007. Future long-term changes in global water resources driven by socio-economic and climate changes. *Hydrological Sciences Journal* 52 (2): 247–275.

Alley, William M., Thomas E. Reilly, and O. Lehn Franke. 1999 (and 2013 update). Sustainability of groundwater resources. USGS Survey Circular 1186. Available at https://pubs.usgs.gov/circ/circ1186/pdf/p1-5.pdf.

Arata. Akira. 2020. Water crisis in India—issues, causes, and consequences. USPC: Union Public Service Commission reprint for the Civil Services of the Government of India. Available at https://upscfreematerials.org/water-crisis-issues-causes-and-consequences/.

Boin, Arjen. 2017. Crisis management. *Encyclopedia Britannica*. Available at www.britannica.com/topic/crisis-management-government.

CDWR (California Department of Water Resources). 2021. California's groundwater update. Available at file:///C:/Users/David/Downloads/calgw2020_highlights.pdf.

Dieter, Cheryl A., Molly A. Maupin, Rodney R. Caldwell, Melissa A. Harris, Tamara I. Ivahnenko, John K. Lovelace, Nancy L. Barber, and Kristin S. Linsey. 2018. Estimated use of water in the United States in 2015. USGS Circular 1441. Available at https://pubs.er.usgs.gov/publication/cir1441.

EIA (U.S. Energy Information Administration). 2021. Hydropower explained. Available at www.eia.gov/energyexplained/hydropower/where-hydropower-is-generated.php.

EPA (Environmental Protection Administration). 2021. A closer look: Temperature and drought in the Southwest. Available at www.epa.gov/climate-indicators/southwest.

Global Water Forum. 2015. Water for thermal power plants: Understanding a piece of the water energy nexus. https://globalwaterforum.org/2015/06/22/water-for-thermal-power-plants-understanding-a-piece-of-the-water-energy-nexus/.

Hanasaki, Naota, S. Kanae, T. Oki, K. Masuda, K. Motoya, N. Shirakawa, Y. Shen, and K. Tanaka. 2008. An integrated model for the assessment of global water resources—Part 1: Model description and input meteorological forcing. *Hydrology and Earth System Sciences* 12 (2): 1007–1025.

Hutson, Susan S., Nancy L. Barber, Joan F. Kenny, Kristin S. Linsey, Deborah S. Lumia, and molly A. Maupin. 2004. Estimated use of water in the United States in 2000. U.S. Geological Survey Circular 1268.

Lall, Upmanu, Tanya Heikkila, Casey Brown, and Tobias Siegfried. 2008. Water in the 21st century: Defining the elements of global crises and potential solutions. *Journal of International Affairs* 61 (2): 1–17.

Maddocks, Andrew, and Paul Reig. 2014. *World's 18 most water-stressed rivers*. Washington, DC: World Resources Institute, May 20. Available at www.wri.org/insights/worlds-18-most-water-stressed-rivers.

NOAA (National Oceanic and Atmospheric Administration). 2019. Drought: August 2019. National Centers for Environmental Information, September 11. Available at www.ncdc.noaa.gov/sotc/drought/201908.

Safeopedia.com. 2017. Water stress. Available at www.safeopedia.com/definition/3092/water-stress.

Trias, Angelo. 2020. Which is which? Differences between crisis, emergencies and disasters. Available at https://portal.edukasyon.ph/blog/which-is-which-differences-between-a-crisis-emergency-and-disaster.

USACE (U.S. Army Corps of Engineers). 2018. National Inventory of Dams (NID). Available at https://nid.sec.usace.army.mil/ords/f?p=105%3A1%3A%3A%3A%3A%3A%3A%3A.

USGS (United States Geological Survey). n.d. Hydroelectric power water use. Available at www.usgs.gov/special-topic/water-science-school/science/hydroelectric-power-water-use?qt-science_center_objects=0#qt-science_center_objects.

USGS (United States Geological Survey). 2018. Groundwater decline and depletion. Available at www.usgs.gov/special-topic/water-science-school/science/groundwater-decline-and-depletion?qt-science_center_objects=0#qt-science_center_objects.

World Atlas. 2022. The eight US states located in the Great Lakes region. Available at https://www.worldatlas.com/articles/the-eight-us-states-located-in-the-great.

CHAPTER 3

Water Problems and Quality Controls

Water-related disasters are the results of a complex confluence of interactions between the oceans, atmosphere, and landmasses that combine to result in either far too much or far too little water in a specific area. When this occurs, these disasters pose both direct impacts such as damage to buildings, crops and infrastructure, and loss of life and property, as well as indirect impacts that include losses in productivity and livelihoods, increased investment risk, indebtedness, and human health impacts. They can occur from too much water appearing too quickly, or they can be the droughts caused by a combination of heat and not enough precipitation. The many storms, floods, and droughts now occurring more often are expected to increase in frequency and severity by 2050 and beyond. The global warming and water-related hazards (also called hydro-hazards) evolve into disasters for the communities they strike. Climate change is increasing the prevalence of water-related hazards, which in turn are contributing to more and more severe water-related disasters. Population growth, combined with migration patterns of people in the United States, poses a significant increase in risk. These pressures expand population concentrations in areas where flooding is increasingly common. Moreover, this rapid population growth is occurring in arid regions where water scarcity is dramatically accelerating. Effects of these phenomena include unplanned urbanization and rapidly increasing suburban population, degradation of ecosystems, resistance to change irrigation practices,

D. E. McNabb and C. R. Swenson, *America's Water Crises*,
https://doi.org/10.1007/978-3-031-27380-3_3

and the continuing inaccurate public perception of risk—all adding to the damage potential from more hydro-disasters. When water disasters strike, they can include floods, landslides, tsunamis, storms, heat waves, cold spells, droughts, and waterborne disease outbreaks, all becoming more frequent and more intense (UNESCO n/d; UN Water n/d). Reducing risk to, and improving the resilience of, water and sanitation services affected during a water disaster will be key to maintaining access during a climatically uncertain future. Water disaster management can help ameliorate the danger.

Factors Shaping a Water Crisis

Taken together, the combined impact of climate change and population growth results in a looming crisis over the amount of water available for human use. As this vital resource is diminished and as demand is the increasing scarcity, the impacts will be felt across the country in an escalating fashion. As scarcity continues, the probability of conflict over access to the declining sources is expected to continue—as well as becoming more confrontational. As a result, the decisions made about the allocation of this resource will be increasingly critical for all aspects of society. Choices between various uses: economic interests, domestic consumption, political priorities, and environmental impacts of creating new sources will become center stage in the drama around water allocation decisions. It is therefore impossible to examine the diminishing water resource without addressing the challenge faced in allocating this scarce critical commodity.

Clean freshwater is, of course, necessary for human life. The same water is also a required commodity for many other ancillary activities. Some, like agricultural production, are contributory uses for sustaining life. Others such as water for recreation and aesthetic uses such as landscaping are not essential but add substantially to the quality of life. Each of the myriad uses for water in society today has advocates who have a very clear stake in assuring that they receive the water, the need and want for their use. As water becomes scarcer, the competition for each existing and new acre foot of freshwater will only increase. As competition increases, the importance of the decision-making process around the allocation of the scarce resource will increase commensurately. Therefore, water allocation decisions will be some of the most important and potentially contentious decisions society makes as climate change and population growth are on a collision course to water scarcity.

Water allocation decisions and the economic and social conflicts inherent in these decisions are certainly not new. A central feature of the population growth and settlement of the Western United States were clashes between competing interests over the right to use available water. Such old west adages as "Whisky's for drinking and water is for fighting over" were born of the deep divides in frontier society over who had access to water. They underscore just how precious the supply of water was. Water was the central most sought after scarce commodities, for without an adequate supply there could be no prosperity. Ranchers, farmers, industrialists, and indigenous peoples all had critical needs, desires, and beliefs centered on the availability of water. Political debates, lengthy court battles, and outright shooting wars have been fought over water. In such cases, the least powerful are typically quite literally "left in the dust" as those with financial and political power achieve their goal of securing for themselves water for their needs or enterprises at the expense of those less powerful. Inherent in the outcomes of such conflicts are the seeds of social discord.

At times, the conflicts are life and death battles where populations are forced to leave areas because they no longer have the supply of water they need to live. In others, the result is a business or enterprise is forced to close or relocate. Not all conflicts are as dire as these, but they are often no less contentious. No matter the severity of the need or depth of the conflict, the importance of a clear and equitable decision-making process is central to the allocation decisions.

In the 1980s, for example, the City of Bellevue Washington was faced with increasing costs for water to serve their growing city. Bellevue had been purchasing water from the City of Seattle, the regional supplier. That water was becoming more expensive, and Seattle had put in place a rate structure that was perceived as inequitable to high growth areas such as Bellevue. These cost increases and equity concerns caused Bellevue elected leaders to seek alternative water sources for their growing city. They chose to evaluate the potential of creating a water supply dam project on the North Fork of the Snoqualmie River not far from their city. The watershed was suitable from a water quality standpoint and the planning commenced in an open public process. Almost immediately the proposed project became controversial. No growth advocates saw an opportunity to slow development and mobilized against the project. Large landowners and business interests responded in kind as they saw the dam as a financial benefit. Environmental interests organized to oppose the dam as it would impound water, thereby wiping out river rapids enjoyed by kayakers, and indigenous Tribes raised concerns over the impact on traditional fishing resources. The public debate over the dam lasted for several years, and in the end, having publicly weighed the arguments on both sides, Bellevue elected leaders abandoned the proposal. This example illustrates how competing interests impact water supply and distribution decisions. This example also illustrates the importance of an open and "fair" public decision-making process where all interested parties can be heard and decision makers can make a well-informed decision.

Conflicts over water frequently have a high social cost. Due to water's essential nature, that is, as an irreplaceable basic human need, those who control the availability of water and the distribution of this scarce resource hold a great deal of social power, be it political, financial, or a combination of both. For this reason, those who control the decision-making power over the allocation of water hold tremendous sway over social and economic life. Historically, water is frequently found at the center of issues related to the desperate treatment of one group by another or those who place their economic interest's paramount to all others. Those who control the distribution of water have considerable power and influence over those who do not. Given the essential nature of water as a substance required to sustain human life, there are few elements that can engender the passion or provide the social or economic control, than the availability of clean freshwater.

Recognizing this inherent power, in the United State water is considered a "public good" and as such, the acquisition, allocation, and distribution of water have been institutionalized in public entities rather than being left solely to private market forces or companies. Where private purveyors of bottled water exist, they do so as a part of a public distribution regime that has at its heart the "public utility" concept for water acquisition and distribution. In a public utility, decisions are, by design, to be made with a greater degree of equity and are an expression of public will consistent with the goals of the jurisdiction in which they are made. Publicly elected individual's guild and oversee the operation are accountable to the broad electorate where the equity and appropriateness of their decisions can be observed, debated, and voted upon. The Bellevue example illustrates this in that the planning work for the proposed water project was publicly available, and all interested groups and individuals were able to organize and present their positions openly, both pro and con. They made their cases, and the publicly accountable elected officials deliberated the matter and decided not to pursue the project in a majority vote of an open public meeting of the City Council.

Where this system breaks down, it is usually a result of the public accountability of the utility being compromised, and/or the critical water acquisition and distribution decisions being obfuscated from public view. The Flint Michigan water crisis illustrates this point. There, the City of Flint's power to control their local operations, including the distribution of water, had been usurped by the State of Michigan through a highly unusual move by the State Legislature and Governor to suspend local elected leader's authority over the operation of the city, including the acquisition and distribution of water. Under this unusual arrangement, the Governor appointed officials to take control of the management of the city, bypassing and excluding the local elected leaders and thereby excluding public discussion and debate over decisions effecting the health and welfare of the citizens of Flint.

This move, made by State officials for the expressed purpose of controlling city finances, resulted in a series of decisions regarding the city's water supply that resulted in unhealthy and highly dangerous water delivered to the citizens through the City's water distribution system. The water was so contaminated when it reached the tap that the citizens were exposed to serious and long-lasting health impacts from the consumption of this water. The water's chemical constituency also degraded the city's distribution piping system resulting in significant damage and considerable future cost. These decisions were made without the local elected officials being able to debate, question, and properly decide the changes before they were made in an open public process, resulting in tragic consequences for the citizens of Flint. Here, the essential accountability so important to decisions about water allocation was bypassed with tragic consequences.

These two examples illustrate how difficult water allocation decisions can be and how important it is that they are made through open public discussion, debate, and questioning by accountable public officials. As difficult as these decisions have been in the past, they will only become more difficult as the water crisis deepens. Federal, state, and local leaders will be faced with the challenge of deciding which constituent interests are satisfied as the available water is distributed among the disparate and frequently competing interests. Questions of economic interest and social equity will require great care and compromise. Decision makers may be asked to decide whether it is equitable to provide water for high water requirement in food production or industrial practices while the community does not have enough water for domestic consumption. Food producers and industrial users, having invested millions of dollars in their production enterprises, will understandably lobby hard for the water they need. Underserved populations with the most compelling need for the water may have less ability to press their case to government decision makers. These difficult decisions on the allocation of water resources are likely to grow more common and more contentious as the water crisis becomes increasingly pressing.

Climate Change Research

The Global Change Research Program (USGCRP) is the Federal Government's management organization for collaborative climate change research programs. The program sponsors collaborative research in 10 federal departments, with research data shared among 13 agencies. It defines climate change as "changes in the global environment (including alterations in climate, land productivity, oceans and other water resources, atmospheric chemistry, and ecological systems) that may alter Earth's capacity to sustain life" (USGCRP 2020). The program endorses, supports, and distributes global research activities and findings that report on the state of the nation' climate assessment every four years. The USGCRP was established in 1990 with passage of the Global Change Research Act, charged with coordinating federal research and investments in understanding the forces shaping the global environment, both human and natural, and their impacts on society, to coordinate research across federal agencies to understand the human-induced and the natural processes that influence the total earth system: the atmosphere, oceans, land, water, ecosystems, and people. USGCRP emphasizes research that can be used to answer critical questions about large-scale forces of change in the earth system, and how the United States and the world can respond to reduce exposure to climate-related risks, protect public health, and strengthen economic and community vitality.

International agencies also sponsor research on climate change. An example of a global climate research study is the UN Water which reports on changes in weather pattern that have affected water accessibility in most of the world. In a 2019 UN report, changes in the global climate were occurring more consistently than expected and with extreme visible consequences from the changes in weather patterns. Also, weather crises are more frequent and no longer following what had been normal and predictable time periods. These rainstorms, very high velocity high windstorms, long droughts, and other destructive weather events associated with higher summer temperatures were apparently influencing the availability of freshwater resources. The more or less predictable availability and distribution data on rainfall, snowmelt, river flows, and groundwater were changing with the weather patterns. More and more prolonged severe droughts are now occurring in many parts of the nation. More extreme floods and storms are occurring in the South and Central sections of the country. These results in changes in water availability and quality are having an impact on public health and growing food security. Also, more weather-related storms and water stress are expected to occur with greater severity as the temperature continued to rise.

Climate-Related Disasters

Climate change impacts are increasingly felt through changing hydrological conditions including changes in snow and ice dynamics. An example of the weather changes water stresses that occur in the U.S. is the June 2021 report by the U.S. Weather Service (USWS): "Multiple rounds of widespread heavy rain, localized flash flooding, and severe thunderstorms are possible from the Central Plains into the Great Lakes today, Friday and Saturday. A dangerous heatwave with record-breaking and triple digit temperatures occurred in the Northwestern U.S. Excessive heat watches and safety warnings were issued" (NWS 2022).

Climate change's estimated influence on the number of regions projected to be water-stressed in increasing. Furthermore, weather changes causing water shortages will occur more often in already water-stressed regions is no longer just a forecast; it is occurring now. In late June 2021, for example, the U.S. Drought Monitor (USDM) reported that Claudette, the third named tropical cyclone of the 2021 Atlantic hurricane season, made landfall in the Central Gulf Coast and moved across the southeast United States. Heavy rainfall resulted in widespread improvement to drought conditions in North Carolina and South Carolina, as well as improved conditions in northern Florida and southern Georgia. Severe thunderstorms, including a Force 3 tornado that hit western suburbs of Chicago, affected parts of northern Illinois, northern Indiana, northern Ohio, and southern Michigan. Meanwhile, relatively dry weeks in both the Northeast and the West caused drought conditions to worsen, for the most part, in both regions.

Although climate change has been shown to be a significant contributor to the decline in surface water availability, it is not the only cause of stresses to the surface water supply. Overpopulation and urbanization are also significant contributors to the scarcity stress of accessible freshwater. In areas where surface water supplies are insufficient to meet the growing demand for water, individuals and water suppliers have turned to groundwater to make up for the losses. This results in resource supply stresses. Elements of a research report on climate change are included in Box 3.1.

Box 3.1 Fourth report of the study of the impact of climate change in the U.S.

The National Climate Assessment:

The Global Change Research Act of 1990 mandates that the U.S. Global Climate Change Research Program (USGCRP) delivers a report to Congress and the President no less than every four years that "1) integrates, evaluates, and interprets the findings of the Program...analyzes the effects of global change on the natural environment, agriculture, energy production and use, land and water resources, transportation, human health and welfare, human social systems, and biological diversity; and 3) analyzes current trends in global change, both human-induced and natural, and projects major trends for the subsequent 25 to 100 years." The Fourth National Climate Assessment (NCA4) fulfills that mandate in two volumes.

Volume II draws on the foundational science described in Volume I; the Climate Science Special Report (CSSR) Volume II focuses on the human welfare, societal, and environmental elements of climate change and variability for 10 regions and 18 national topics, with particular attention paid to observed and projected risks, impacts, consideration of risk reduction, and implications under different mitigation pathways. Where possible, NCA4 Volume II provides examples of actions underway in communities across the United States to reduce the risks associated with climate change, increase resilience, and improve livelihoods.

Source NCAR (2018).

Water Access Limits

The depletion of the nation's critical aquifers is growing faster than they can be replenished. Groundwater levels depend on regular recharging of rain, melting snow and other surface water sources. When there is no rain or when it comes as a flood, little of the water can be caught and used to replenish the aquifer, the water table drops. When withdrawals of groundwater exceed replenishment, water levels below ground also drop. Many aquifers, especially those which don't have abundant recharge, are affected by the amount of water being pumped out of local wells. Disagreements over the accessibility to the declining sources of freshwater for municipal water supplies, agriculture, and power production and for supporting aquatic life have evolved into an increase in the amount and the greater intensity of conflicts over limited groundwater resources.

Approximately 75% of all groundwater pumping is occurring in just seven nations: In decreasing amounts, they are India, the United States, China, Pakistan, Iran, Mexico, and Saudi Arabia. In a global effort to bring this fact to the public's attention, the United Nations declared the focus of 2020 world water day to be of the growing groundwater crisis. This book describes with the water crisis now underway and some of the steps municipal water utilities are taking to deal with the emerging groundwater crisis.

In a report from U.S. Jet Propulsion Laboratory (JPL) researchers, the world is already too late to avert a climate change crisis and its influence in the resulting global warming, but there is still hope that groundwater crisis can be successfully averted. "While reversing climate change and its impact on groundwater resources is no longer a possibility for humanity, managing our way through the global groundwater crisis is" (Famiglietti 2014, 947).

Changes in Climate Patterns

Weather patterns in the United States naturally vary slowly from decade to decade, with the occurrence of extreme events ranging at different times from year to year. Water supplies also vary from year to year. Predictions of increasingly warmer winter weather from 2020 to 2050 are that the world will continue to become warmer, with some parts of the country becoming drier and other parts wetter. More severe winter weather events are also predicted to occur. Annual predictions for seasonal weather are collected annually by the Central Intelligence Agency (CIA) from nations around the globe (Table 3.1).

Table 3.1 U.S. participation in global change research programs

Organization	*Selected focus*
Department of Agriculture (USDA)	USDA assesses climate change effects on natural and economic systems associated with productive lands; develops cultivars, cropping systems, and management practices to improve drought tolerance and build resilience to climate variability
Department of Commerce (DOC)	DOC's National Oceanic and Atmospheric Administration's (NOAA) mission is to understand and predict changes in climate, weather, oceans, and coasts; the National Institute of Standards and Technology (NIST) works with other agencies to develop or extend measurement standards, methodologies, and technologies for greenhouse gas emission inventories and climate research
Department of Energy (DOE)	DOE's Office of Science supports fundamental research to address key uncertainties in regional to global-scale earth system change arising from the interactions and interdependencies of the atmospheric, terrestrial, cryospheric, oceanic, and human-energy components of the earth system
Department of Health and Human Services (HHS)	HHS supports USGCRP by conducting fundamental and applied research on linkages between climate variability and change and health, translating scientific advances into decision support tools for public health professionals, conducting monitoring and surveillance of climate-related health outcomes, and the public health community in communication about climate change
Department of the Interior (DOI)	USGS scientists work with other agencies to provide policymakers and resource managers with scientifically valid information and an understanding of global change and its impacts with the ultimate goal of helping the nation understand, adapt to, and mitigate global change
Department of Transportation (DOT)	The Department of Transportation (DOT) coordinates with USGCRP and its participating agencies to inform transportation system mitigation and resilience solutions. DOT initiatives to improve the resilience of the U.S. transportation sector include: the Federal Highway Administration (FHWA), the Maritime Administration (MARAD) Ports Team, and the Office of the Assistant Secretary for Research and Technology (OST-R)
Environmental Protection Agency (EPA)	The core purpose of the Environmental Protection Agency's (EPA's) global change research program is to develop scientific information that supports policymakers, stakeholders, and society at large as they respond to climate change and associated impacts on human health, ecosystems, and socioeconomic systems

(continued)

Table 3.1 (continued)

Organization	*Selected focus*
National Aeronautics and Space Administration (NASA)	NASA's global change activities have four integrated foci: satellite observations, research and analysis, applications, and technology development. Satellites provide critical global atmosphere, ocean, land, sea ice, and ecosystem measures
National Science Foundation (NSF)	The National Science Foundation (NSF) addresses global change issues by support of instrumentation and facilities, developing new analytical methods, and enables cross-disciplinary collaboration with global change programs for understanding of physical, chemical, biological, human systems, and interactions
Smithsonian Institution (SI)	Smithsonian Institution's global change research is primarily conducted at the National Air and Space Museum, the National Museum of Natural History, the National Zoological Park, the Smithsonian Astrophysical Observatory, the Smithsonian Environmental Research Center, and the Smithsonian Tropical Research Institute

Source US National Park Service (2021)

An example of a typical winter forecast for the United States suggests a warm wet winter weather pattern for the nation, with most regions experiencing weather that is "mostly temperate, but tropical in Hawaii and Florida, arctic in Alaska, semiarid in the High Plains west of the Mississippi River, and arid climate in the Great Basin of the Southwest, the low winter temperatures in the northwest are warmed somewhat in January and February by warm chinook winds from the eastern slopes of the Rocky Mountains" (CIA 2022).

Projected Temperature Changes

Projected global warming changes for the U.S. West range from about 3.6 to 9 °F (2 to 5 °C) over the next century. Regional estimates for areas such as California indicate that the upper value may be as high as 12.6 °F (7 °C) for some areas (Dettinger and Anderson 2015). These areas are home to what are among the fastest growing populations in the nation. Minimum temperatures in the Basin have increased more than maximum temperatures and variability in interannual temperatures has declined. As a result, the probability of very warm years increased and very cold years declined. An estimated change in stream flows in the multi-state Great

Basins continued change in the seasonal water flows of increased winter flow, reduced and earlier spring peaks, and reduced summer and fall flows (Chambers 2008).

The temperature of the world has been increasing since the 1800s and is expected to continue to warm over this century and beyond. This conclusion is based on weather scientists' understanding of how the climate system works and on computer models designed to simulate earth's climate. Results from a wide range of climate model simulations suggest that our planet's average temperature could be between 2 and 9.7 °F (1.1 to 5.4 °C) warmer in 2100 than it is today (Herring 2021). In 2012, the Union of Concerned Scientists (UCS) published the following selected list of possible changes to the water supply expected to occur due to climate change (UCS 2011).

- Decline in drinking water—both quantity and quality—is expected for these reasons:

  Municipal sewer systems may overflow during extreme rainfall events, gushing untreated sewage into drinking water supplies.

  - Loss of mountain snowpack and earlier spring snowmelt spurred by higher temperatures reduce the availability of drinking water downstream.
  - The shrinking of mountain glaciers threatens drinking water supplies for millions of people.
  - Sea-level rise can lead to saltwater intrusion into groundwater drinking supplies, especially in low-lying, gently sloping coastal areas.

- Decline in irrigation supplies. Loss of mountain snowpack reduces the amount of water available for irrigation downstream, while earlier spring snowmelt affects the timing. Saltwater intrusion may contaminate the supply from groundwater.
- Disruptions to power supply. Lower lake and river levels may threaten the capacity of hydroelectric plants, while higher temperatures may mean that water is too warm to cool coal and nuclear power plants, leading to power brownouts.
- Increased drought in dry areas. In drier regions, evapotranspiration may produce periods of drought. The U.S. Southwest is getting drier, and relatively wet can experience long, dry conditions between extreme precipitation events.

- Expansion of dry areas. Scientists expect the amount of land affected by drought to grow by mid-century, with water resources in affected areas to decline as much as 30%.
- Extreme precipitation is likely when a storm passes through a warmer atmosphere holding more water. In warmer months, it takes the form of torrential rainstorms; in winter, blizzards are more likely.

Global Temperature Change

The global temperature has been increasing since the 1800s and will continue to warm over this century and beyond. This conclusion is based on weather scientists' understanding of how the climate system works and on computer models designed to simulate earth's climate. Results from a wide range of climate model simulations suggest that our planet's average temperature could be between 2 and 9.7 °F (1.1 to 5.4 °C) warmer in 2100 than it is today (Herring 2021). In 2012, the Union of Concerned Scientists (UCS) published the following selected list of possible changes to the water supply expected to occur due to climate change (UCS 2011).

- Decline in drinking water—both quantity and quality—is expected for these reasons:

  Municipal sewer systems may overflow during extreme rainfall events, gushing untreated sewage into drinking water supplies.

  - Loss of mountain snowpack and earlier spring snowmelt spurred by higher temperatures reduce the availability of drinking water downstream.
  - The shrinking of mountain glaciers threatens drinking water supplies for millions of people.
  - Sea-level rise can lead to saltwater intrusion into groundwater drinking supplies, especially in low-lying, gently sloping coastal areas.

- Decline in irrigation supplies. Loss of mountain snowpack reduces the amount of water available for irrigation downstream, while earlier spring snowmelt affects the timing. Saltwater intrusion may contaminate the supply from groundwater.

- Disruptions to power supply. Lower lake and river levels may threaten the capacity of hydroelectric plants, while higher temperatures may mean that water is too warm to cool coal and nuclear power plants, leading to power brownouts.
- Increased drought in dry areas. In drier regions, evapotranspiration may produce periods of drought. The U.S. Southwest is getting drier, and relatively wet can experience long, dry conditions between extreme precipitation events.
- Expansion of dry areas. Scientists expect the amount of land affected by drought to grow by mid-century, with water resources in affected areas to decline as much as 30%.
- Extreme precipitation is likely when a storm passes through a warmer atmosphere holding more water. In warmer months, it takes the form of torrential rainstorms; in winter, blizzards are more likely.

Adapting to Climate Change

Climate change has substantial impacts on both water resources demand and availability (UN-W). It is critical for water managers to understand the processes driving these changes, the sequences of the changes, their impact of their systems, and their occurrence at different locations and when they will occur in the system. These changes are likely to be an increasingly powerful driver of water availability, acting with other drivers that are already having a serious impact on its quality and availability. Increased water-related risks associated with the changes in frequency and intensity of extreme events, such as droughts, floods, storm surges, and landslides, will put additional strain on water resources management and increase uncertainty about quantity and quality of water supplies. These risks will continue regardless of mitigation measures applied over the coming decades.

Cities need to find ways to adapt to the changes that are expected and to render its water infrastructure and services more resilient in coping with new conditions and extreme weather patterns. Climate change is a complex problem that has increased the need for an integrated, multisectoral, and multidisciplinary response. Apart from the normal water domain, decision makers in other spheres must use and consume water efficiently.

The necessary adaptation to climate change is linked to water and its role in sustainable development (UN Water *Climate Change Adaptation:*

The Pivotal Role of Water). The need for adaptation to climate change is urgent. To respond to better able to cope with the effects of climate change on water supplies, it will be necessary to adapt land and water management practices that can create resilience to climate change and enhance water security. Innovative practices and implementation of strategies are needed for adaptation as well as for mitigation. Adaptation to climate change is urgent. Water plays a pivotal role in it, but the political world has yet to recognize this notion. Adaptation measures in water management are often underrepresented in national plans or in international investment portfolios. Therefore, significant investments and policy shifts are needed and should be guided by the following principles:

- Mainstream adaptations within the broader development context;
- Strengthen governance and improve water management;
- Improve and share knowledge and information on climate and adaptation measures, and invest in data collection;
- Build long-term resilience through stronger institutions and invest in infrastructure and in well-functioning ecosystems;
- Invest in cost-effective and adaptive water management as well as technology transfer; and
- Leverage additional funds through both increased national budgetary allocations and innovative funding mechanisms for adaptation in water management.

Application of these principles will require joint efforts and local-to-global collaboration among sectoral, multisectoral, as well as multidisciplinary institutions. Responding to the challenges of climate change impacts on water resources requires adaptation strategies at the local, regional, national, and global levels.

Water Cycle Changes

Extreme weather events have become more frequent and intense in many regions, resulting in a substantial increase of water-related hazards. At the same time, demographic changes are exposing more people to increased flooding, cyclones, and droughts. The impacts of recent major flooding, which have resulted in many deaths and cost billions of dollars in damages, are an indication of what could lie ahead from increased climate variability. At the opposite end, the more intense droughts experienced in the past decade, which have affected an increasing number of people, have been linked to higher temperatures and decreased precipitation. Moreover, in its Fourth Assessment Report, the Intergovernmental Panel on Climate Change (IPCC) has concluded that there is a 90% probability that the extent of drought-affected areas will increase.

Conclusion

Climate change affects the water cycle directly and, through it, the quantity and quality of water resources available to meet the needs of societies and ecosystems. Climate change can result in an increased intensity in precipitation, causing greater peak runoffs but less groundwater recharge. Receding glaciers, melting permafrost and changes in precipitation from snow to rain are likely to affect seasonal flows. Longer dry periods are likely to reduce groundwater recharge, lower minimum flows in rivers, and affect water availability for agriculture, drinking water manufacturing, and energy production. With less water available for thermal power plant cooling, production of electricity is likely to be reduced. Global warming is also melting glacial ice, resulting in rising ocean levels. Their rise is already having serious effects on coastal aquifers and the supply of water to some coastal cities. Extensive forest fires like what has been occurring in California are resulting in large-scale deforestation, which, in turn, is increasing soil erosion and depriving the soil's ability to absorb rainfall for recharging declining aquifer water.

The effect of global warming on water temperatures is expected to have substantial effects on energy flow and on the recycling of water. The composition and quality of water in rivers and lakes are likely to be affected owing to changing precipitation and temperature resulting from climate change. At the same time, changes in precipitation intensity and frequency influence non-point water supply source pollution, making the management of wastewater and water pollution more demanding and urgent. Climate change will directly affect the demand for water. For instance, changes in demands will derive from industrial and increases in the numbers of households requiring more water or from thermoelectric production and agriculture irrigation.

Throughout much of the coterminous U.S., water acquisition and distribution decisions will become ever more complex and common as scarcity increases. Public entities accountable to the broad populations they serve will be increasingly called upon to assure that the competing interests for water be heard and their positions aired and debated in open public processes, as allocation and distribution decisions frequently engender a high degree of impact on social and economic equity effecting all aspects of society.

References

Chambers, Jeanne C. 2008. Climate change and the Great Basin. Reno, NV: USDA Forest Service, Rocky Mountain Research Station, Reno, NV.

CIA (Central Intelligence Agency). 2022. Filed listing: Climate. Available at www.cia.gov/the-world-factbook/field/climate/.

Dettinger, Michael D., and Michael L. Anderson. 2015. Storage in California's reservoirs and snowpacks in this time of drought. *San Francisco Estuary and Watershed Science* 13 (2). Available at https://doi.org/10.15447/shews.2015v13issrart1.

Famiglietti, James S. 2014. The global groundwater crisis: Groundwater depletion the world over poses a far greater threat to global water security than is currently acknowledged. *Natural Climate Change* 4 (2114): 945–948.

Herring, David. 2021. Climate change: Global temperature projections. NOAA: Climate.gov. Available at www.climate.gov/news-features/understanding-climate/climate-change-global-temperature-projections.

NWS (National Weather Service). 2022. Oregon water supply summary as of February 8, 2022. Available at www.weather.gov/media/pqr/WaterSupplyOutlook.pdf.

UCS (Union of Concerned Scientists). 2011. Climate hot map: Global effects around the world. https://climatehotmap.org/global-warming-effects/water-supply.html.

USGCRP (U.S. Global Change Research Program). 2020. *Our changing planet: The U.S. global change research program for fiscal year 2020*. Washington, DC, USA. https://doi.org/10.7930/ocpfy2020.

PART II

Shape of the Water Resource

CHAPTER 4

Drought's Role in a Water Crisis

As the climate in the U.S. continues to get warmer, in large sections of the country, it is also getting dryer. When it stays dry too long, a drought can ensue. Drought has become a common contributor to water stress wherever it occurs. Because drought typically develops slower than a storm event and is often unnoticed until it is difficult to plan for and has diverse and indirect consequences, drought is called "the Creeping Disaster."

Drought in the Southwest is not a new event. People have lived through periods of extreme drought often. Similar conditions have occurred before. The Dust Bowl, the periods of severe drought that occurred during the decade of the 1930s started in 1930 and lasted for close to a decade. The damaged that resulted exacerbated the damage that occurred during the Great Depression. Thousands of acres of farmland topsoil were blown away during the combination of extended drought, unusually high temperatures, poor agricultural practices, and the effects of periods of wind erosion all contributed to what became America's Dust Bowl (McElvaine 1984). Millions of people lost their livelihood and were forced to start life again, away from the semiarid mid-western and southern plains. Parts of Texas and Oklahoma were particularly hard hurt.

The connection between climate warming and drought to life that depends on freshwater resource supplies has long been known. The United States government began recording annual changes in average temperatures as early as the late 1800s. The record since then clearly

D. E. McNabb and C. R. Swenson, *America's Water Crises*,
https://doi.org/10.1007/978-3-031-27380-3_4

shows that the U.S. has been becoming warmer and drier every year. Moreover, the report shows that the changes are occurring at a faster rate each year, even as some sections of the Continental U.S. are getting wetter. These changes are having significant effects on the nation's surface and groundwater resources. For that reason, this chapter begins with a brief review of how water and a drought are related and what a drought can do to a water resource.

Defining Drought

There are many ways to define the term "drought." The USGS defines drought as "a period of drier-than-normal conditions that results in water-related problems. When rainfall is less than normal for several weeks, months, or years, the flow of streams and rivers declines, water levels in lakes and reservoirs fall, and the depth to water in wells increases. If dry weather persists and water-supply problems develop, the dry period can become a drought" (USGS n/d).

Among the possible descriptions of drought, three types commonly encountered are hydrological drought, meteorological drought, and agricultural drought. Hydrological drought is defined as "a lack of water in the hydrological system, manifesting itself in abnormally low streamflow in rivers and abnormally low levels in lakes, reservoirs, and groundwater. It is part of the bigger drought phenomenon that denotes a recurrent natural hazard" (Van Loon 2015). When drought impacts the water supply in streams, reservoirs, and ground water, it is known as hydrological drought. Periods of drought can lead to inadequate water supply, threatening the health, safety, and welfare of communities. According to the National Drought Mitigation Center, hydrological drought results from precipitation shortfalls on the surface or subsurface water supply. This, in turn may alter streamflow, lower reservoir and lake levels, and rapidly deplete groundwater resources.

Meteorological drought is an unofficial term that is usually based on long-term precipitation departures from normal or traditional measures, but, according to NOAA (2019), there is no consensus regarding the amount, the type, or what is considered normal precipitation, or the time period that the minimum duration of the lack of precipitation that makes a dry spell an official drought.

The type of drought common in farming centers is agricultural drought. This drought occurs when there is insufficient moisture in the soil to meet the needs of a particular crop at a particular time. Water needs can vary significantly among types of crops. Among the most water-needing crops are some rice paddy varieties, some olive varieties, and some almonds; nuts like peanuts, almonds, and groundnuts can fall somewhere between moderate to high water use; tomatoes, cabbage and lettuce, and cucumber, are among some of the least water intensive crops to grow per unit of weight produced. For example, among the crops that may use the most water to produce in California, are pasture (clover, rye, other grasses), nuts (almonds and pistachios), alfalfa, citrus and subtropical fruits (grapefruit, lemons, oranges, dates, avocados, olives, jojoba), and sugar beets (Pera 2021). A deficit of rainfall over areas during critical periods of the growth cycle can result in destroyed or underdeveloped crops with planted crops that are likely to receive greatly depleted yields. Agricultural drought is typically evident after meteorological drought but before a hydrological drought. According to the U.S. Department of Agriculture, the products' most drought-affected crops are corn, soybeans, hay, cattle, and winter wheat.

Defining Drought Severity

When rainfall is less than normal for several weeks, months, or years, the flow of streams and rivers decline, water levels in lakes and reservoirs fall, and the depth to water in wells increases. If dry weather persists and surface water supply problems what was once just a dry period becomes a drought. As rainfall occurs less than normal for several weeks, months, or years, the flow of streams and rivers declines, water levels in lakes and reservoirs fall, and the depth to water in wells increases. The drought can then significantly impact the region's water resources. While the drought is occurring and for some time afterward, users of groundwater may have to turn to groundwater to supplement or replace to lost surface water. This means less water is available to recharge the aquifer. Moreover, where some recharging is possible, it may take hundreds of years to complete the process.

The two common ways to identify the levels of dryness and drought are the U.S Drought Monitor (USDM) and the Palmer Drought Severity Index (PDSI). The U.S. Drought Monitor uses five classifications: abnormally dry (D0), showing areas that may be going into or are coming out of drought, and four levels of drought: moderate (D1), severe (D2), extreme (D3), and exceptional (D4). The USDM is produced jointly by

Table 4.1 USDM drought index and PDSI range of drought severity scores

USDM	*USDM severity label*	*Palmer drought severity index*
D0	Abnormally Dry	1.0 to 1.9
D1	Moderate Drought	2.0 to 2.9
D2	Severe Drought	3.0 to 3.9
D3	Extreme Drought	4.0 to 4.9
D4	Exceptional Drought	5.0

Sources USDM (2022)

the National Drought Mitigation Center (NDMC) at the University of Nebraska-Lincoln, the National Oceanic and Atmospheric Administration (NOAA), and the U.S. Department of Agriculture (USDM 2022).

Values of the DM and Palmer Index drought severity index are shown in Table 4.1. Percent of area under drought by month are in Table 4.2. The Palmer Index is a way weather researchers collect the data and for comparing changes in the weather. The data for the scale is collated for some 20,000 or more locations in the 48 states of the Continental United States (CONUS) and at sea. Alaska and Hawaii, and U.S. possessions and territories are also monitored for drought severity.

The PDSI uses temperature and precipitation data to estimate relative dryness. It is a standardized index that generally spans −10 (dry) to +

Table 4.2 Percent of area in drought for CONUS in 2020

Month	*USDM CONUS*	*USDM All US*	*Palmer CONUS*
January	11.0	9.2	4.4
February	11.5	9.6	9.4
March	14.5	12.2	11.9
April	14.8	12.3	12.7
May	19.9	16.7	12.4
June	25.2	21.4	17.9
July	32.7	27.4	20.6
August	39.4	33.3	30.8
September	42.6	35.8	35.1
October	45.1	37.9	35.2
November	48.0	40.2	37.3
December	49.0	41.0	36.7

Source NCEI (2022) (NOAA)

10 (wet). Maps of operational agencies like NOAA typically show a range of –4 to + 4, but more extreme values are possible (DAI 2020). For example, USDM drought conditions for states in early 2022 are included in federal drought reports. They can change from week to week with changes in the local state's weather.

Defining "Normal" Patterns

The CONUS temperatures and precipitation patterns from 1961 to 1990 are generally considered to have been "normal." However, as the century-long slow warming of the atmosphere was accelerating as the Twentieth Century was closing during the 1990s, to the point where climate change was finally recognized as a major threat to the health of the world's freshwater supplies; 1999 was the third warmest year since climate records were kept in 1980s; it was also the 22nd driest year of the century. Dry conditions developed early that year, with a drought well established by April in the Northeast and the Ohio Valley. By late summer, wildfires developed in many parts of the West, and short-term dry spells hit much of the southern Plains and Southeast. It was not until early September that the remnants of Hurricane Dennis brought much-needed rain to parts of the mid-Atlantic and Northeast states.

The early drought evolved into a different crisis when two hurricanes struck the eastern seaboard. Hurricane Floyd was the strongest storm of the 1999 Atlantic Hurricane Season. Millions of coastal residents from Florida through North Carolina evacuated ahead of the storm. Floyd's death toll reached 36 in North Carolina and 57 across the United States as a whole, with total damage estimates near $6.5 billion.

During the summer of 1999, the United States experienced an intensifying drought and heat wave. The east coast was the area hardest hit by the drought, with record and near-record short-term precipitation deficits occurring on a local and regional scale resulting in agricultural losses and drought emergencies being declared in several states. According to NOAA's Rapid Response Project, the USDA estimates that the drought could cost eastern farmers as much as $1.1 billion in lost income. In addition, wildfires had developed in many parts of the Western U.S. by late summer, and short-term moisture deficits were plaguing much of the southern Plains and Southeast.

Drought in the Twenty-First Century

The new century began with drought affecting a little more than one-third of the CONUS, with the portion of severe drought somewhat larger than it had been in 1999. Compared to other droughts in the ending decade, the 2000 drought was as extreme as all the major droughts that had occurred during 1960s. The American Southeast suffered through an extremely dry summer and fall. From December 1999 to September 2000, portions of the South experienced the driest season ever recorded. Alabama and Louisiana had the driest July to September ever experienced; Texas and Oklahoma had the driest two months from August to September since records had been kept. By the end of September, severe to extreme drought stretched from the Southeastern U.S. to the Southern Plains of the Midwest, up along the Rocky Mountains into the far West. Relief came in some areas in September in the form of Tropical Storm Helene and Hurricane Gordon. Overall, the 12 months from October 1999 to September 2000 was the driest in the region's 106-year records. In the Deep South, the driest was November to September. Box 4.1, provided by the USPS, provides a clue to why climate change influences the occurrence and severity of drought in the United States and the world.

Box 4.1 Drought and the impact of climate change

"Defining drought may seem easy. If an area receives less rain or snow than expected over the course of a year, it can be classified as being in drought. The severity of drought increases over time depending on how long an area remains arid. However, there's more to the story than solely if there isn't enough rain or snow."

"Droughts don't just affect water stored in wetlands, lakes, and rivers, but also water below ground stored in aquifers and in the soil. When this groundwater gets used up, the dry ground can act like a sponge, sucking surface water straight in. The surface water-groundwater relationship gets even more complicated with snowpack. If snow melts too early in the year, water can move through the environment too quickly, causing the ground to dry up and become 'thirsty' too soon. So even if there is 'enough' water, the timing of the water may dictate whether an area is in a drought."

> "Climate change has further altered the natural pattern of droughts, making them more frequent, longer, and more severe. Since 2000, the western United States is experiencing some of the driest conditions on record. The southwestern U.S., in particular, is going through an unprecedented period of extreme drought. This will have lasting impacts on the environment and those who rely on it."
>
> *Source* USGS

Mix of Hot and Wet

The weather in 2003 was up and down with drought. The year started with a very dry January in the Southwest and Southeastern regions, bringing moderate to extreme drought close to a third of the CONUS. Much of the West and parts of the Northern Plains and Great Lakes regions were abnormally dry. Although the heat remained, the damage was reduced when the region received widespread summer precipitation, enough to reduce the drought coverage to about 20% by June before it began again, with four summer months of abnormally dry and warmer weather. By October, roughly 41% of the region was again coping with moderate to extreme drought. Winter storms helped to reduce the high drought to a little over 30% at the end of the year.

The relief for much of the northwest did not benefit from the short storm respite, however. At its peak in early summer of 2004, about 60% of the west from the Rocky Mountains to the West Coast was suffering under moderate to extreme drought conditions. For the regions west of the Mississippi River, the 1999 to 2004 drought in the Western U.S. was one of the most severe droughts of the 100 years. The five years of drought led the U.S. Department of Agriculture to declare the drought disaster status in all or large parts of Arkansas, Illinois, Iowa, Kansas, Missouri, Texas, and Wisconsin. Only some of the Southwest escaped the extreme drought, as early rain and snow in September relieved the Southwest, made it possible for just 19% of this section to miss the more extreme drought conditions.

By 2010, large sections of the Southeast, and the West, the states of the Northern Great Lakes, region and south-Central Texas had to endure very dry conditions for the year. For the next five years, large sections of CONUS endured moderate to extreme drought conditions expanded and contracted from season to season and year to year. Regions with peak drought moderate to severe drought hovered from the 20 to mid-30% range, with other sections well below these ranges.

2010 was the first in a five-year period of respite for much of the drought-ridden west and Central U.S. The weather/precipitation year began with drought in the West, sections of the Southern High Plains, and the Great Lakes regions began the year with moderate to extreme drought with as low as 4% at the start of the summer and ending about 16%. In large areas of the country, the weather was warm and wet, with 20% even experiencing record cold during February, May, and December. Among areas with the warmest weather included the Mid-Atlantic.

Temperature Still Rising

The five years from 2017 to 2021 saw an increasingly warmer climate for most of the CONUS. This was counterbalanced somewhat with very wet episodes in some months for some regions. The result was a year of both wet and dry extremes at the start of the year for more than 30% of the sections with others with very wet extremes. From January to October from 10 to 20% of the country benefited from greater than average precipitation. When years averaged for the all but the months of May, September, October, and December, 20% or more of the central and western regions suffered through very warm conditions that produced high evaporation rates for open surface water reservoirs. Over the year, the Northern High Plains were very dry during the summer and, as summer turned to fall, the drought and dryness areas expanded outward. By the end of the summer, the drought in the states in the center covered nearly 25% of the country.

2018 began with extensive drought in the Southwest, the High Plains, the Mid-Mississippi Valley, and the South and East Coast, covering nearly 28% of the CONUS. After a dry January, the percentage of moderate to extreme drought area rapidly expanded to exceed four months with 38.5%. Relief came with several months of rain in an early spring, bringing the percentage to roughly 26%, with the drought all but disappearing in the East and Southeast. However, a dry and hot summer again raised the percentage to more than 35%. Except for continuing extreme drought in the Southwest and Four-Corners regions, winter was a continuation of hot and dry/warm and wet up-and-down weather, with the year ending with 21.9% still under drought.

Although 2019 began with another dry period, the full year saw the year end with the record for the second-highest year in the 1895 2019 full period. The very wet conditions continued for the first six months of the year, with a return to dry drought conditions in the Southwest and the south Plains. A very wet November and December helped lower the high drought percentage to 11.2% remaining in the Pacific Northwest, Four-Corners and Southern Plains.

2020 was a warm year with precipitation averaging in the middle of the 100-plus year record. Table 4.2 lists temperatures in 2020. Overall, the year was the fifth warmest year and the 63rd wettest year over the period. Despite the 12-month averages records, the year continued in the pattern of variable extremes. The weather was abnormally wet in the Southeast, with North Carolina recording the second hottest year on record and 10 states across the Southeast and East Coast having the second warmest year on record. Only 4.4% of CONUS was in the moderate to extreme drought category at the beginning of the year; this grew each month, closing the year 36.7%.

Approximately 5% of CONUS was very dry during every month of the year except December. August was the hottest month, averaging nearly 39% of the country with extreme high temperatures. For all of CONUS, drought increased each month (Table 4.2).

Monthly changes in the percent of area in drought are shown in Table 4.2, with values measured using the USDM approach for CONUS only, the USDM method when all U.S. areas are included and by the Palmer Drought Index. According to the USDM method, by December 2020 nearly half of the contiguous U.S. was under drought conditions.

Varied Extremes Again

2021 was an unusually warm year during most of a year for CONUS. At least 10% of the states were fit in the very warm classification during February, April, and May. About a third of states were very warm in March and nearly sixty of the 48 states (58.3%) were very warm in June. August precipitation helped drop this to 30.8% with the remaining year ranging from 37.9% in October to 62.7% in December. The heat amplified the evapotranspiration from plans, with the loss of reservoir water during the last half of the year becoming a big problem in the West, the Great Lakes, Mid-Atlantic, and Southern Florida regions.

Areas where the weather fluctuated significantly during the year included:

- The Southwest: Dry during the first half of the year and very dry during the summer and the month of December.
- California and Nevada: Dry from February to September and November, then shifting to wet in October and December.
- Northern High Plains: Similar to California, dry during spring and summer, wet in October and December.
- The Southern High Plains: Wet from May through August, turning to dry from September and November and December.

Drought soon followed the hot and dry weather. On August 17, nearly 27% of the CONUS was experiencing extreme to exceptional drought—the largest area of the largest level of drought since 2000. The West from the Rocky Mountains to the Pacific Ocean suffered a mid-year drought characterized by minimal participation and persistent high temperatures. Overall, the 2021 water year (October 2020 to September 2021) was the fourth driest year on record. The two years from June 2019 to June 2021 was the driest 24 month period since records were kept in 1895. The Southwest, which had been in drought condition for the previous 20 years, was still from moderate to exceptional drought in early 2022. New Mexico is one of Southwest states most severely affected by the prolonged drought. Climate scientists have reported that the first two decades of the current century in the Southwest were the driest period in at least 1,200 years. And it is continuing. On May 1, 2022 residents of northern New Mexico villages were forced to evacuate as fierce winds drove what was the largest active U.S. wildfire racing toward their drought-parched mountain valley. Winds gusting over 40 miles per hour blew embers a mile ahead of the blaze, where they started new fires.

- Increased drought severity lengthens recovery time of groundwater.
- Time-lag between rain and groundwater drought in unconfined aquifers can be large.

- Drought intensity controls the time-lag when groundwater is deep.
- Land–atmosphere interactions control the time-lag when groundwater is shallow.

Challenges Ahead

As the forgoing illustrates, the increasing frequency and severity of drought in the United States and around the world is the result of the rapid climate change. Our planet is warming, and prolonged drought threatens huge portions of the country. This presents significant challenges for leaders and decision-makers in government. Both elected officials and appointed professional managers and administrators must make decisions on where, how, and when to implement water conservation practices. Some of the challenges ahead for these leaders as they face increased drought are examined in this chapter.

The states most impacted by drought are those in the southwest. Leaders there are not unfamiliar with water scarcity of course, as they are in the most arid regions of the country. Each of these states has independently developed water source methodologies, and water allocation schemes based on the amount of water they expect to have to support future growth and development. Further, each state has collaborated with surrounding states in planning for future water use and allocation. Watersheds and groundwater aquifers cross state boundaries necessitating significant formal interstate water planning agreements and compacts regarding the distribution of the scarce resource.

In this complex arrangement, the state's policy goals that often range around population growth, agricultural need, and economic development are based on assumptions leaders make regarding the availability of water. Elected and appointed leaders in each state have established water allocation plans based on assumptions made from data extrapolations of the recorded history of precipitation and groundwater tables in their locations. These assumptions then form the basis of their planning for population growth, land in agricultural production, and future economic development. Overlaying these economic priorities against the water that the assumptions predict they will have, is a foundational element in their planning. These assumptions undergird their preparedness for growth and enable them to reassure the public that there will be sufficient water to quench the thirst of current and future residents, farmers and ranchers, and industrial and technological growth and expansion.

In these dry states, each economic interest has a well-organized and politically influential constituency that press their financial priorities, and attendant requisite water needs, to the elected leaders who control government action. Government actions and decisions regarding water availability often constitute the lifeblood of their states' economic and financial interests as well as the political futures of individual lawmakers. Water availability is a critical element in all economic development decisions made by state leaders. Quite simply, without water there is no economic activity and as water resources are increasingly stressed from drought induced by climate change, it will be increasingly difficult to satisfy each potential user's desire for water. To illustrate this challenge, it is easy to envision a scenario where state government leaders are presented with information from hydrologists and climate scientists that explain and outline how rapid climate change has caused their previous assumptions about water availability to become outdated. These experts may then argue persuasively that they no longer believe that there is the amount of available water that they had previously predicted. This then sets up a scenario where the state leaders heeding this advice should change their assumptions about water availability and alter their growth plans downward to match the new lower water availability projections.

Should elected leaders come to believe and act on these new assumptions about water availability, they might be faced with the prospect that there is no adequate water to support the population growth and/or economic development of the state they have touted, campaigned on, planned for, and tethered their political fortunes. This scenario sets up a very difficult set of challenges: Logic would suggest that the leaders act to withhold land-use approvals and permits for new residential subdivisions to control the rate of growth, enact programs to support agricultural and grazing interests to switch to low water crops and practices, and urge industry to reduce water consumption, all to reduce future water demand. However, actions such as moratoriums on development, for example, are likely to be at odds with the policy priorities of the state and its leaders. Building moratoriums tends to have an instant and lasting negative impact on economic activity and growth in an area. Any governor or state legislature that attempts this would be subject to enormous pressure from the construction industry and economic interests in their states whose financial well-being depends on building homes or creating economic activity.

Similarly, should these leaders press agricultural interests to alter their crops or cease certain traditional practices such as flood irrigation, they would be met with fierce opposition. Also, high-tech industries such as computer chip manufacturing plants that use large amounts of water would likely resist any limits on availability and cool their interests in locating or expanding plants in the state. All of these interests have considerable sway over who gets elected and the policies they enact and priorities they follow once in office. The economic interests around the availability of water are immense and powerful, placing government decision-makers in the very difficult role of determining who will and importantly who will not receive the water they covet.

Conclusion

Understanding groundwater, surface water, and the integrated nature of the hydrologic system enables resource managers and policymakers to better prepare for and respond to drought (USGS n/d). Groundwater overdraft (pumping out water faster than it can be replenished or recharged) is a major problem in areas of the Western United States. It has long been a major problem in California, so much so that the California legislature passed a law requiring the end of groundwater overdraft by 2040 (Azura et al. 2019). The potential effects of reducing used agriculture groundwater with the negative impacts of warmer and dryer climate on economic and environmental factors. Mixing reclaimed water to the regional resources plan as a source for use in a prolonged drought is also recommended.

In reported findings of a 2021 study of data from 266 wells, each with at least 10 years of operating, across 19 basins across the contiguous United States by Schreiner-McGraw and Hoori Ajami found that it takes an average of "three years for shallow aquifers to recover the storage lost during the drought. This recovery time increases with higher drought severity and is influenced by the time-lag between the initiation (termination) of a meteorological drought and initiation (termination) of a groundwater drought". Schreiner-McGraw and Hoori Ajami summarized initial findings in the following list:

The connection between drought and the increase in the withdrawals of groundwater resources for agriculture irrigation are common wherever it is available. In recent years, the long-term negative impact on groundwater resources become known. Often, the nature of existing water laws enables this to occur without aquifer recharge despite the known adverse effects on future supplies. In 2018, a report on the connection on groundwater declines drought by the USGS emphasized that all future groundwater levels are dependent on recharge from infiltration of precipitation. When a long-term drought occurs, a decline of water levels also occurs. Where surface water is limited along with precipitation, groundwater resources decline, particularly so where aquifers are typically the preferred resource. The amount of water being pumped from local wells in normal periods, can be expected to increase during a drought.

As the USGS indicated in 2016, "When rainfall is less than normal for several weeks, months or years, the flow of streams and rivers declines, water levels in lakes and reservoirs fall, and the depth of water in wells declines. If dry weather persists and water-supply problems develop, the dry period can become a drought" and that can be a prolonged crisis for the nation (USGS 2016).

Elected and appointed government leaders will be increasingly challenged to make water resource allocation decisions between competing interest in their states or localities. The economic interests of the state and competing high priority interests will place decision-makers at odd with important constituencies as drought, induced by climate change, increases in severity and duration.

References

Dai, Aiguo and National Center for Atmospheric Research Staff (eds.). 2020 (Last modified 12 December 2019). The climate data guide: Palmer Drought Severity Index (PDSI). Available at https://climatedataguide.ucar.edu/climate-data/palmer-drought-severity-index-pdsi.

NCEI (National Centers for Environmental Information). 2022. U.S. Climate Regions NOAA release. Available at https://www.ncei.noaa.gov/monitoring-references/maps/us-climate-regions.

NOAA (National Oceanic and Atmospheric Administration). 2019. Drought: August 2019. National Centers for Environmental Information, September 11, 2019. Available at https://www.ncdc.noaa.gov/sotc/drought/201908.

McElvaine, Robert S. 1984. *The Great Depression: America, 1929–1941*. New York: Times Books.

Pera, Matt. 2021. These are the California crops that use the most water. *The Press Democrat*, June 24. Available at www.pressdemocrat.com/article/specialsections/these-are-the-california-crops-that-use-the-most-water/.

USDM (United States Drought Monitor). 2022. Measuring drought. Available at https://droughtmonitor.unl.edu/About/WhatistheUSDM.aspx.

USGS (United States Geological Survey). 2016. Groundwater and drought. Available at https://water.usgs.gov/ogw/drought/.

Van Loon, Anne F. 2015. Hydrological drought explained. WiresWater, April 15. Available at https://doi.org/10.1002/wat2.1085.

CHAPTER 5

Surface Water: The Fading Everyday Resource

Surface water is the water in the nation's rivers, streams, creeks, lakes, wetlands, ponds, reservoirs, and all other collections of water that touch upon the surface of the earth. Close to 74% or 237,000 million gallons per day of water from all surface water sources comes from these resources. came from surface-water sources. About 70% is freshwater; the rest can be saline or any variety of non-fresh sources. All types of surface water are open to a wide variety of pollutants and other harmful treatments. When that water is no longer available for a beneficial use, it may be said to be in, or about to be, subject to a water crisis. In an effort to cope with surface water crises, the USGS maintains an extensive body of research and collected information about the size, structure, and sustainability of surface water in different watersheds (hydrologic units). The watershed is an area of land that drains all the streams and rainfall to a common outlet such as the outflow of a reservoir, mouth of a bay, or any point along a stream channel.

Surface water bodies are managed at the watershed level. Watersheds, or drainage basins, are an area where all rain and melting snowmelt water drains to a central point like a lake, river, or other body of water. Watersheds, which can be as small as a footprint or large enough to include all the land that drains water into rivers, may grow as water from smaller watersheds flow into others, creating larger and larger watershed areas. The largest are the ones that drain into final outlets where it enters

D. E. McNabb and C. R. Swenson, *America's Water Crises*,
https://doi.org/10.1007/978-3-031-27380-3_5

a sea or an ocean. There are 18 major watersheds in the coterminous United States. Each major watershed includes a variety of smaller units. In watershed management, they are also known as drainage basins or catchments. Ridges and hills that separate two watersheds are that watershed's *drainage divide*. The watershed includes all surface water—lakes, streams, reservoirs, and wetlands—and all the underlying groundwater. It all depends on the outflow point; all of the land that drains water to the outflow point is the watershed for that outflow location. Watersheds are important because the streamflow and the water quality of a river are affected by activities, human-induced or not, occurring on the land area above the river-outflow. When water from all outside resources and flooded water bodies enters the watershed too quickly for the land to absorb it, flooding can occur. Floods can result from rapid melting of winter snows, severe thunderstorms, tropical storms, and other precipitation events. In the United States, flooding causes billions of dollars in damages and takes dozens of lives every year.

Using water for human consumption in these watersheds are not permanent as the level of pollutants are too extensive. In addition to seasonal changes, climate change researchers point to the climate change as likely causing a decline in the total quantity of these resources. In some areas, projections on the increased demand for surface water due to population increases are less strong. This is because of increases in water use efficiency appear to be offsetting projected increases in demand in agriculture irrigation and other uses (Averyt et al. 2013). Additional efficiency will be needed however, because the problem is the decline is surface water availability is already underway, and decreasing supplies are expected to continue, particularly in the Southwestern United States and the central High Plains regions.

Surface Water Crises

What can go wrong with the nation's surface water resource? A simple answer to this question, differs by where the question is asked. In much of the Western half of the coterminous United States, the answer might be we are running out of water. If the question is asked in the eastern half of the country, the answer may be more and more often, we are having more water than we can deal with. Everywhere, for many years the answer could have been the crisis is that our surface water is becoming too polluted to be used. In June of 2022, the question for the Western

United States was most pressing. That's because associated with the water loss was the world's climate is getting warmer and dryer, resulting in a mega-drought lasting more than a dozen years. The source of much of the western surface water lies in a series of continental mountain ranges, where precipitation that fell as snow was now falling as rain before the needed spring and summer seasons. When precipitation falls as snow, it is held and typically released gradually as spring and summer temperatures rise, enabling run off capture in dams and reservoirs. When precipitation arrives as rain, it tends to run off rapidly and therefore avoids capture by the system of catchments made to impound snowmelt.

In June of 2022, U.S. Bureau of Reclamation Commissioner C. C. Touton reported conditions of the agency's water storage operations in the Colorado River Basin, California Great Basin, Columbia Pacific Northwest, Missouri River Basin, the Arkansas, Gulf and North Platte River basins, the New Mexico Basin, and the basins in the Southwest. Elements of that report and other sources are included in the following pages.

One of the most important rivers that much of the Western United States depend on is the Colorado River. Prolonged drought over the course of the river has resulted in the inability of the U.S. to comply with negotiated agreements to supply guaranteed amounts of water across the border to Mexico, the ultimate end of the river. Box 5.1 focuses on the resulting demand by Mexico to renegotiate the Colorado River amounts water negotiated in 2007.

Box 5.1 Prolonged drought and U.S.–Mexico water sharing agreements

Drought in the U.S. Southwest also effects agriculture across the southern border. As early as 1848, sharing water with Mexico began what is now one of the longest standing bilateral cooperation between Mexico and the U. S. The agreements reached in 1906 and 1944 formalized water cooperation between the two countries by regulating water flows and cross-border deliveries of the Colorado and Rio Grande rivers. The Mexican Water Treaty of 1944 committed the U.S. to deliver 1.5 million acre-feet of water to Mexico on an annual basis, plus an additional 200,000 acre-feet under surplus conditions. The treaty is overseen by the International Boundary and Water Commission.

Colorado River water is delivered to Mexico at Morelos Dam, located 1.1 miles downstream from where the California-Baja California land boundary intersects the river in northwestern Mexico and Yuma County, Arizona. While this agreement usually functions regularly, lower water volumes in border rivers, with periods of intense drought, have resulted in strained relations. According to a series of USGS education reports, about 10% of the total flow of the Colorado river makes it across the border. Most of this remaining amount is used for irrigated agriculture and water for the cities of Mexicali, Tecate and Tijuana in Mexico. Satellite phots of the river show that hardly any of the flow reaches the Gulf of Mexico (USGS: *Endpoint of the Colorado River, Mexico*) Drought-related disagreements over the river flow agreements were aired in 2020 and 2021, resulting in plans to revisit the drought-reduced shared water in 2023. Meanwhile, in Mexico, the government has declared a water emergency in all northern states.

Sources USGS public domain and water education reports.

The Upper Colorado River Basin has a drainage area covering portions of Colorado, Wyoming, Utah, Arizona and New Mexico. The Continental Divide marks the eastern boundary of the basin whereas the western boundary is defined by the Wasatch Mountains. The Wind River and Wyoming Ranges from the northern border and the southern portion includes the San Juan Basin. From 1984 to 2012, total streamflow deliveries from the upper basin's outlet at Lees Ferry, Arizona, to the Lower Colorado River Basin averaged 10.3 million acre-feet/year (maf/yr.).

Baseflow accounted for nearly a third of this (2.8 maf/yr). Baseflow depends on spring and summer snowmelt. However, increasingly without the mountain snow that feeds the rivers and streams. Climate change means the world is getting hotter, changing weather patterns and increasing surface water evaporation. Baseflow deliveries to the Lower Colorado River Basin may decline overall by the end of the twenty-first century despite potential increases in precipitation and baseflow in some areas.

Disappearing Wetlands

Much of the nation's wetlands have or are fast disappearing to make room for more housing and industry needed by the nation's growing population. Much of the existing methods used for agriculture is poorly or inefficiently used. Crops that need large amounts of water are used in areas where freshwater resources are fast disappearing. Agriculture consumes more water than any other source and wastes much of that through inefficiencies. Climate change is altering patterns of weather and water throughout all of the United States, causing shortages and droughts in some areas and extreme storms and floods in others.

Managers of public water and wastewater system operators and administrators of public surface water sources must contend with two supply pressures that can significantly curtail the amount and quality of these critical natural resources. The first is the attitude of the users of freshwater that they have an absolute right to use water in any way and to what degree they desire. Second, discarding everything, from raw sewer to deadly poisons, and failing to segregate and reuse water that has not been dangerously degraded and a notion that we can let others deal with any mess we leave in the rivers or lakes. A third is the often-heard reaction to the increasing cost of treated (finished) water is the statement that "water is everywhere, a natural gift and not made. Therefore, it should be free." However, it isn't free. It is costly to collect, treat, transmit, and, once used, be cleansed after use and before it is returned to nature. A common attitude is, water, and especially surface water, is almost everywhere and free.

Polluted Water

Water pollution, inefficient water use, and population growth are contributing to a freshwater crisis. Clearly, the United States uses a lot of water and about two-thirds of that water is taken from lakes, rivers, and streams. About 70% of all the freshwater used in the United States comes from surface water sources. The other 30% is from groundwater. Less than three-tenths of 1% of total water use across the United States is recycled water, with Florida contributing the largest amount. Surface water is any body of water above ground, including streams, rivers, lakes, wetlands, reservoirs, and creeks. Surface water is used for many purposes, especially irrigation and public uses, supplying people with drinking water and for everyday uses. Only one percent of all the freshwater is surface water, including the water held behind dams and other reservoirs. It includes the rain and snow that do not infiltrate into the ground or return to the atmosphere. Of the surface water total, 87% is stored in lakes and ponds, 11% in swamps, and just 2% in rivers. The ocean is also considered to be surface water but is not included in any amounts of saline surface water that is used.

Rivers, lakes, and aquifers are drying up or becoming too polluted to use. More than half the world's wetlands have disappeared, drained, filled-in, or carelessly polluted. Agriculture consumes more water than any other source and wastes much of that through inefficiencies. Climate change is altering patterns of weather and precipitation from the Atlantic to the Pacific coast. Droughts and extreme weather events are drying rivers and lakes and killing crops and destroying wetlands. In the process, they are causing wildfires in some areas and floods in others.

Surface water is natural water flowing on many different surfaces of the earth. It includes water in rivers, lakes, streams, ponds and reservoirs, lagoons and brackish water. Lagoons, a shallow channel or pond near or communicating with a larger body of water, or a surface water body, a surface water body managed for processing sewage, or surface water pond for storage of a tainted industrial liquid, and saline ocean water are also considered as surface water but are not included as freshwater. To some degree, one or more variety of pollution is affecting all these freshwater sources. According to the World Health Organization (WHO), water pollution occurs when the composition of water is changed to the extent that makes it unusable as freshwater. When surface water comes in contact with hazardous materials, chemicals, physical mixing with saline or some brackish water, it becomes polluted.

Sources of surface water pollution are generally divided into two classes according to their origin: Point source pollution and non-point pollution. Point source pollution includes all types of contaminants that enter a water source from an identifiable source. These can be discharges from an industrial operation and oil spills, a livestock feed lot, a sewage treatment plant, a municipal storm drain, and mixed wastewater and surface runoff water that often occur during super storms, the discharge of high temperature water from a power plant, water from mining operations, and other sources. Non-point source pollution is any variety of pollution from dispersed, non-identifiable sources. Examples include pollution from large farming areas, the stormwater and melting snow and ice runoff in cities, shoreline debris, and a variety of individual, non-identifiable polluters. Non-point source pollution is the major cause of water pollution in the United States. Moreover, it is difficult to regulate because there is no single, identifiable origin (Linquip Team 2021).

Uncontaminated surface water is an important natural resource for many reasons, led by water for irrigated agriculture, for thermoelectric power generation, and for public supply that includes water for all the cities and towns in the nation. Of all these uses, the largest amount of polluted surface water is withdrawn for agriculture, followed by water used in for thermoelectric power generation. According to the U.S. Department of Agriculture (USDA), irrigation is a major user of both surface water and groundwater in the United States, accounting for approximately 80% of the nation's water use and over 90% in many Western States. The power industry uses water to cool electricity-generating equipment. In 2015, of all the water used in the U.S. about 322,000 million gallons per day was fresh and saline water for cooling power plant production and industrial purposes. From 70 to 74% of the water used for power generation are from fresh surface water and saline sources. Only 30% of the remaining water needs by the power industry is met by groundwater. However, continuing droughts in the Western U.S. have resulted in increased reliance on this limited supply.

An ongoing problem with surface water continues to be pollution that occurs from the use of treated freshwater and then returning it to the source without first being cleansed and purified. Despite the advances in protecting surface water resources that have occurred since passage of the revised Clean Water Act in 1972, significant pollution continues to be a problem in many parts of the United States. The Clean Water Act (CWA) established the basic structure for regulating discharges of pollutants into the waters of the United States and regulating quality standards for surface waters. The basis of the CWA was enacted in 1948 as the Federal Water Pollution Control Act, then reorganized and expanded in 1972 as the Clean Water Act.

The EPA spends large sums on safe water efforts. In 2020, the EPA awarded $1.6 billion nationwide in new federal grant funding for use by drought-ridden Southwestern states with the Clean Water State Revolving Fund (CWSRF), including $155,631,000 to assist the Pacific Southwest Region. The funding was for a wide range of water infrastructure projects, including modernizing aging wastewater facilities, implementing water reuse, recycling, and addressing stormwater problems. The EPA also awarded $1.07 billion across the country in new federal grant funding with the Drinking Water State Revolving Fund (DWSRF), including $191,189,000 for the Pacific Southwest Region. This funding to be used for loans that help drinking water systems install treatment for contaminants, improve distribution systems by removing lead service lines and improve system resiliency to natural disasters such as floods.

Types of Surface Water Pollutants

The many sources of sources that are the main types of pollutants that contaminate both surface and ground water. One such list identifies the following types:

- Poly and perfluoroalkyl substances
- Industrial waste of all types
- Global warming resulting in heat pollution
- Mining activities and abandoned mines, including surface ore treatment pits
- Urban development, including construction waste
- Leakage from landfills
- Leakage from sewer lines and treatment spills
- Accidental oil spills and leakage
- Underground storage leakage
- Burning of fossil fuels
- Radioactive waste
- Sewage and wastewater treatment facility spills
- Agricultural activities, including food preparation wastes
- Marine dumping
- Transportation, including exhaust carbon and fuel spills
- Construction activities, including demolitions

Stormwater Pollution

Stormwater is the broad term used to identify large quantities or rapid surge of water from rainfall, ice or snowmelt, or retention-source release. The uncontrolled surge of water can be destructive to property and take human life. As the water or melting ice flows across streets, parking lots, yards, fields, sidewalks, and factories, it carries pollutants into surface

Table 5.1 Example stormwater management practices

Practice description	
Detention basins	Basin designed to empty after a storm or snowmelt, common in rapidly developing urban areas
Retention basins/ponds	Similar to retention basins but designed for longer retention; allows for treatment before release
Infiltration basins	Similar to retention ponds but where space is limited; runoff retained for soaking into permeable soil
Infiltration trenches	Used where space is limited
Sand filters	Collects sediment and pollutants from runoff; filtered outflow collected for non-consumption use
Grassed swales	Vegetated channels instead of permanent curbs or gutters. Natural filtration removes pollutants
Vegetated filter strip	Vegetated filtered strips adjacent to roadway allows flow across vegetation
Wetlands	Retention basin and pond filtration modified to permit area use for low storm event water
Porous pavements	Modified asphalt pavements to trap and contain water on site
Water quality inlet	Modified treatment inlets to control some solids, oil and grease
Nonstructural practices	Examples included street sweeping to clear stormwater flow, parking limitations, educational programs, etc

Source Hearney et al. (2002) and McNabb (2017)

water streams and lakes, which can result in the destruction of aquatic life, fish and wildlife habitats, and food crops. In combined sewer and storm drain systems, stormwater mixes with wastewater and may be treated at a water resource recovery facility. However, the growing numbers and severity of storms has resulted in discharge of untreated sewage during severe storms and rapid snow and ice melt runoff. To better cope with these changes, stormwater management has become a requirement for cities throughout the United States. Population increases have resulted in increases in the loss of natural land cover and the expansion of impervious surfaces, preventing rainwater from soaking in or infiltrating the soil. Stormwater is managed in many different ways, some of which are listed in Table 5.1.

Box 5.2 EPA's 2020 report on sources of drinking water pollution

Various chemicals, microbes, and radionuclides can contaminate surface water and aquifers. Disinfection of drinking water has dramatically reduced the prevalence of waterborne diseases (such as typhoid, cholera, and hepatitis) in the United States. Other processes may also be used to treat drinking water depending on the characteristics of and contaminants in the source water.

Common sources of drinking water contaminants include:

- Industry and agriculture. Organic solvents, petroleum products, and heavy metals from disposal sites or storage facilities can migrate into aquifers. Pesticides and fertilizers can be carried into lakes and streams by rainfall runoff or snowmelt or can percolate into aquifers.
- Human and animal waste. Human wastes from sewage and septic systems can carry harmful microbes into drinking water sources, as can wastes from animal feedlots and wildlife. Major contaminants include Giardia, Cryptosporidium, and E. coli.
- Treatment and distribution. While treatment can remove many contaminants, it can also leave behind byproducts (such as trihalomethanes) that may themselves be harmful. Water can become contaminated after it enters the distribution system, from a breach in the piping system or from corrosion of plumbing materials made from lead or copper.
- Natural sources. Some ground water is unsuitable for drinking because the local underground conditions include high levels of certain contaminants. For example, as ground water travels through rock and soil, it can pick up naturally occurring arsenic, other heavy metals, or radionuclides.

Source USEPA 2020 Update.

The EPA's National Pollutant Discharge Elimination System (NPDES) stormwater program aids communities in regulating some stormwater discharges from three potential sources: municipal separate storm sewer systems, construction activities, and industrial activities.

Most states are authorized to implement the stormwater NPDES permitting program. EPA remains the permitting authority in a few states, territories, and on most land in Indian Country.

Population growth and the development of urban/urbanized areas are major contributors to the amount of pollutants in the runoff as well as the volume and rate of runoff from impervious surfaces. Together, they can cause changes in hydrology and water quality that result in habitat modification and loss, increased flooding, decreased aquatic biological diversity, and increased sedimentation and erosion.

Unfit for Human Consumption

An increasing amount of surface water withdrawn from sources in the United States is unfit for human consumption. Box 5.1 is an excerpt from the EPA's 2018 report on water pollution in the U.S. Research has long indicated that emerging organic pollutants (EOPs) in surface water cab adversely affect human health and wildlife. Published data has reported that between 2010 and 2016, the concentration of pesticides in the U.S. surface water was higher than in the European Union and the amount found in freshwater in China.

Freshwater withdrawn from surface water sources accounts for more than 60% of all the water delivered to American homes. But a significant pool of that water is in peril, according to the most recent surveys on national water quality from the U.S. EPA nearly half of our rivers and streams and more than one-third of our lakes are polluted and unfit for swimming, fishing, and drinking. Nutrient pollution, which includes nitrates and phosphates, is the leading type of contamination in these freshwater sources. While plants and animals need these nutrients to grow, they have become a major pollutant due to farm waste and fertilizer runoff. Municipal and industrial waste discharges also contribute their fair share of toxins. Other pollutants include all the random junk that industry and individuals dump directly into waterways.

Surface Water Pollution

When demands for surface water in a watershed exceed natural supplies, the source is said to be stressed, the watershed must then withdraw from reservoir storage, water imported from other systems, or groundwater to meet current water demands. Irrigated agriculture and thermoelectric power system cooling are major demand-side driver of water stress in the U.S. Water stress studies have found that surface water withdrawals for cooling power plants indicating that a single power plant, for example, has the potential to stress water supplies at the watershed level. Watersheds in the Western U.S. are particularly sensitive to low flow events and projected long-term shifts in flow driven by climate change.

In the late 1990s, forest researchers at the U.S., Forest Service's Southern Research Station (SRS) in North Carolina scientists participating in several national-scale assessments of climate change science and climate-related impacts. In 2005, they developed the Water Supply Stress Index (WaSSI), a hydrologic model for examining how long-term climatic changes interacting with human factors could influence water availability. WaSSI has been used to assess the impacts of changes in temperature and precipitation, land use, and human population on the relationship between water supply and demand in the southern region of the United States. Modified versions of WaSSI have been used to assess stress in other regions such as the Eastern Forest Environmental Threat Assessment Center (EFETAC) that used the stress model with monthly water supplies and carbon sequestration for watersheds and the effects of global change on water, carbon, and biodiversity at both local and continental scales across the continental United States. Another team has used the index to study regional trends and different sector demands (Averyt et al. 2013).

A Washington State University-led research team has received a collaborative National Science Foundation grant to improve understanding of how contaminants, nutrients, and naturally occurring minerals interact with the environment as they move downstream in a river. The researchers, including collaborators from New Mexico State University, University of New Mexico, and the University of Illinois at Urbana-Champaign, will develop a database of river experiments around the world as they work to improve the mathematical modeling of the transport processes. Rivers can be thought of as the kidney or liver of the planet, filtering contaminants and moving important weathered materials and nutrients, as well as contaminants through the environment, says Tim Ginn, professor in the Department of Civil and Environmental Engineering who is leading the NSF project. A wide variety of chemicals and minerals end up in river water, including runoff from agricultural fertilizers and effluent from treated wastewater that often includes remnants of common pharmaceuticals—everything from caffeine to ibuprofen. Rivers also move solutes from natural processes, such as the chemical weathering of rocks, to the ocean. As these chemicals move through river systems, they react with the environment and can have significant impacts on ecosystems.

Agriculture Pollutants

The sector contributing the largest amount of pollution to our supplies of surface water is agriculture. It is followed closely by the next two contributors, thermoelectric power generation and cities and other urban centers. Combined, these human activities are placing a significant amount of freshwater in peril. According to the U.S. Environmental Protection Agency, pollution by nutrients, which includes nitrates and phosphates, is the leading type of contamination in the freshwater sources. While plants and animals need these nutrients to grow, inefficient irrigation methods and farm waste runoff and fertilizer, become pollutants in follow-on surface water. Municipal and industrial waste discharges can contribute their fair share of toxins as well. All the random junk that industry and individuals dump directly into waterways continue to pollute the surface water resource.

Agricultural water is used in irrigation, pesticide and fertilizer applications, crop cooling and frost control. According to the United States Geological Survey (USGS), water used for irrigation accounts for nearly 65% of the world's freshwater withdrawals after the water used for thermoelectric power is excluded. There are 330 million acres of land used for agricultural purposes in the United States that produce an abundance of food and other products. Poor water quality can affect the quality of food crops and lead to illness in those who consume them. For example, the water may contain germs that cause human disease. Irrigating crops with contaminated water can then lead to contaminated food products which lead to illness when eaten. In addition to food crops, water quality can be affected by animal farms and barnyards and feedlots (CDC 2016). Table 5.2 lists the amounts of water used in agriculture applications.

Surface water runoff from local watersheds can often be collected and stored in a pond and then used to supply agricultural water needs. Individual home or farmstead water supplies seldom utilize surface water because the water quality is not satisfactory and requires some level of treatment before it is suitable for consumption. Surface runoff can often be collected and stored for irrigation during periods of lower than normal precipitation.

Discharged Heat Pollution

Eighty-nine percent of electricity in the United States is produced with thermally driven water-cooled energy conversion cycles. Thermoelectric power plants withdraw a tremendous amount of water, but only a small%age is evaporated. The evaporative or consumptive use is approximately 2.5% or 3,310 million gal per day (MGD). Hydroelectric plants use falling water rather than heat to drive generators, produce approximately 9% of the nation's electricity. Evaporative water loss from the hydropower reservoir surfaces results in some water loss from (USGS n/d).

Thermoelectric power plants account for 40% of the freshwater withdrawn every year in the U.S., roughly the same as the agriculture sector. Power plants require water for the steam cycle, ash handling, flue gas desulfurization systems, and others. About 90% of the total, are used for cooling purposes. Although most of the cooling water is not consumed, it is returned to the water source. However, these huge volumes of water withdrawn by the power sector have an impact on the ecosystem and on the water resources of a region (Martin 2015).

A 2003 study by researchers at the U.S. Department of Energy's National Renewable Energy Laboratory (NREL) reported that the water

Table 5.2 Agricultural water needs in gallons and acre-feet

Type of water use	*Daily water needed (gallons/day)*	*AnnuawWater needed (Acre-feet/year)*
Personal Use (per person)	50	0.060
Dairy cow	35	0.040
Cattle	20	0.022
Horse	12	0.013
Chickens (per 100 head)	9	0.010
Swine	1.5	0.002
Spraying	Variable	0.015 per acre
Irrigation (humid climate)	Variable	0.5 per acre
Irrigation (arid climate)	Variable	2.0 per acre

Source Pennsylvania State University Extension

consumed during evaporative cooling in thermoelectric power plants was close to one-half gallon (0.47 gal (1.8 liters) of freshwater evaporated per kilowatt hour (kWh) of electricity produced. Hydroelectric plants evaporate an average of 18 gallons (68 L) of freshwater per kWh used by the consumer. The national weighted average for thermoelectric and hydroelectric water use is 2.0 gal (7.6 L) of evaporated water per kWh of electricity consumed at the point of end use (Torcellini et al. 2003).

In 2019, 89% of electricity in the United States was produced with thermally driven water-cooled energy conversion cycles. Thermoelectric power cooling water sources include fresh and saline water from both surface water and groundwater sources; 48% of all the fresh surface water withdrawals is for thermoelectric use. Reclaimed wastewater is a supplemental source of water for thermoelectric power, especially in areas where additional water sources are needed for plant operations. Fuels used in this process include coal, natural gas, and nuclear fuels. Total withdrawals for thermoelectric power for 2015 were 133,000 Mgal/d, nearly 100% of which was withdrawn from surface water sources, predominantly freshwater. Total withdrawals for thermoelectric power accounted for 41% of total water withdrawals, 34% of total freshwater withdrawals, and 48% of fresh surface water withdrawals for all uses. Thermoelectric power withdrawals are used in either once through or recirculating systems.

Subsequent water withdrawals for a recirculating system are used to replace water lost to evaporation, blowdown, drift, and leakage (USGS n/d). Thermoelectric power plants withdraw a very large amount of freshwater, with a small%age of the water lost through evaporation. The evaporative (consumptive) use consumes approximately 2.5% of the total withdrawn or 3,310 million gallons per day (MGD). In addition, hydroelectric plants produce approximately 9% of the nation's electricity.

An example of the surface damage occurred in Michigan in 2020. The Edenville Dam on the Tittabawassee and Tobacco rivers failed on May 19 when torrential rains sent water over the surface of the dam that resulted in a catastrophic 500-year flood. The flooding reservoir water then caused the subsequent failure of the downriver Sanford Dam and catastrophic flooding in Midland County. Dow Chemical, headquartered in Midland, Michigan, was in the path of the flood. The surface water for all that distance was subject to a variety of human and animal wastes and industrial poisons. Dam failures spread these poisons long distances and into cities that are miles downstream from the dam failure.

Withdrawals of surface waters have immediate effects on the rate of streamflow at the point of withdrawal. However, the adverse effect of groundwater on surface water may take years or even decades to show up. The effects of groundwater withdrawal include the timing, rates, and locations of streamflow depletions in ways different than that of surface water on groundwater. As shown in the pollution caused by failure of the Edenville Dam in Michigan decades of discharging chemicals into and on the banks of the Tittabawassee and Saginaw rivers. To protect the nation against disasters that follow the breaching of a dam, the Federal Government carries out a comprehensive inventory of dams and their potential for failure.

Conclusion

The Clean Water Rule is the Federal EPA's guide to maintaining clean and fit natural water resource. The agency, in its explanation for adopting the rule, explains that the health of rivers, lakes, bays, and coastal waters depend on the streams and wetlands where they begin. Streams and wetlands provide many benefits to communities by trapping floodwaters, recharging groundwater supplies, filtering pollution, and providing habitat for fish and wildlife. People depend on clean water for their health: Our cherished way of life depends on clean water. Healthy ecosystems provide wildlife habitat and places to fish, paddle, surf, and swim. Our economy depends on clean water: manufacturing, farming, tourism, recreation, energy production, and other economic sectors need clean water to function and flourish (undated www.epa/cleanwaterule).

Surface water pollution from tainted sources is a major problem for city and state water agencies. Stormwater runoff from non-point source pollution, for example, is one of the most significant threats to aquatic ecosystems in the United States. As water runs over and through the watershed, it picks up and carries contaminants and soil. If untreated, these pollutants wash directly into waterways carried by runoff from rain and snowmelt increasing pollution and adding phosphorous, thereby altering and degrading the ecosystems of the receiving waters. In states where wildfires are occurring, the runoff from the fire-blackened areas has resulted in huge amounts of polluted surface water.

References

Averyt, Kristen, James Meldrum, Peter V. Caldwell, Ge Sun, Steven G. McNulty, and Annette Huber-Lee. 2013. Sectoral contributions to surface water stress in the coterminous United States. *Environmental Research Letters* 8(2-013). Available at https://doi.org/10.1088/1748-9326/8/3/035046/pdf.

CDC (Center for Disease Control). 2016. Types of agriculture and water use. Available at https://www.cdc.gov/healthywater/other/agricultural/types.html.

Linquip. 2021. 10 types of water pollution in 2021. Available at www.linquip.com/blog/types-of-water-pollution/.

Hearney, James P., David J. Sample, and Leonard Wright. 2002. Costs of urban stormwater control. U.S. EPA, National Risk Management Research Laboratory. Available at https://play.google.com/store/books/details/Costs_of_Urban_Stormwater_Control?id=q3EFWwJJGrMC&gl=US.

Martin, Anna Delgado. 2015. Water for thermal power plants: Understanding a piece of the water energy nexus. Available at https://globalwaterforum.org/2015/06/22/water-for-thermal-power-plants-understanding-a-piece-of-the-water-energy-nexus/.

McNabb, David E. 2017. *Water resource management: sustainability in an era of climate change*. London: Palgrave-Springer.

Torcellini, Paul, Nicholas Long, and Ron Judkoff. 2003. Consumptive water use for U.S. power production. National Renewable Energy Laboratory (NREL TP-550-33905). U.S. Department of Energy.

CHAPTER 6

Groundwater: The Disappearing Resource

Groundwater is the water in the water underground in the cracks and spaces in soil, sand, and rocks. It is stored underground in the confining layers and gaps in the soil. Where possible, it moves slowly through aquifer composition formed from the soil, sand, and rock sediment in the soil. In most of the land, it is freshwater that is readily purified for most uses. However, in coastal areas and where inland seas once existed, it can be saline and unfit for domestic uses without extensive desalination. Groundwater is thus a large reservoir of usable freshwater, containing more than 30 times more surface water in rivers and lakes combined. Most groundwater originates from rain or snowmelt, which infiltrates the ground and moves downward until it reaches the saturated zone, where it collects in spaces. Other sources of groundwater include seepage from surface water (lakes, rivers, reservoirs, and swamps), surface water deliberately pumped into the ground, irrigation, and underground wastewater treatment systems (septic tanks). Recharge areas are locations where surface water infiltrates the ground rather than running into rivers or evaporating. Wetlands, for example, are excellent recharge areas (Doršner 2018).

Groundwater aquifers are underground areas of porous rock that holds water and allows water to flow and be withdrawn to provide water for agriculture, industrial, municipal, and personal use. When the water is withdrawn to an excess, it becomes depleted such that it can be many

D. E. McNabb and C. R. Swenson, *America's Water Crises*, https://doi.org/10.1007/978-3-031-27380-3_6

Table 6.1 List of important U.S. groundwater aquifers and basins

Colorado Plateau System
California Central Valley Aquifer
California Coastal Basins
Coastal Lowlands Aquifer System
Columbia Plateau Regional Aquifer System
Denver Basin Aquifer
Floridan Aquifer System
Glacial Aquifer System
Great Basin Carbonate and Alluvial Aquifer System
Great Lakes Basin—Lake Michigan Basin
Hawaii Volcanic-Rock Aquifers
Middle Rio Grande Basin
North and South Carolina Atlantic Coastal Plain Aquifer System
Northern Atlantic Coastal Plain Aquifer System
Northwest Volcanic Aquifer System
Ozark Plateaus Aquifer System
Pennsylvanian and Mississippian Aquifer and Appalachian Plateau Systems
Southwest Alluvial Basins.
Transboundary Aquifer
Williston and Powder River Basins

Source USGS

years or even centuries before they are recharged. Aquifers can be consolidated or unconsolidated. Unconsolidated aquifers consist primarily of sand and gravel that allow water to move through, whereas the major consolidated aquifers are limestone and dolomite (carbonates) and sandstones that can bar water flow. Unconsolidated aquifers are porous and permeable enough to accept, store and allowed to flow through in usable quantities (Table 6.1).

Aquitards are the confining layers and consist of materials with low permeability, which inhibit groundwater movement. Examples are a layer of clay or of a consolidated shale layer. An aquitard over an aquifer may limit recharge to the aquifer but may also protect the aquifer from surface contamination. Where an aquitard directly overlies an aquifer, the water in the aquifer is confined because the aquitard prevents or restricts upward movement of water.

Artesian aquifers are water stored in confined aquifers and where underground pressure exceeds atmospheric pressure, allowing the water to overflow the surface soil. Wells in confined aquifers have water levels that rise above the water-bearing formation until the local hydrostatic pressure in the well is equal to the atmospheric pressure. These water sources may or may not be where water flows out of the wells without any mechanical force behind it.

Water in an unconfined aquifer exists under atmospheric pressure; and wells that are completed in such aquifers have water levels that correspond to the local water table. An unconfined aquifer is known as a table aquifer water. The static water level is the level of water in a water well that is not influenced by pumping. The potential surface height is the level to which water will rise in tightly cased wells. A water table map shows spatially as groundwater is pumped from water wells, where there is usually a localized drop in the water table around the well which is known as a cone of depression. When there are a large number of wells that have been pumping water for a long time, the regional water table can drop significantly. This is called groundwater mining, which can force the drilling of deeper, more expensive wells that commonly encounter more saline groundwater. Rivers, lakes, and artificial lakes (reservoirs) can also be depleted due to overuse. Some over-withdrawn rivers and groundwater sources are dry in some years.

Another water resource problem associated with groundwater mining is saltwater intrusion, where over-pumping of freshwater aquifers near ocean coastlines causes saltwater to enter freshwater zones. The drop of the water table around a cone of depression in an unconfined aquifer can change the direction of regional groundwater flow, which could send nearby pollution toward the pumping well instead of away from it. Finally, problems of subsidence (gradual sinking of the land surface over a large area) and sinkholes (rapid sinking of the land surface over a small area) can develop due to a drop in the water table distribution of water levels in wells in an unconfined aquifer.

Aquafer Under Extreme Stress

In the United States, some 60% of irrigation relies on groundwater. As in 2010, water withdrawals in four States—California, Texas, Idaho, and Florida—accounted for more than one-quarter of all fresh and saline water withdrawn in the United States in 2015. California accounted for 9% of the total freshwater withdrawals for all categories of use; Texas withdrew 7% of the total; Idaho, 6%; Florida 5%; and Arkansas, 4%.

Throughout the nation, more groundwater is withdrawn annually for agriculture than is returned to the aquifers. The result is a *crisis of groundwater depletion*.

Many areas of the United States are experiencing groundwater depletion. Aquifer overexploitation is an ongoing threat to the continuation of crop production. Groundwater depletion in the irrigated High Plains of the Midwestern U.S. and California Central Valley accounts for 50% of all the groundwater depletion now underway in the United States. Other aquifers where very high depletion is occurring include the Mississippi Embayment, the Pacific Northwest's Columbia Plateau, and the Southwestern Desert Aquifer, where the deeper groundwater are now accessed by high-capacity large-diameter drilled wells and turbine pumps (Sharp 2022).

The Columbia Basin Aquifer System

The Columbia Plateau Regional Aquifer System (CPRAS) covers an area of about 44,000 square miles of the Columbia Plateau and extends along the Washington/Oregon/Idaho border (Fig. 3.20). A series of aquifers over basalt that formed some 17 billion years ago, and which today form the Columbia River Basalt Group (CRBG). More than 350 lava flows have been identified and individual flows range in thickness from 10 to more than 300 feet; interbeds occur in places where sediment accumulated between flows.

A 2015 USGS study of the Basin's water tables was by Kahle and Vaccaro identified extensive water withdrawals in large sections of this three-state system. Groundwater levels in these sections have declined over a range of more than 10,000 square miles (about 23% of the CPRAS). The declines are a result of pumping at rates that exceed groundwater recharge. Areas with large and widespread declines are located in the northern part of the study area (referred to as the Odessa Subarea), in parts of the Yakima River basin in Washington, in the Pullman-Moscow area in Washington and Idaho, and in parts of the Umatilla River basin in Oregon. These declines are in areas known to rely heavily on groundwater for irrigation and other uses. In contrast, about 5% of the CPRAS has experienced groundwater level rises due to the delivery and application of surface water for irrigation within the large Bureau of Reclamation irrigation projects.

Groundwater supply management in the CPRAS is crucial for maintaining the supply of water resources due to the high water demand for agriculture, economic development, and ecological needs and the extensive competition for what is a limited and declining resource. Water-resource issues that have implications for future groundwater availability include:

- Widespread water-level declines associated with the development of groundwater resources for irrigation and other uses;
- Decrease in base flow to rivers and associated effects on water temperature and quality;
- Limited availability of non-appropriated surface water;
- Potential capture of surface water, which was appropriated through senior water rights by pumpage of groundwater, which was appropriated through junior water rights; and
- Current and projected effects of climate change and variability on increasing pumping demand, groundwater recharge, base flow in rivers, and ultimately, sustainable groundwater yields.

Ongoing activities in the region for enhancement of fisheries and obtaining additional water for agricultural, municipal, and domestic use are affected by groundwater withdrawals and rules implemented under the Endangered Species Act for numerous stocks of salmonids. The study addressed some of these groundwater availability issues by improving the understanding of the hydrogeologic system, the status and trends of the groundwater system, the general relation between groundwater and surface waters, current water use, and the water budget for the CPRAS.

Groundwater-Surface Water Interactions

Groundwater does not always remain deep underground. There is a significant relationship between groundwater and surface streams and lakes. The two are interconnected in two ways: Streams and rivers recharge aquifers and in turn, groundwater replenishes surface water bodies. Studies by the USGS have found that groundwater's contribution to annual streamflow volume can be as large as 90% in some locations and times. When dams fail, they most often spread polluted water and toxic chemicals for miles past the failure. Both soils and aquifers are recipients of that pollution (Table 6.2).

Table 6.2 States with estimates of the most irrigated agriculture fields

Rank	*State*	*Irrigated acres (averages)*	*Percent of irrigated estimate*
1	Nebraska	8.3 million	14.9
2	California	7.9 million	14.1
3	Washington	6.4 million	9.3
4	Arkansas	4.8 million	8.6
5	Texas	4.5 million	8.0
6	Idaho	3.4 million	6.0
7	Kansas	3.1 million	5.2
8	Colorado	2.6 million	4.5
9	Montana	2.0 million	3.4
10	Mississippi	1.8 million	3.0
11	Oregon	1.7 million	2.9
12	Florida	2.1 million	2.7
13	Wyoming	1.4 million	2.6

Sources Variety of published state and federal reports and range of dates 2005–2020

The Overused Aquifers

Agriculture is the major user of ground and surface water in the United States, accounting for approximately 80% of the nation's consumptive water use and more than 90% in many western states. Consumptive water use is the amount of water used by the vegetation in a given area. It is not directly returned to the soil. It is measured by amount of water consumed by a crop for its growth, the building of plant tissues plus evaporation from soils and precipitation that is not absorbed in that growth. Consumptive use varies with temperature, humidity, wind speed, topography, sunlight hours, method of irrigation, moisture availability.

Efficient irrigation systems and water management practices can help farm profitability in an era of increasingly limited and more costly water supplies. Improved on-farm water management, combined with institutional measures improving watershed-scale water-management, including the use of conserved water rights, dry-year water banks, option water markets, and regulated irrigation withdrawals, may also reduce the impact of irrigated production on offsite water quality while conserving water for growing nonagricultural demands. Irrigation withdrawals of freshwater were 118 billion gallons per day (BMG) in 2015, an increase of 2% from 2010 when it was 116 BGD. These amounts continue to be approximately equal to withdrawals estimated in the 1960s. While the effectiveness of agricultural conservation programs in supporting water conservation and environmental policy goals may vary with local hydrologic conditions, they have succeeded in controlling the rate of water consumption used for agricultural irrigation. The USDA's Economic Research Service research, data, and information program studies agriculture water allocation, conservation, and management issues associated with water scarcity challenges facing irrigated agriculture in a changing water environment and monitors the type, size, and location of irrigated farms and the legal and institutional measures governing water use (USDA 2019).

Agricultural irrigation includes water used before, during, and after growing seasons to suppress dust, prepare fields, apply chemicals, control weeds, remove salt from root zones, protect crops from frost and heat, and harvest crops. Water use for agricultural practices varies throughout the year and depends on many factors, such as weather patterns, other land uses within the watershed, crop type, evolving technologies and practices, and cost. The variability of agricultural water use may continue as weather and climate patterns shift, thus changing water availability and demand in some areas (EPA 2018).

Types of Irrigation Systems

The major types of irrigation systems used throughout the world are (CDC 2016):

- Rain-fed farming: the natural application of water to the soil through direct rainfall.
- Surface irrigation: furrow or border irrigation; basin (similar to flood irrigation) as in rice paddies. Water is distributed over and across land by gravity or natural flow, no mechanical pump involved.
- Flood irrigation: Water is turned into a field without any flow control such as furrows, boarders, or corrugations. This is the least efficient, least uniform, and least effective method of irrigation, with much water wasted.
- Localized irrigation: Water is distributed under low pressure, through a piped network and applied to each plant.

- Drip irrigation: Localized irrigation; drops of water are delivered at or near the root of plants. In this type of irrigation, evaporation and runoff are minimized; one of the most efficient user of water.
- Sprinkler irrigation: Water is distributed by overhead high-pressure sprinklers or guns from a Central location in the field or from sprinklers on moving platforms. More efficient than surface irrigation; less efficient than drip irrigation.
- Center pivot irrigation: Water is distributed by a system of sprinklers that move on wheeled towers in a circular pattern. This system is common in flat areas of the United States.
- Lateral move irrigation: The sprinklers move a certain distance across the field and then need to have the water hose reconnected for the next distance. This system tends to be less expensive but requires more labor.
- Sub-irrigation: Water is distributed across land by raising the water table, through a system of pumping stations, canals, gates, and ditches. This type of irrigation is most effective in areas with high water tables.
- Manual irrigation: Water is distributed across land through manual labor and watering cans in a labor intensive system.

Aquifer Depletion

Aquifer depletion describes the falling water tables as a result of sustained groundwater pumping without sufficient recharging. Groundwater is the sole source of water supply for much of the world, including significant parts of the United States. Groundwater supplies the water for half of the population of the United States and nearly all the rural population, where self-supplied wells are the main water source. Supplies of well water for crop irrigation results in more than 50 million gallons every day withdrawn from already overdrawn aquifers.

Water for agriculture that is withdrawn faster than nature can resupply it is taking irreplaceable water from wells in most of the high plains of the United States, the Middle East, Asia, South America, and Asia. As world population and urbanization continues to grow, water withdrawals are increasing at an even faster rate. For many years, groundwater depletion has been a concern in the California Central Valley, the American Southwest, and the High Plains from North Dakota to Texas. However, increases in groundwater withdrawals have overstressed aquifers in many areas of the nation, not just in arid regions. The long drought in California has resulted in many new wells dug and the agriculture water consumption rate shifted from 40% groundwater to 60% groundwater in the state. California Central Valley farm well drillers that used to strike good water at 500 feet deep must now drill down 1,000 feet or more, at a cost of more than $300,000 for a single well (Dimick 2014).

California Aquifer Example

The crises occurring in California's Central Valley Aquifer System is an example of what over withdrawals of groundwater can do to a groundwater system. Discovery of gold in the Sierra Nevada and the subsequent proliferation of hydraulic mining operations provided the impetus for the construction of a surface-water diversion system that consisted of hundreds of miles of canals used to transport water to where it was needed for gold-washing operations. This was the beginning of the valley's modern-day aqueduct system, which has become vital to the agricultural economy.

Fertile soil, favorable climate, abundant water, and rapid population growth in the Central Valley encouraged the development of agriculture, which soon became one of the major industries of California. Surface water satisfied most irrigation needs until the late nineteenth century, when a rapid increase in irrigated acreage produced a demand for water that exceeded the surface-water supply, and groundwater supplementation became necessary; the drought of 1880 was a major stimulus for groundwater development. Wells were used to supplement less dependable surface-water supplies and to provide water where surface-water diversion canals had not been constructed. Shallow groundwater was obtained easily in 1880, and artesian pressure was sufficient to produce flowing wells in much of the valley. After 1900, groundwater gradually became a more significant part of the total irrigation supply and, eventually, the large number of wells reduced artesian pressure to such an extent that it became necessary to install pumps in order to obtain water.

The invention of the deep-well turbine pump around 1930 allowed withdrawals from greater depths, which encouraged further development of groundwater resources for irrigation. Withdrawals increased sharply during the 1940s and 1950s and averaged about 11.5 million acre-feet per year by the 1960s and 1970s, which was approximately 20% of the total irrigation withdrawals for the United States at that time. Withdrawals reached a maximum of 15 million acre-feet per year during 1977, a drought year.

During the 1960s and 1970s, withdrawals greatly exceeded recharge, and water levels declined precipitously, as much as 400 feet in places. The declines caused a major reduction in the amount of groundwater in storage and resulted in widespread land subsidence, mainly in the western and southern parts of the San Joaquin Valley.

Increased rainfall and construction of additional surface-water delivery systems halted most of the serious water-level declines after 1977, and water levels recovered to pre-1960 levels. The network of aqueducts in the Central Valley is usually sufficient to provide one-half or more of the water needed for irrigation in years of average or above-average precipitation. In dry years, however, reliance on groundwater supplies is greater, and aquifers are again subject to withdrawals in excess of recharge during a severe drought.

The Central Valley has become one of the most important agricultural areas in the world. No single region of comparable size in the United States produces more fruits, vegetables, and nuts. More than 7 million acres are currently under irrigation. During 1985, crop irrigation accounted for 96% of the surface water and 89% of the groundwater withdrawn in the Valley.

System Climate and Drainage

The climate of the Valley is Mediterranean and Steppe, characterized by hot summers and mild winters, thus allowing for a year-around growing season; at least one crop is under cultivation at all times. About 85% of the precipitation falls from November to April. Most of the precipitation that falls on the valley floor evaporates before it can infiltrate downward to become recharged. Much of the moisture that moves inland from the Pacific Ocean is intercepted by the Coast Ranges, so that annual precipitation in the valley is relatively low. Annual precipitation decreases from north to south, with an average of about 23 inches in the northern part of the Sacramento Valley, to about 6 inches in the southern part of the San Joaquin Valley. Rainfall amounts vary greatly from year to year. Annual precipitation is exceeded by potential evapotranspiration throughout the entire valley, which causes a net annual moisture deficit.

In contrast, the mountains that surround the Central Valley intercept moisture from eastward-moving weather systems and have, except in severe drought years, an annual surplus of moisture in the form of rain and snow. Precipitation can exceed 80 inches annually in the Sierra Nevada range. Annual runoff from rainfall and snowmelt is approximately 32 million acre-feet. Most of the runoff originates in the Cascade Range and the northern Sierra Nevada. This water flows to the valley in perennial streams and provides nearly all the average annual 12 inches of recharge the valley aquifer system receives. Runoff from the Coast Ranges is principally on the western slopes to the Pacific Ocean.

Structure of California Aquifer System

The California Central Valley system is shaped into five subbasins that are, from North to South, the Sacramento Valley system, the Delta and Eastside Streams, the San Joaquin Basin, and the Tulare Basin. The Sacramento Valley is in a trough of about 400 miles long and from 20 to 70 miles wide, including more than 20,000 square miles of exceptional agricultural land. The trough is filled to great depths by marine and continental sediments, which are the result of millions of years of inundation by the ocean and erosion of the rocks that form the surrounding mountains. Sand and gravel beds in this great thickness of basin-fill material form an important aquifer system. From north to south, the aquifer system is divided into the Sacramento Valley, the Sacramento San Joaquin Delta, and the San Joaquin Valley subregions, on the basis of different characteristics of surface water basins. The Valley is considered one of the most important agricultural areas in the world. No single region of comparable size in the United States produces more fruits, vegetables, and nuts. In 1995, more than 7 million acres were under irrigation. During 1985, crop irrigation accounted for 96% of the surface water and 89% of the groundwater withdrawn in the Valley.

Columbia and Colorado Plateau Examples

The Northwest Volcanic Aquifer area lies in the middle of a broad volcanic region of the Northwest, referred to in other sources as the Northwest Volcanic Province. Both the Columbia Plateau Aquifer and the Colorado Plateau Aquifer systems are included in this northwest Volcanic Aquifers range. This large region includes multiple hydrogeologic provinces such as the Columbia Plateau, the Cascade Range, the Modoc Plateau, the High Lava Plains of Oregon, the volcanic Snake River Plain, and the northwest corner of the Basin and Range Physiographic Province. The boundary between the Northwest Volcanic Province and the Basin and Range Province is not well-defined. For the USGS, it is available for study purposes as the transition in northwestern Nevada from the volcanic dominated areas to the predominant basin and range structures of the Great Basin to the south. Most regional groundwater flow in the Northwest Volcanic Province occurs in volcanic deposits (largely lava flows) or volcanic-derived basin-filling sediments (USGS n/d).

Both the Columbian Plateau and the Colorado Plateau aquifers are located adjacent to mountain ranges and benefit from centuries of natural groundwater charging from melting snow and ice. The Columbia Plateau covers ground east of the Cascades and west of the Rocky Mountains. This aquifer system covers an area estimated to range from 44,000 than 60,000 square miles within the Columbia River Basin in Washington, Oregon, and Idaho. More than 10 million acres support a $6 billion per year agricultural industry, leading the nation in production of apples and nine other commodities. Groundwater availability in the aquifers of the area is a critical water resource management issue because this water demand for agriculture, economic development, and ecological needs is high (USGS 2010). Water supplied from the Columbia River system and the basalt and overlying unconsolidated aquifers is essential to nearly 2 million acres of irrigated land. The system of aquifers is also an important water source for meeting the needs of growing rural and urban populations and ecological needs in the region. Groundwater availability is also a primary consideration for balancing the use of surface water and groundwater supplies throughout the three-state region.

- Increased drought severity lengthens recovery time of groundwater.
- Time-lag between rain and groundwater drought in unconfined aquifers can be large.
- Drought intensity controls the time-lag when groundwater is deep.
- Land–atmosphere interactions control the time-lag when groundwater is shallow.

Columbia Plateau Aquifer System

Most groundwater used in the Columbia Plateau is stored in layers of basalt collectively known as the Columbia River Basalt Group, a complex aquifer system made up of hundreds of individual basalt flows. Groundwater pumping has removed water from storage in the basalt aquifers and resulted in maximum water-level declines of 300 feet and declines of more than 100 feet over extensive areas, placing important agricultural sectors at risk. Groundwater depletion in some areas of the aquifer that have resulted in negative environmental conditions, including streamflow in many subbasins that is inadequate for fish listed under the Endangered Species Act. Declining groundwater levels have measurably reduced streamflows and have contributed to loss of wetlands and degradation of aquatic habitat. Groundwater availability issues in the basin include: (1) widespread water-level declines caused by pumping, (2) reduction in base flow to rivers and associated effects on temperature and water quality, and (3) effects of global climate change on recharge, base flow, and groundwater availability.

The Columbia River Plateau supports a $5 billion/year agricultural industry, leading the nation in production of apples and nine other commodities. Groundwater availability in the Columbia River basalts and the basin-fill sediments of the Columbia Plateau Regional Aquifer System (CPRAS) is a critical water-resource management issue in the Basin where the water demand for agriculture, economic development, and ecological needs is high. Groundwater availability is a primary consideration for balancing the conjunctive use of surface water and groundwater supplies throughout the CPRAS. Groundwater depletion also has contributed to adverse environmental impacts. Declining groundwater levels have significantly reduced stream flows and have contributed to loss of wetlands and degradation of aquatic habitat.

Current streamflow in many subbasins is considered inadequate for several fish breeds that are listed under the Endangered Species Act. Conversely, in some areas of the CPRAS excess recharge from surface water irrigation over the past century has resulted in additional groundwater in both the sediments and basalts that is potentially available for use. The excess "artificial" recharge has created wetlands now dependent on continued irrigation and has contributed to landslides and excess sediment loads in the Columbia River at the Hanford Reach National Monument.

Groundwater levels in the CPRAS have been altered by extensive irrigation practices in the region. Diverted surface water or pumped from streams or reservoirs for irrigation has had a diverse impact on the region's groundwater. Surface water diverted to agriculture use have increased the recharging of the aquifer in some locals, causing some water levels to rise. Groundwater level rises of as much as 300 feet have been recorded in some locations, resulting in waterlogging in places. On the other hand, extensive withdrawals of groundwater in other areas of the water system have resulted in water-level declines of as much as 300 feet in Oregon and 150 feet in Washington (USGS 2000). Isolated aquifers in the eastern part of the Columbia Plateau system in Idaho have been generally limited to domestic, limited commercial and small public supplies in lowland areas. Groundwater resource development in much of this section is sparse with rugged lands and small populations.

Groundwater Management

The states supplied from this aquifer system play a significant role in the management of water resources within their boundaries. Most states water management processes developed with the contribution of various federal agencies. For example, in Washington state, The Washington Department of Ecology (Ecology) has primary responsibility for managing the state's water resources. Ecology gathers data and information needed to support agency water management decisions and long-term planning through a variety of environmental monitoring activities. As part of these activities, Ecology maintains a network of groundwater monitoring stations in each of the agency's four regions. Information from these networks is compiled annually to assess both the status in current year conditions and longer-term groundwater level/storage trends in state aquifers. The report summarizes the results of the most recent evaluation in spring 2016, which turned out to be an unprecedented drought year.

A goal of the periodic measures of groundwater depth is to see how well aquifers bounced back or were affected by drought conditions (WAECY 2016). Groundwater measurements both the Western and Eastern sections of the state were found to have mixed aquifer status, as exemplified in the report section in Box 6.1.

This system of associated aquifers extends by different amounts from a central location where the Four Corners of Utah, Colorado, Arizona, and New Mexico meet. The surface water that spring from this region are over-allocated in serving more than 35 million people and 4.5 million acres of mostly irrigated farmland.

The Colorado Plateau Aquifer System

The 2021 principle Aquifer Studies report by Degan and Musgrove produced the most current information about the Colorado Plateaus aquifers. Earlier studies have found that the system extends to about 141,000 square miles that include sections of southwestern Wyoming, eastern Utah, western Colorado, northwestern New Mexico, and northeastern Arizona. Groundwater from this system supplies drinking water for many of the 1.2 million people living in the region. The aquifer system ranks 28th in the nation as a source of groundwater for public supply, with about 102 million gallons per day pumped for this use.

Many of the aquifers in this system are regularly tapped for agricultural irrigation, which in earlier studies was reported to be about 81.9 million gallons (Maupin and Barber 2005; Kingsbury et al. 2021). Land use overlying the aquifers is composed primarily (97%) of natural land cover consisting of regional wild plants that grow over the ground and protects it from erosion, with a few small areas of cultivated pasture and agricultural land (1.9%). Urban and other developed land accounts for just one percent. The two major urban areas overlying the aquifer in 2021 are Flagstaff, Arizona, with a metropolitan area population of about 145,382, and Grand Junction, Colorado, with a metropolitan area population of about 156,728 (NPS 2021).

The Colorado Plateaus aquifer system consists of four stratigraphically distinct regional aquifers: the Uinta-Animas, Mesa Verde, Dakota–Glen Canyon, and Coconino–De Chelly aquifers. These are made up of several local aquifers in different formations, primarily sandstones and conglomerates, siltstone, carbonate rocks, and shale layers. Fractures, faults, and solution cavities in carbonate rocks locally increase transmissivity. The sedimentary rocks making up the Colorado Plateaus aquifers are thousands of feet thick, offset by faults, and variably uplifted or subsided. The geologic structure and topography (mountains and deeply incised canyons) add to the complexity of the aquifers (Kingsbury et al. 2021).

Groundwater quality in the Colorado Plateaus aquifers was evaluated by the USGS in 2017 by sampling 60 public supply wells. Sampling sites were spatially distributed across the extent of the aquifers: Arizona (15 wells), New Mexico (10 wells), Utah (19 wells), Colorado (11 wells), and Wyoming (5 wells). Parts of the region are sparsely populated, and public-supply wells are not evenly distributed. Wells were completed at depths from 100 to 3,636 feet with a median of 565 feet, and open intervals were

10 to about 1,790 feet. Water-quality data collected from the spatially distributed wells are representative of water quality in the study area.

Box 6.1 Washington groundwater level status, Spring 2016

Each spring, Ecology Water Resources measures groundwater levels in selected wells across the state. These measurements provide a point-in-time snapshot of groundwater conditions prior to the start of the annual irrigation season. In spring of 2016, Ecology measured 212 wells. Seventy-four percent (or 157) of these wells were located in the greater Yakima and Columbia basins of Eastern Washington. The remainder were located in Western Washington.

Sixty two percent of the Ecology measured wells in Western Washington (34 of 55) had sufficient data to calculate summary statistics. Thirty of these wells (or 88%) entered spring 2016 with normal-to-above-normal groundwater levels, and 20 set new minimum (high) spring water level depths this year. The above-normal groundwater levels is related to the past winter's above normal rainfall. Four wells on the Sequim-Dungeness Peninsula exhibited below normal water levels this spring, despite this winter's above normal rainfall. All of these wells have a history of longer-term water level decline, however.

Approximately 71% of the wells in Eastern Washington (112 of 157) had sufficient data to calculate summary statistics. Forty-five of these wells (or 40%) exhibited normal-to-above-normal groundwater levels relative to typical spring conditions. The remaining 67 wells exhibited below normal groundwater levels, and 42 wells declined sufficiently to set new spring maximum water level depth values. Most of the Eastern Washington wells that set new maximum low water level values this spring are in areas of significant groundwater pumping for irrigation. A majority of these well also have a long history of water level declines.

Source WAECY (2016)

The Colorado Aquifer in Utah

The Colorado Plateau aquifers underlie a broad area of regional uplift in southeastern and south-Central Utah. They are characterized by essentially flat-lying sedimentary rocks, the age of which ranges from 540 to 66 million years old. The rocks are flat, lying in most places due to a lack of major faulting and folding in most of the region. The aquifers vary in rock type, degree of cementation, and hydraulic characteristics. The space (porosity) between sand grains, vesicles in volcanic rocks such as what characterize this part of Utah is often low due to compaction and layer cementation during consolidation. As a result, secondary porosity—the joints, fractures, and bedding planes—transmits the majority of groundwater flow. Aquifers of the Colorado Plateau are often interbedded with relatively impermeable fine-grained rocks like shale or siltstone, causing confined conditions. In confined conditions, the water is under pressure so that the water level in a sealed well would rise above the top of the aquifer layer (USGS 2021).

USGS research suggests that small amounts of groundwater can be obtained from wells throughout most of Utah, but large amounts that are of suitable chemical quality for irrigation, public supply, or industrial use generally can be obtained only in specific areas. Most wells in Utah yield water from unconsolidated basin-fill deposits, and most are in intermountain basins that have been partly filled with rock materials eroded from adjacent mountains.

A small percentage of wells in Utah yield water from consolidated-rock (bedrock) aquifers. Consolidated rocks that have the highest water yields are basalt, which contains interconnected vesicular openings (volcanic rocks that contain holes formed by gas bubbles in lava), fractures, or permeable weathered zones at the tops of lava flows. Also porous is limestone that contains fractures or other openings enlarged by water, and sandstone that contains open fractures. Most wells that yield water from consolidated-rock aquifers are in the eastern and southern parts of the state in areas where water cannot be obtained readily from unconsolidated deposits.

The Aquifer in Colorado

The use of groundwater in Colorado for public supply and domestic and industrial purposes began before 1900. Groundwater resources currently supply approximately 18% of the state's needs and groundwater development is continuing at a fast pace. Historic patterns of groundwater development reflect the state's climate and population distribution patterns. Future patterns of groundwater use will undoubtedly have to respond to evolving climate and population changes.

Large quantities of groundwater occur in deep basins formed during several of the major geologic events that define the evolution of the geologic landscape of Colorado. These basins typically contain multiple geologic formations made up of sand, gravel, and mudstone shed off nearby uplifts. "Stratigraphic setting," or the way different underground layers are stacked, can be complex with some formations changing characteristics rather quickly with distance away from an active mountain range front. In Colorado three episodes of basin, development created important hydrogeologic settings. The earliest was the Pennsylvanian to Permian Ancestral Rocky Mountain event. Next came the mountain building event, which created the foundation for many of the mountain ranges present today. Basins formed during two events form large-scale aquifer systems recognized today. The third, and most recent, episode of basin development is the development of deep basins during extension of the Rio Grande Rift. These basins form many of the deep valleys included in the mountainous region.

The Northwest Volcanic Aquifer Study Area lies in the middle of a broad volcanic region of the Northwest. This larger region includes multiple hydrogeologic provinces such as the Columbia Plateau, the Cascade Range, the Modoc Plateau, the High Lava Plains of Oregon, the volcanic Snake River Plain (east to and including parts of the Yellowstone Plateau volcanic field), and the northwest corner of the Basin and Range Physiographic Province. The boundary between the Northwest Volcanic Province and the Basin and Range Province is not well-defined and is used by USDS researchers to indicate the transition in northwestern Nevada from the volcanic-dominated areas to the predominant basin and range structures of the Great Basin in aquifers located to the south of this region.

Conclusion

Since 1950, the USGS has collected and analyzed water use data for the United States and its territories. That data is revised every 5 years. The only data available in 2022 was that of the 2015 measurement report.

Then, water use by all U.S. sectors was 322 billion gallons of water per day. Climate change and population increases have placed surface water supplies in stress in many locations. In 2015, the three sectors with the greatest dependence upon water from all sources were irrigation (118 billion gallons per day), thermoelectric power (133 Bgal/day), and public supply (39 Bgal/day), cumulatively accounting for 90% of all water used by all sectors. Surface water resources for these and other users are declining in many locations. As a result, to continue to meet the needs of those who depend on water have had to increase their use of groundwater.

There is no question that groundwater is one of the nation's most important natural resources. Regular testing and evaluating the withdrawals and remaining levels of the resource in important underground water levels, federal and state water supply monitors had determined that this source supplies about 37% of the water that county and city water departments supply to households and businesses (public supply). About 42% of the water used for irrigation comes from groundwater. In addition to this municipal supply, groundwater provides drinking water for more than 90% of the rural population. It is also the sole supply of freshwater for some major cities.

Withdrawals of groundwater are expected to rise as the population increases and available sites for surface reservoirs become more limited. When the aquifers that hold most of the accessible groundwater become depleted, those that depend on this source must either find replacement resources or change the way they depend of the source. However, in many regions of the contiguous United States, the water levels in many aquifers is declining faster than it can be recharged.

References

CDC (Center for Disease Control). 2016. Types of agriculture and water use. Available at https://www.cdc.gov/healthywater/other/agricultural/types.html.

Dimick, Dennis. 2014. If you think the water crisis can't get any worse, wait until the aquifers are drained. Available at https://news.nationalgeographic.com/news/2014/08/140819-groundwater-california-drought-aquifers-hidden-crisis/.

Doršner, Kamals (Modified by Matthew R. Fisher). 2018. Water supply problems and solutions, Chapter 7. *Essentials of Environmental Science*. Available at https://openoregon.pressbooks.pub/envirobiology/chapter/7-2-water-supply-problems-and-solutions/.

EPA (Environmental Protection Administration). 2018. Agricultural water use. *EPA EnviroAtlas*. Available at https://enviroatlas.epa.gov/enviroatlas/DataFactSheets/pdf/ESN/Agriculturalwateruse.pdf.

Kingsbury, James A., Laura M. Bexfield, Terri Arnold, MaryLynn Musgrove, Melinda L. Erickson, James R. Degnan, Anthony J. Tesoriero, Bruce D. Lindsey, and Kenneth Belitz. 2021. Groundwater-quality and select quality-control data from the National Water-Quality Assessment Project, January 2017 through December 2019. USGS Data Series 1136. Available at https://pubs.usgs.gov/ds/1136/ds1136.pdf.

Maupin, Molly A., and Nancy L. Barber. 2005. Estimated withdrawals from principal aquifers in the United States. USGS Circular 1279. Available at https://pubs.usgs.gov/circ/2005/1279/ (online).

National Park Service (NPS). 2021. Water Resources on the Colorado Platau. Availale at https://www.nps.gov/articles/000/water-on-the-colorado-plateau.htm.

Sharp, Jay. 2022. Water resources in the Southwest. *DesertUSA*. Available at www.desertusa.com/desert-water/water-in-the-southwest.html.

USDA (United States Department of Agriculture). 2019. Irrigation & water use. USDA Economic Research Service Circular. Available at www.ers.usda.gov/topics/farm-practices-management/irrigation-water-use/.

USGS (United States Geological Survey). 2000. Ground Water Atlas of the United States: California, Nevada (HA 730-B). James A. Miller, ed. Available at https://pubs.usgs.gov/ha/730/report.pdf.

USGS (United States Geogical Survey). 2010. Scientific Investigations Report 2010–5040. Available at https://pubs.usgs.gov/sir/2010/5040/section5.html.

WAECY (Washington State Department of Ecology). 2016. Washington groundwater level status and trends—Spring 2016. Available at https://Vermont-WUDR-Workplan.pdf.

CHAPTER 7

Stored Water: Failing Dams and Other Storage Crises

Surface water stored in a lake or in a river behind a dam or contained in in concrete or steel tank, or frozen in mountain snowpack is an asset stored for later use. However, when it is water that is stored for a particular intended use, the potential for a crisis to occur becomes possible. The storage of water is not without danger. For example, problems can occur when water is used some way and then discharged uncontrollably into a surface water system downstream. In another scenario, when a storage tank or distribution pipe fails, released water that was once an asset takes on the potential for an uncontrolled threat. The result of any such uncontrolled release of rushing water can be catastrophic, causing damage to life and property. Both of these disasters occur often in the United States, and when they do, they have an impact beyond the banks of a brief but uncontrolled flow. The water from a flooding reservoir or distribution system can leave water that pollutes surface and groundwater in its wake.

In most states large public water districts maintain some type of outdoor water storage facilities. California, for example, controls one of the largest water storage systems, with almost 1,300 reservoirs throughout the state. However, only approximately 200 are reserved for water storage, and many of the larger ones are critical components of the Central Valley and State Water Project distribution facilities. California's reservoir system depends upon the annual snowpack for supplying the water needed to supply the state's large spring and summer supply

D. E. McNabb and C. R. Swenson, *America's Water Crises*,
https://doi.org/10.1007/978-3-031-27380-3_7

of water for its agriculture industry. Reservoirs fed by annual snow and rainfall supplies are also in annual demand for the thermoelectric sector. According to the USGS, there are nearly 1,300 reservoirs throughout the state, but only approximately 200 of them are considered storage reservoirs, and many of the larger ones are critical components of the Central Valley and State Water Project facilities. Storage reservoirs capture winter precipitation for use in California's dry summer months. In addition to its large water storage and distribution system, California also depends on water that is annually stored in the statewide snowpack. As the mountain snow and ice melts slowly over the course of the summer, this supply provides surface water and groundwater storage that augments the state's water supply.

Stored Water Loss

Surface water stored in a lake or in a river behind a dam or contained in concrete or steel tank is an asset held for later use. In approximately a dozen states, winter snow becomes another important water storehouse. Despite the importance of this open storehouse of water, large amounts of freshwater are lost every year. In some cases, water loss is due to failure of the dams that hold the water back, or tanks that hold stored water. In these circumstances flooding can occur in the land immediately surrounding the tank or storage pond. When a raised local storage pond or a storage tank or distribution line fails, the released water asset becomes a destructive uncontrolled threat. The result can be catastrophic damage to life and property in the path of the released wall of rushing water. Both of these disasters occur with some frequency in the United States. The flooding reservoir or distribution water also leaves polluted surface water and groundwater in its wake.

Because the volume of a local surface water fluctuates with changes in the seasons, supplies of surface water rise and fall. Local storage facilities are needed to supplement declining supplies to last across the drier periods. Conservation and efficiency efforts help somewhat in compensating for declining supplies in city water, but water withdrawals in very large amounts that is occurring in many arid locations due to climate change surpass the ability of many jurisdictions to hold enough water. Demand for reliable adequate supply of water requires increasingly large amounts of stored supplies to compensate for seasonal decline beyond available surface water supplies. This is exacerbated by the annual withdrawals of surface water that continue to support irrigated agriculture and thermoelectric production that compete with the domestic needs of a growing population.

Evaporation of Stored Water

Most stored water is maintained in large open reservoirs. Evaporation is a common problem with the large reservoirs behind large dams in the Western United States. In wet areas, such as the American Southeast, the water that evaporates tends to fall again as rain nearby. However, in hot, dry areas such as most of the Southwest, evaporation can result in a huge loss of water. The level of reservoirs in desert areas can drop 5 feet (1.5 meters) in a single year due to evaporation alone. Moreover, in growing areas such as the Phoenix region, water is transported through an extensive array of open canals where evaporation causes significant water loss in this dry hot climate. Some modest efforts for controlling evaporation have been tried with various types of floating shade covers, but there is little agreement that covers are a cost-effective method of evaporation reduction.

Stored water can contribute to water stress in a variety of ways. In warmer climates, evaporation can result in the loss of large amounts of freshwater. Beyond evaporation, both flowing and stored surface water can see algae become a major pollutant. Nutrient such as agriculture fertilizers and animal waste, for example, can result in a variety of types of algae, some of which are harmful to humans and farm animals. Harmful algal blooms are overgrowths of algae that are a major environmental problem in all 50 states. Red tides, blue-green algae, and cyanobacteria are examples of harmful algal blooms that can have severe impacts on human health, aquatic ecosystems, and the economy. Some produce dangerous toxins in fresh or marine water, but even nontoxic blooms hurt the environment and local economies (EPA 2022d). Harmful algal blooms can:

- Produce extremely dangerous toxins that can sicken or kill people and animals
- Create dead zones in the water
- Raise treatment costs for drinking water
- Hurt industries that depend on clean water.

Box 7.1 US Fish and Wildlife service description for dam removals

According to the US Fish and Wildlife service, dams are used extensively throughout the United States for a variety of purposes, including navigation, flood control, and power generation. While well-designed and properly managed dams can provide many benefits, they drastically alter natural river communities. The natural flow of water and sediment is impeded and populations of native fish, mussels, and other aquatic animals are damaged. In some cases where the costs of maintaining a dam outweigh its benefits, a decision is made to remove, or decommission, it. Dam removal is a planning process that can often take several years and be expensive. However, the safety and environmental benefits are priceless. In most instances, rivers heal quickly, and it's hard to realize that a dam once blocked the river. Decommissioning provides an opportunity to restore a river's health and return it to its natural, free-flowing state.

To take our rivers for granted is to risk the quality and quantity of our water and to endanger some of our most beautiful natural places. When a dam no longer serves its intended purpose, removing it provides an opportunity for us to return a river to its original state, where natural systems are allowed to work without barriers.

Source US FWS n/d.

In a move to control the dangers that can occur with toxic algae blooms, NOAA established the National Phytoplankton Monitoring Network (PMN) as a public algae awareness organization. The network functions as a community-based body of volunteers monitoring marine phytoplankton and harmful algal blooms (HABs). Recognizing the interrelationships between humans and coastal ecosystems, PMN provides volunteer citizen scientists with meaningful opportunities for hands-on science engagement. Moreover, with the sharing of information and research findings, the network enhances the nation's ability to respond to and manage the growing threat posed by HABs by collecting important data for species composition and distribution in coastal waters and creating working relationships between volunteers and professional marine biotoxin researchers (NCCOS 2017).

An example of a surface disaster occurred in Michigan in 2020. The Edenville Dam on the Tittabawassee and Tobacco rivers failed in May 19 when torrential rains sent water over the surface of the dam that resulted in a catastrophic 500-year flood. The flooding reservoir water then caused the subsequent failure of the downriver Sanford Dam and catastrophic flooding in Midland County. Dow Chemical, headquartered in Midland, Michigan, was in the path of the flood. The surface water for all that distance was subject to a variety of human and animal wastes and industrial poisons. Dam failures spread these poisons long distances and into cities that are mile downstream from the dam failure.

Withdrawals of surface waters have immediate effects on the rate of streamflow at the point of withdrawal. However, the adverse effect of groundwater on surface water may take years or even decades to show up. The effects of groundwater withdrawal include the timing, rates, and locations of streamflow depletions in ways different than that of surface water on groundwater, as shown in the pollution cause by failure of the Edenville Dam in Michigan decades of discharging chemicals into and on the banks of the Tittabawassee and Saginaw rivers. To protect the nation against disasters that follow the breaching of a dam, the Federal Government caries out a comprehensive inventory of the number of dams and their potential for failure.

Impact on Stored Water Quality

The fisheries unit of the National Oceanic and Atmospheric Administration (NOAA) is an office that works to preserve water quality. This agency works in all regions of the country to preserve fisheries habitat by assuring the health of rivers. The purpose of many dams in the Pacific Northwest is to collect and store water for uses such as hydropower and irrigation. Water diverted from the river results in lower natural flows and less habitat for fish downstream. In addition, changes occur in the quality of water when it is stored behind a dam. The withdrawals from these dam reservoirs during prolonged periods of drought in many Western U.S. states has resulted in high stress of the region's water stored in the resources. The stress can come from any of the following natural phenomena (NOAA Fisheries n/d).

- Flow timing: Fish are adapted to a particular natural pattern of flows in a river. In the northwest, most rivers have peak flows in the spring and lower flows in summer. The way dams use water often results in changes to these natural patterns. Salmon flows are particularly affected.
- Ramping: Sudden release of water or reduced flows cause river levels below the dam to rise or fall suddenly, potentially stranding fish. Water is stored in the reservoir during periods of low power demand and then released later to generate electricity when demand is high.
- Temperature: Water held in reservoirs tends to heat up, increasing the temperature of the river. Salmon and steelhead prefer cool water. However, water on the bottom of the reservoir remains cold through the summer. If cool water is taken from the bottom (some dams cannot do this), it may actually be used to help keep the river cool through the summer.
- Total dissolved gas: Falling water, such as from a dam spillway, may mix nitrogen from the atmosphere into the water. Water can only hold a certain amount of nitrogen, and when this is exceeded, bubbles are formed. These bubbles can also form inside the bodies of fish which are swimming in the water, causing injury or death.
- Oxygen: Water released from the bottom of the reservoir is often low in oxygen, causing problems for fish downstream. However, water falling over a spillway may actually mix more oxygen into the water.
- Toxic substances: Sediment settles out in reservoirs behind dams. Many toxic substances may be trapped in these sediments such as pesticides, or heavy metals from mine tailings. If these sediments are disturbed, these substances may be released into the water.
- Eutrophication: Eutrophication is the process that occurs when the water in a lake or a pond becomes enriched with nutrients, resulting in extensive growth of simple plant life. The bloom of algae and plankton in a water body is an indication of this process. The process often results in the deterioration of water quality and the depletion of dissolved oxygen in the water source. Eutrophic waters can eventually become "dead zones" that are incapable of supporting life. The process that occurred in the Great Lakes was described by the USGS in 1965: Several changes commonly associated with eutrophication in small lakes have been observed in the Great Lakes. These

changes apparently reflect accelerated eutrophication in the Great Lakes due to man's activity (such as run off of farm chemical fertilizers). The plankton has changed somewhat in Lake Michigan, and the plankton, benthos, and fish populations of Lake Erie are greatly different today from those of the past (USGS 1965).

- Reservoirs often produce large amounts of algae and other plants. The result is higher levels of nutrients in the river downstream. Too many nutrients can cause problems such as low oxygen levels or excessive growth of algae.

NOAA Fisheries, also known as the National Marine Fisheries Service, is an office of the National Oceanic and Atmospheric Administration within the Department of Commerce, and has five regional offices, six science centers, and more than 20 laboratories around the United States and U.S. territories.

Snowpack: Nature's Storehouse

The amount of precipitation that falls as snow and the resulting high elevation snowpack that provides water to lower lands through the spring and summer plays a key role in the water cycle in the western contiguous United States. The snowpack that remains stores the water until needed in spring and early summer. The annual snowmelt is the water stored in the mountain ranges during the winter and released as runoff in spring and summer. Millions of people in the Western U.S. depend on the melting of mountain snowpack for hydropower and cooling for thermopower, irrigation, and drinking water. In most western river basins, snowpack is a larger component of water storage than are constructed reservoirs. As the snowfall and snowmelt events continue to occur earlier and in lower amounts, the sectors that depend on it suffer.

Moreover, with continued climate warming, the reductions in snowpack, and the shifts times and amounts of snowmelt, are expected in future (EPA 2021f). As much as 75% of water supplies in the western states are from snowmelt. In some locations, during certain times of the year water from snowmelt can be responsible for almost all of the streamflow in a river. An example is the South Platte River in Colorado and Nebraska. Historically, the South Platte River was essentially "turned off" after the supply of water coming from melting snow was exhausted in late spring. Today, though seepage of irrigation water from ditches and fields replenishes the alluvial aquifer (water-bearing deposit of sand and gravel left behind by a river) during spring and summer, the aquifer slowly drains during fall and winter by discharging groundwater to the South Platte River.

Long-term monitoring by federal and state climate and agricultural agencies closely follow timing and changes in amounts of snowfall. The extent and retention of the annual snowpack is a critical source of surface water in the CONUS Northwest and Southwest The time of its melting influences downstream surface water resources as a contributor to groundwater levels for the agricultural and public freshwater supplies for these regions. The amount of annual snow and longevity of the resulting snowpack is changing in response to climate change by melting sooner and faster, thereby reducing the surface water supplies when most needed during the hot, dry periods. The ability to survive through extended summer droughts that are occurring more often and continuing for longer periods are influenced by changes in the length of the snow season and by the occurrence and length in the amount of snow that accumulates through the winter. The snowpack disappearance dates in southern Alaska, the Cascades and Sierra Nevada ranges, across the lower Midwest, and along parts of the Appalachians are changing more rapidly than in regions like the Rocky Mountains and Upper Midwest (Stevens 2021).

Monitoring Snowpack

The U.S. Department of Agriculture, with other federal, state, and local organization researchers, have measured snowpack since the early 1900s. Measurements are collected for eleven CONUS states and for Alaska. Records are kept for states in three CONUS regions: West Coast: (Washington, Oregon, and California); Mountain: (Montana and Wyoming, Idaho, and Nevada); and Southwest: (Utah, Colorado, Arizona, and New Mexico).

In the early years of data collection, researchers manually measured snow water (the amount of water equivalent in the snow), However, since 1980, measurements at some locations are now collected with automated instruments as part of the snow telemetry (SNOTEL) network. The changes in mountain snowpack can affect agriculture, winter recreation, and tourism in some areas, as well as forest and wildlife. Additionally, warming and earlier snowmelt accelerate the start of the wildfire season and promote more wildfire activity in the Western United States and Alaska. Among the states most dependent upon annual snowpack are Montana, Wyoming, Utah, and Colorado. The following EPA-published data on the nearly 70-year history of monitoring snowpack conditions showed a continued decline in the size and term before melting of snowpack in the United States:

- From 1955 to 2020, April snowpack declined at 86% of the sites measured. The average change across all sites amounts to about a 19% decline.
- Large and consistent decreases in April snowpack were observed throughout the Western United States. Decreases have been especially prominent in Washington, Oregon, northern California, and the northern Rockies.
- While some stations have experienced increases in April snowpack, all 12 states included in this indicator experienced a decrease in snowpack on average from 1955 to 2020. In the Northwest (Idaho, Oregon, Washington), all but four of the widespread system of stations saw decreases in snowpack over the period of record.
- About 81% of sites experienced a shift toward earlier peak snowpack. This earlier trend is most worrisome in Southwestern states like Colorado, New Mexico, and Utah. Across all stations, peak snowpack has shifted earlier by an average of nearly eight days since 1982, based on the long-term average rate of change.

Over the 45 years from 1955, the spring snowpack in the 10 mountain states in the West has declined by nearly 20% on average. These 10 states and their 2020 average snowfall in inches are included in Table 7.1. The decline is even greater in some locations. From 1955 to 2020, April snowpack declined at 86% of the sites measured, decreases have been especially prominent in Washington, Oregon, northern California, and the northern Rockies. All of these states have seen decreases in snowpack over the period of record (Michon and Lindsey 2022; EPA data).

Colorado and Wyoming are among the contiguous U.S. states receiving the highest annual snowfall over the 2010–2021 period. Three other states were in the top 20 group of states with highest amounts of snow for the same period: Utah, Montana, and Idaho. California had the least snowfall that year, just 3.76 inches. Ranks within the continuous

Table 7.1 The 8 continuous mountain states 2018–2019 winter snowfall (inches)

Rank (of 48)	*State*	*Yearly snowfall (inches)*
4	Colorado	67.3
9	Wyoming	51.0
12	Utah	40.99
16	Montana	37.28
18	Idaho	36.49
25	New Mexico	22.74
29	Nevada	18.83
39	Arizona	4.1

Source World Population Review: Snowiest states

Table 7.2 Winter snowfall (inches) of the 3 west coast states, 2018–2019

Rank (of 48)	*State*	*Yearly snowfall (inches)*
31	Washington	15.57
35	Oregon	10.91
42	California	3.76

Source World Population Review: Snowiest states

eight Mount States and snowfall inches for the year are in Table 7.1. Similar data for the three west coast states with important snowpacks are included in Table 7.2.

Despite the low amounts of snow in the states that depend on snow-packs, the surface water supplies supported by snowmelt are important. The snowfalls and water content retained in snowpacks retained provide spring and summer surface water and are a major source of water for groundwater aquifer recharging.

Snowmelt in Mountain States

In August of 2022, the 2021–2022 winter snowpack in Colorado was facing a continuation of the early annual melt. While the amount of snow received in Colorado's mountains over the winter was nearly normal for that time of year, the statewide snowpack levels were about 91% of average USDA officials say that number is starting to drop as snow in some areas starts to melt early with warmer-than-average spring temperatures. Colorado's runoff season is vital to replenishing the state's rivers and reservoirs, most of which are currently sitting at below-normal levels after multiple years of climate change-driven drought. Colorado will likely face another year of below-normal water supplies despite nearly normal snowpack totals (Sakas 2022).

In Montana, the major demand for freshwater is for hydropower use. The source of most of that water is the melting surface water from snowpack in headwater streams that drain three of North America's largest watersheds: the Columbia, Missouri-Mississippi, and Canada's Saskatchewan rivers. Montana's dependence on snowpack to meet the state's surface water need was not a significant problem until the 1980s, when the declines were deemed noteworthy and subject to closer monitoring. Since records were first kept in the 1930s, the annual snowpack in Montana has declined regularly in the mountains west and east of the Rocky Mountains. The rate of decline has increased since the 1980s along with increasing temperatures throughout the state.

Divided in two parts by the Continental Divide, Montana has had what in effect are two climates. Western Montana is mountainous timberland, and the eastern section is rolling prairie land. Each section has its own weather patterns. In the western section, as the climate gets warmer, less precipitation now falls as snow, and more snowmelts during winter, reducing the availability of snowmelt for use in the eastern agriculture lands. Over the past century, most of the state has warmed about two degrees (F). Heat waves are becoming more common, and snow is melting earlier in spring. Rising temperatures and changes in rainfall are likely to have both positive and negative effects on Montana's farms and ranches, and the net effect is unknown. Rising temperatures and recent droughts have killed many trees by drying out soils, increasing the risk of forest fires, or enabling outbreaks of forest insects. In the coming decades, the changing climate is likely to decrease the availability of water in Montana, affecting agricultural yields, and further increasing the risk of wildfires. The headwaters of three surface water rivers begin in Western Montana watershed, becoming the Columbia, Missouri and Mississippi rivers, and the Saskatchewan River in Canada.

Wyoming's Snowpack Resources

Wyoming's water resources include surface water, groundwater, ice precipitation (rain, sleet, hail, and snow), snow, glaciers, and water vapor in the atmosphere. These water resources are constitutionally the property of the state, which specifies that the state must provide a set volume of flow or a percentage of the total amount of river water that flows to other states in the region (WSGS 2022).

Wyoming's water resources must be enough to meet the water needs of electric power generation, agriculture (irrigation and livestock watering), domestic supply, public water systems, lawn/garden watering, fish hatching/rearing, and environmental purposes including groundwater recharging, recreation, and industry.

Snowmelt is a significant contributor to the water resources of Wyoming. Changes in its timing affect agriculture in the state and contribute to earlier and more frequent wildfires, ecosystem changes, and other impacts. A NASA study of a basin in northwestern Wyoming revealed that the snowmelt season in the area is now ending on average about sixteen days earlier than it did from the 1970s through the 1990s. Dorothy Hall, lead author of the study, concluded that earlier snowmelt is probably happening all along the Wind River Range. This is a mountain range of the Rocky Mountains in northwestern Wyoming. This section of the Rockies runs roughly northwest to southeast for approximately 100 miles, with the Continental Divide following the crest of the range. The study on data from several stream gauges was analyzed and recorded for a previous study. All gauges showed lower current stream flows in springtime in the 2000s than in the 1970s, 1980s, and 1990s. There was a warming trend in spring and summer nighttime temperatures over those decades (Viñas 2015).

Dam Failures and Water Loss

Dams are primarily built for retention of surface water. Water thus held behind the dam is then available for productive purpose while functioning as a bar to flooding. A dam stops or limits the flow of water or underground streams by making use of obstructions that are constructed of earth or concrete.

The U.S. Army Corps of Engineers (USACE) operates and maintains approximately 740 dams and associated structures nationwide that provide significant, multiple benefits to the nation—its people, businesses, critical infrastructure, and the environment. These benefits include flood risk management, navigation, water supply, hydropower, environmental stewardship, fish and wildlife conservation, and recreation (USACE 2021). The Bureau of Reclamation (USBR) manages, develops, and protects water and related resources. This agency is the nation's largest wholesale water supplier, operating 338 reservoirs with a total storage capacity of 140 million acre-feet (an acre-foot, 325,851 gallons of water [USBR 2022]).

Congress first authorized the U.S. Army Corps of Engineers (USACE) to inventory dams in the United States with the National Dam Inspection Act of 1972 and published a National Inventory of Dams (NID). The NID, first published in 1975 with a few updates as resources permitted over the next ten years, continues to be a record of all qualifying dams in the United States. The Water Resources Development Act of 1986 authorized US Army Corps of Engineers to maintain and periodically publish an updated NID, with re-authorization and a dedicated funding source provided under the Water Resources Development Act of 1996. USACE also began close collaboration with FEMA and state regulatory offices to obtain more accurate and complete information. The National Dam Safety and Security Act of 2002 and the Dam Safety Act of 2006 reauthorized the National Dam Safety Program and included the maintenance and update of the NID by USACE. The NID was reauthorized as part of the Water Resources Reform and Development Act of 2014 and the Water Resources Development Act of 2018 (FEMA 2020). The inventory is a jointly managed program with water resource agencies in 49 states. As of 2020, Alabama was the only state without any dam safety legislation or formal dam safety program. The NID consists of dams meeting at least one of the following criteria:

- High hazard potential classification—loss of human life is likely if the dam fails.
- Significant hazard potential classification—no probable loss of human life but can cause economic loss, environmental damage, disruption of lifeline facilities, or impact other concerns.
- Equal or exceed 25 feet in height and exceed 15 acre-feet in storage.
- Equal or exceed 50 acre-feet storage and exceed 6 feet in height.

The inventory database tracks dams that could be life threatening or cause economic or environmental damage. Of all the dams listed by the NID, about 15,621 dams in the U.S. (close to 17 percent of all dams) were identified as having high-hazard potential. In the 2017 Infrastructure Report Card prepared by the American Society of Civil Engineers (ASCE), America's dams collectively received a D rating for safety. No record of the total number of small dams that have failed in the U.S. since 2005 was found, but one or more exists in every state. From January 2005 to June 2013, state dam safety programs reported 173 dam failures and 587 incidents that without preventative action would likely have resulted in dam failure. The most disastrous impact of water loss occurs when a dam fails. A dam failure or dam burst is a catastrophic type of failure characterized by the sudden, rapid, and uncontrolled release of impounded water or the likelihood of such an uncontrolled release. The NID inventory lists the purposes for which the dams were constricted. Purposes included irrigation, flood control, navigation, fire protection, fish and wildlife and debris control, water supply, and power generation. Only 15% of the nearly 90,000 dams identified power generation as one of their purposes (Lee et al. 2018).

The majority of America's dams were built for agriculture or flood control or a combination of both purposes and did not include capacity for power generation. Many of these dams were developed by the U.S. Department of Agriculture and were constructed with a 50 life span. The DOE identified 600 dams built for agriculture rather than power generation, which have reached that point and must be reconstructed from the ground up. In addition, another 1,500 dams need repairs so that they can continue to provide flood control, municipal water supplies, recreational activities, water for livestock, and wildlife habitat.

Box 7.2 USACE description of benefits from existing dams on the Snake River

"The Army Corps of Engineers' Walla Walla District owns and operates the four lower Snake River dams, all of which are multiple-use facilities that provide navigation, hydropower, recreation, and fish and wildlife conservation benefits. Because of their locations, size and ability to help meet peak power loads, these four dams do much more than generate energy--they are key to keeping the system reliable and helping to meet its multiple uses—including supporting wind energy. The Snake River dams lie east of the other federal generators, so they provide a significant technical contribution to transmission grid reliability.

The Lower Snake River system of locks and dams deliver a significant economic benefit to the nation. Barging on the inland

Columbia Snake River System moves, on average, approximately 10 million tons of cargo valued at over $3 billion each year. Forty percent of the Nation's wheat transits through this system. While building four dams on the lower Snake River between the 1950s and 1980s, the U.S. Army Corps of Engineers constructed adult fish ladders to allow passage of upstream-migrating adult salmon and steelhead (salmonids) on their way to their natal spawning areas. Since construction, the Corps has made both facility and operational improvements to assist adult salmonid migration as needed."

Source USACE (2010).

Example Dam Failures

Dams of the many rivers and streams in the eastern states have existed since the eighteenth century. The South Fork Dam was an earth- and rock-fill dam constructed on South Fork Creek, a feeder of the Conemaugh River about eight miles east of Johnstown, Pennsylvania. In dry years during the 1830s, the canal was not enough water in the canal for the canal boats to pass through. The problem was solved in 1852 with construction of a dam and a reservoir on the south Fork Creek that could function as a feeder for the canal. The dam was approximately 72 feet high, 918 feet long, 10 feet wide at its crest, and 220 feet wide at its base. As a result of poor maintenance, the outlet works' culvert collapsed and a portion of the dam washed out in 1862.

With the arrival of the railroad, the canal was no longer needed, and the canal and dam were left unused. In 1879 the breached dam and adjacent land were sold to Benjamin Ruff who planned but did not complete needed repairs to the dam. Under Ruff's ownership the area became the South Fork Fishing and Hunting Club. The South Fork Dam was inadequately repaired, with the top of the dam lowered and widened to permit carriages to pass over.

Dam Oversight and Management Complexity

In June of 2021, Washington State Gov. Jay Inslee and U.S. Sen. Patty Murray of Washington sought a report whether reasonable means for replacing the benefits would be provided by the four Lower Snake River Dams (LSRD Options 2022), and whether those benefits were sufficient enough to justify breaching of the dams as part of a comprehensive salmon recovery strategy for the Pacific Northwest. A draft of the 119-page report was released in the summer of 2022. A draft report was requested to provide a platform for public input through the remainder of their process. The report describes the range of services and associated benefits currently provided by the dams and the actions that have been considered to replace or improve upon these services and benefits if the dams were to be breached, along with expected results of replacement actions and anticipated costs. Following public input, tribal consultation, and other means of engagement, this report will be updated and released in final form. Then the governor and senator, considering all the information provided, will make their recommendations (LSRD Options 2022). Findings from this process are to be used by Gov. Inslee and Sen. Murray to determine their recommendations for the LSRD.

Washington Congressman Dan Newhouse, representing the region, objected to the report, saying it ignored the interests of the agriculture sector. Environmentalists have argued that the four dams must be removed to promote salmon recovery, but agricultural stakeholders say those arguments don't consider the larger environmental threats to the fish, or the broader impacts to the environment and economy. The dams provide benefits from hydroelectric production and enable multi-state benefits from barge shipment of the region's extensive wheat production and other products.

The report added that the continuing warming of the environment, change in the Lower Snake River and the Columbia River System, is inevitable. Several experts interviewed referenced the situation in the Klamath River Basin on the border of Oregon and California where four dams are slated to be breached in 2023. Somewhat similar to the process that has unfolded for the LSRD, the basin engaged in a series of agreements between all stakeholders on the four privately owned Klamath dams that aimed to balance environmental, agricultural, tribal, and fishery needs from the early 2000s through the mid-2010s. Just like with the agreements that were reached on the LSRD, multiple parties believed that these agreements for the Klamath River did not provide sufficient resources for ESA-listed species and did not affect the implementation of the ESA. Ultimately the parties came together on the Klamath Basin Settlement Agreement in 2016 which agreed upon the removal of the four projects

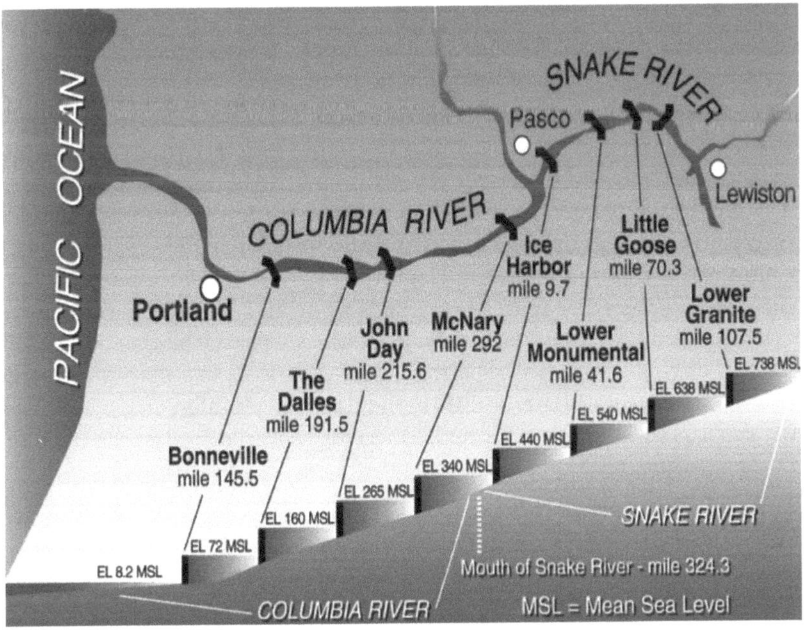

Fig. 7.1 Map of the Snake and Columbia rivers with dam sites and elevation identified (*Source* U.S. Wheat Association, June 14, 2022)

through the traditional Federal Energy Regulatory Commission approval process. Figure 7.1 shows the location of the four dams on the Snake River between Paso Washington and Lewiston, Idaho.

Recent Dam Failures

Hope Hills Dam

The first Hope Mills Dam was a rock-crib dam built in 1839 by the Rockfish Mills Co. for the powering of four cotton mills nearby. In 1865, Yankee troops burned the cotton mills but spared the dam from destruction. Two mills were rebuilt after the Civil War and the surrounding area including the dam was named *Hope Mills*. On May 9, 1923, this first dam was breached by a flood and construction of an earthen-embankment dam began the same year. The second dam costs $27,093 and was completed in early 1924 (Table 7.3).

Rockfish Mills Co. went bankrupt in 1930 and was not purchased until 1940, when the Dixie Farms company took over. Dixie Yarns eventually donated the dam to the town in 1984. In 1993, a two-lane Lakeview Road bridge was installed over the dam and at the same time a leak in the dam was repaired. In 1994, an unused turbine shaft from the old cotton mill site was uncovered and filled with cement. In addition, walls behind the dam's spillway were reinforced with rock after concerns were voiced about their strength. Heavy rains associated with Hurricane Floyd in 1999 forced the opening of the floodgates to save the dam. In 2001, a 6-foot (1.8 m) crack was repaired where engineers had determined that it was not structural. Later in 2001, the dam passed safety inspection, but two small holes and some eroded concretes were repaired in 2002.

Table 7.3 Sample of recent failures of U.S. dams

Dam	*Location*	*Year(s) built*	*Year(s) failed*	*River*	*Damage*
Spenser Dam	Nebraska	1927	1935, 1960–1966, 2019	Niobrara River	1 fatality; dam to be removed
Big Bay Dam	Mississippi	1990–1991	2004	Leading to Pearl River	104 structures destroyed; no deaths
Hadlock Pond Dam	New York	1986	2005		No injuries; est. $4 million loss
Lake Delhi Dam	Iowa	1922–1929	2010	Maquoketa River	No fatalities; est. $1 million loss
Oroville Dam	California	1968	2017	Feather River	No injuries; est. $1.1 billion loss
Edenville Dam Sandford Dam power dams	Michigan	1924 1925	2020	Tittabawassee River	No fatalities reported; est. $250 million damage

Sources Wikipedia and assorted reports

In late May 2003, heavy rains again caused the embankment dam to give way. About 40 homes and 1600 people downstream were evacuated as water from Hope Mills Lake rushed down the creek causing $2.1 million in danger before making its way to the Cape Fear River. One of the floodgates on the dam had failed to open because of the pressure exerted by the flood waters, which helped push water over the top of the dam.

Construction on a new 14-million dollar Hope Mills Dam began in March 2007. The new dam was to include a concrete labyrinth spillway that would expand the dam's ability to discharge flood waters. The new dam would be 600 feet (180 meters) longer and would incorporate a fish ladder. Construction of the new Hope Mills Dam was complete in June 2008, and the lake was again full by August.

Water passing through a sinkhole in the dam's foundation resulted in another failure in 2010. On Wednesday, June 16, 2010, a controlled release began after engineers noticed silt coming out of the lower side of the dam. Town officials later told residents the lake may have to be lowered as much as nine feet to inspect the dam. Overnight a sinkhole developed at the base of the dam, and the lake water drained underneath the foundation. By Thursday morning (June 17, 2010) almost all of the water in the lake was gone. Repairs for the dam were expected to begin in August 2012 and end October 2013, but they never occurred. A lawsuit was filed by the town in mid-October 2012, complaining that the companies involved were delaying the repairs which had yet to begin. In April 2013 Mayor Jackie Warner requested a plan to remove the dam, but the town sued the builders for $10 million in May in order to pay for repairs. In July 2014, the town received a $9.4-million settlement from the builders.

Spenser Dam

Spencer Dam was constructed for hydropower in 1927 on the Niobrara River in northern Nebraska. The dam included a 3-MW powerhouse, a 400-foot-long concrete spillway, and a long earth-filled embankment dam. This dam was owned and operated by the Nebraska Public Power District. The failure occurred on March 14, 2019, during a major flood and ice run on the river. The failure resulted in the death of one resident in the flood path. An assessment of the failure was carried out by the

Association of State Dam Safety Officials, with the final report published in 2020.

Three major ice run events had damaged the dam in the past. The first occurred in 1935 and resulting damage occurred when the dam's embankment was reached by an ice run. Similar ice runs followed in 1960 and 1966, heavily damaging the structure. Due to the passage of time, turnover of employees, and destruction of records in the 1966 ice run, both the owner and the regulator (Nebraska Department of Natural Resources Dam Safety program) had little knowledge of the dam's vulnerability to ice runs.

On the evening of March 13, dam operators opened all four of the dam's gates on the spillways crest to their maximum six-foot opening. They then tried to open all dam bays, but ice made it impossible to open them all. Meanwhile, upstream one or more ice jams burst, sending ice rubble and flood water toward the dam. The rubble clogged the open spillway gates and other openings, allowing the reservoir to rise to its crest. More ice pushed through the upstream brick wall of the powerhouse, and then rose over the blocked spillway and caused a breach of the dam to occur. A second breach followed, sending the flood through a house and other buildings downstream of the dam, destroying the buildings and causing the death of the one resident. The ice flood continued downstream, overflowing a highway and damaging several bridges. The final report concluded the ice storm made it impossible for the workers present to do anything to halt the ice storm and subsequent damage and death.

Big Bay Dam

The 51.3-feet-tall Big Bay Dam was an earth-filled dam located 11 miles west of Purvis, Mississippi. The dam had several main purposes, led by a means of eliminating overflows from Bay Creek's periodic storm flooding. A second purpose was for recreation. At the time of its failure, it was also considered for hydroelectric power generation. On March 12, 2004, the dam cement and brick exterior embankment had undergone partial failure at a Big Bay Dam which was an earthen dam located 11 miles west of Purvis, Mississippi, in Lamar County. On March 12, 2004, the Big Bay Dam embankment failed by piping in the vicinity of the principal spillway.

While there were some important warning signs of a developing problem over many years of operation, the majority of breach formation occurred in a period of only a few hours. At about 11:00 am on the day of the failure, discharge exiting the ground surface was described as about a ½-inch-diameter flow bubbling about ½ inch above the ground surface. An hour or so later, the discharge was muddy and spraying 30–40 feet into the air. The erosion into the downstream slope of the dam progressed quickly. The engineer noted that the crest of the dam had breached and uncontrolled release of the lake pool began at approximately 12:25 pm.

On May 8, 2007, the Mississippi Department of Environmental Quality (MDEQ) approved the application to rebuild the dam. Work on the new concrete dam began in June of 2007 and was completed in 2009.

Hadlock Pond Dam

On Saturday, July 2, 2005, at about 5:30 p.m., Hadlock Pond Dam in Fort Ann, New York, failed suddenly, releasing about 520 million gallons through the Fort Ann community. The flood flows from the dam failure caused extensive damage downstream of the dam, totaling about $10 million in value, but emergency warning helped to prevent any casualties. Constructed in 1896, the original dam was of earthen construction, or rock fill. It had a height of 29 feet, with a width of 850 feet.

The dam was reconstructed in 2005, only to collapse months later, flooding and damaging nearby properties. The pond is owned by the town of Fort Ann and is primarily used for recreational purposes. The failure occurred as the reservoir was being refilled following the construction of a new spillway and emergency spillway. Internal erosion of the embankment soils near the emergency spillway created an enlarged flow conduit through the embankment soils, which progressively breached the entire dam width.

Lake Delhi Dam

Lake Delhi Dam was originally constructed by the Interstate Power Company between 1922 and 1929 primarily for hydroelectric power generation. The dam site is approximately 1.4 miles south of Delhi, Iowa on the Maquoketa River, and impounds the nine-mile-long Lake Delhi. Reaching a maximum height of 59 feet, the dam consisted of four sections: a 60-foot-long concrete-armored earthen embankment, a 61-foot-long reinforced concrete powerhouse, an 86-foot-long concrete spillway with three control gates, and a 495-foot-long earth embankment section with a concrete core wall.

The recorded rainfall and flooding impacted the state of Iowa during July of 2010. As a result, the area draining to Lake Delhi Dam received approximately 10 inches of rain over the course of 12 hours. This caused overtopping of the dam and its eventual failure on July 24, 2010. A large number of downstream residents possible with no loss of life.

Several deficiencies in the design and maintenance of the dam were discovered and determined to have led to its failure. The dam, which had maintained successful operations for over 90 years prior to the failure, was inspected by an Iowa Department of Natural Resources dam safety representative every five years. During the 2009 inspection, repairs to one of the three spillway gates, which was inoperable at the time, were advised. The Lake Delhi Recreation Association (LDRA), which owned and operated the dam, agreed to complete the repairs by the end of the calendar year. Though repairs were initiated, it was clear that at the time of failure in 2010, the gate remained inoperable. The inability to open one of the three spillway gates during the heavy rainfall of July 2010 only increased the potential for overtopping. Failure followed.

Box 7.3 Heavy rains result in failures of two dams

"Over a 48-hour period, May 16-18, 2020, heavy rainfalls, ranging locally from 6-8 inches, hit mid-Michigan, concentrating in Arenac, Gladwin, Iosco, and Midland Counties. Subsequent rainfall from the evening of May 18, 2020, through the afternoon of May 19, 2020, placed additional stress on many dams located on the Tobacco River and Tittabawassee River systems, including the Chappel and Beaverton Dams (Tobacco River) and the Secord, Smallwood, Edenville, and Sanford Dams (Tittabawassee River). Due to this heavy rainfall combined with the already saturated state of the water system, the Tittabawassee River had surpassed flood stage in many areas.

Around 5:30 p.m. on May 19, 2020, a portion of the Edenville Dam's earthen embankment failed, causing an uncontrolled release of impounded water to rush downstream toward Edenville, Sanford Lake, and Sanford Dam. The level of Sanford Lake rose quickly over the next two hours, and around 7:45 p.m. the Sanford Dam was overtopped by floodwaters and failed. The combined failures sent a torrent of water rushing down the already swollen Tittabawassee River, through the village of Sanford, and toward the cities of Midland and Saginaw, where the Tittabawassee River joins the Saginaw River and ultimately outlets to Saginaw Bay. The Tittabawassee River crested in Midland midday on May 20, 2020, at 35 feet.

Although more than 11,000 people were evacuated and 2,500 structures were damaged by the floods, no major injuries or fatalities were reported. Visual observations indicate that the flood wave generated by the failure of the two dams started around 20 feet high, dissipated to around 4 feet at the Midland United States Geological Survey gauge, and continued downstream. Preliminary damage estimates are more than $250 million."

Source EGLE (2020).

Oroville Dam

Oroville Dam is an earth-fill embankment dam on the Feather River in the foothills east of the northern California Central Valley at 770 feet high; it is the tallest dam in the U.S. and serves mainly for water supply, hydroelectricity generation, and flood control. The dam impounds Lake Oroville, the second largest fabricated lake in the state, capable of storing more than 3.5 million acre-feet (1.1 trillion U.S. gallons).

Embankment dams come in two principle varieties: Earth-fill embankment dams and rock-fill embankment dams. An earth-fill embankment dam is made by constructing a foundation wall that is embedded into the rock below the dam to prevent water flowing beneath it and then adding a core of impermeable clay. Above this the remaining structure is built from earth, normally from the surrounding area (Breeze 2014).

Oroville Dam is a key element in the California State Water Project (SWP), one of two major projects that set up California's statewide water system. Construction began in 1961, and despite numerous difficulties encountered during its construction, including multiple floods and a major train wreck on the rail line used to transport materials to the dam site, the embankment was topped out in 1967 and the entire project was ready for use in 1968. The dam began to generate electricity shortly afterward.

Since its completion in 1968, the Oroville Dam has allocated the flow of the Feather River into the State Water Project's California Aqueduct, adding a major supply of water for irrigation in the San Joaquin Valley and municipal and industrial water to coastal Southern California. In February 2017, the dam's main and emergency spillways threatened to fail, leading to the evacuation of 188,000 people living near the dam. After deterioration of the main spillway largely stabilized and the water level of the dam's reservoir dropped below the top of the emergency spillway, the evacuation order was withdrawn.

Edenville and Sandford Dams

The purpose for the Edenville Dam was to provide water level control for the purpose of hydroelectric power generation and received its original license from FERC in 1998. Installation of the Sandford Dam downstream from the Edenville Dam occurred in 1925 for flood control and the production of hydroelectric power. On May 19, 2020, an extreme rain event caused the Tittabawassee and Tobacco rivers to swell and resulted in the breach of the Edenville Dam. When the Edenville dam failed, the resulting flood also caused the Sandford dam to fail.

A final report on the May 19, 2020 Edenville and Sandford dam failures in Midland and Gladwin counties, Michigan, was prepared by Michigan Department of Environment, Great Lakes, and Energy (EGLE), the state's responsible agency. The following data are taken from that report. "Over a 48-hour period, May 16–18, 2020, heavy rainfalls, ranging locally from 6–8 inches, hit mid-Michigan, concentrating in Arenac, Gladwin, Iosco, and Midland Counties. Subsequent rainfall from the evening of May 18, 2020, through the afternoon of May 19, 2020, placed additional stress on many dams located on the Tobacco River and Tittabawassee River systems, including the Chappel and Beaverton Dams (Tobacco River) and the Secord, Smallwood, Edenville, and Sanford Dams (Tittabawassee River). Due to this heavy rainfall combined with the already saturated state of the water system, the Tittabawassee River had surpassed flood stage in many areas.

The history of the failure began approximately 5:30 p.m. on May 19, 2020, with a portion of the Edenville Dam's earthen embankment failed, causing an uncontrolled release of impounded water to rush downstream toward Edenville, Sanford Lake, and Sanford Dam. The level of Sanford Lake rose quickly over the next two hours, and around 7:45 p.m., the Sanford Dam was overtopped by floodwaters, causing the dam to fail. The combined failures sent a torrent of water rushing down the already swollen Tittabawassee River, through the village of Sanford, and toward the cities of Midland and Saginaw, where the Tittabawassee River joins the Saginaw River and ultimately outlets to Saginaw Bay. The Tittabawassee River crested in Midland midday on May 20, 2020, at 35 feet."

More than 11,000 people were evacuated and 2,500 structures were damaged by the floods, and no major injuries or fatalities were reported. Visual observations indicate that the flood wave generated by the failure of the two dams started around 20 feet high, dissipated to around 4 feet at the Midland United States Geological Survey gauge, and continued downstream. Preliminary damage estimates are more than $250 million.

Conclusion

Despite the growing need for storage facilities for coping with changes to precipitation patterns and prolonged periods of drought, the move to eliminate the number of older dams continues in the United States. In 2022, a total of 1,797 dams had been removed in the U.S. since 1912. Many of the removals were carried out to prevent potential breaching and prevent the flooding that would follow. Others were removed because their original purpose no longer existed. The states with the most dam removals in 2020 were Ohio (11), Massachusetts (6), and New York (6). An estimated total of 69 dams were removed in 2020, despite the COVID-19 pandemic. Box 7.2 includes one federal agency's explanation of why removal of many long-standing dams is justified through many different programs and agencies. It is difficult to track down exactly how much money is available in each of these programs each year, and it is especially difficult to figure out how much has been spent on projects that include dam removal.

Not all agencies support the dam removal movement. The U.S. Army Corps of Engineers, builder and operator of approximately 740 dams and associated structures in the United States and in Puerto Rico, including four dams on the Lower Snake River. Removal of the four dams has been a controversial proposal for many reasons. The Snake River is a major tributary to the Columbia River., which it joins in eastern Washington State. The Snake passes through Washington, Idaho, Oregon, Wyoming, Utah, and Nevada. In the 1960s and early 1970s, the Corps of Engineers owns and operates four large dams on the Snake River: Ice Harbor, Lower Monumental, Little Goose, and Lower Granite. All four dams are described as "multiple-use facilities that provide navigation, hydropower, recreation, and fish and wildlife conservation benefits. The Snake River dams lie east of the other federal generators, so they provide a significant technical contribution to transmission grid reliability" (USACE 2010). A 2005 Army Core of Engineers description of the benefits received from the dams are included in Box 7.2.

Total surface water withdrawals for thermoelectric power accounted for 41% of total water withdrawals, 34% of total freshwater withdrawals, and 48% of fresh surface water withdrawals for all uses. Water withdrawn can be from water stored behind a dam or from a natural water course such as a river or lake. Water returned can result is heat pollution and other damage to fish. Failure of dams used for hydropower is described in the Association of State Dam Safety Officials (ASDSO) documenting 250 dam failures since 2010, plus more than 500 other incidents that were caught and fixed just ahead of a failure. Most were small and had limited impact, but some were catastrophic, including a dam break in Nebraska. A few recent dam failures in the U.S., beginning with the Youngstown, Pennsylvania failure that resulted in the loss of 2,200 lives are described in the following pages. The selected stories of some of the failures occurred between 2003 and 2020.

References

Breeze, Paul. 2014. *Power generation technologies*, 2nd ed. Amsterdam, Netherlands: Elsevier.

EGLE (Michigan Department of Environment, Great Lakes and Energy). 2020. Preliminary report on the Edenville Dam failure and response efforts. Lansing, MI: Environmental Assistance Center. Available at www.michigan.gov/documents/egle/egle-EdenvilleDamPreliminaryReport_700997_7.pdf.

EPA (Environmental Protection Administration). 2021a. Water reuse research Available at www.epa.gov/water-research/water-reuse-research.

EPA (Environmental Protection Administration). 2021b. Building sustainable water infrastructure. Available at www.epa.gov/sustainable-water-infrastructure/building-sustainable-water-infrastructure.

EPA (Environmental Protection Administration). 2021c. A closer look: Temperature and drought in the Southwest. Available at www.epa.gov/climate-indicators/southwest.

EPA (Environmental Protection Administration). 2021d. Climate change indicators: Weather and climate. Available at www.epa.gov/climate-indicators/weather-climate.

EPA (Environmental Protection Administration). 2021e. Climate change indicators: Snowpack. Available at www.epa.gov/climate-indicators/climate-change-indicators-snowpack.

EPA (Environmental Protection Administration). 2021f. A closer look: Land loss along the Atlantic Coast. Available at www.epa.gov/climate-indicators/atlantic-coast.

EPA (Environmental Protection Administration). 2022a. History of the Clean Water Act. Available at www.epa.gov/laws-regulations/history-clean-water-act.

EPA (Environmental Protection Administration). 2022b. Basic information about water reuse. Available at www.epa.gov/waterreuse/basic-information-about-water-reuse.

EPA (Environmental Protection Administration). 2022c. What is green infrastructure? Available at www.epa.gov/green-infrastructure/what-green-infrastructure.

EPA (Environmental Protection Administration). 2022d. Harmful algal blooms. Available at www.epa.gov/nutrientpollution/harmful-algal-blooms.

EPA (Environmental Protection Administration). 2022e. What is radon? Available at www.epa.gov/radon/what-radon.

EPA (Environmental Protection Administration). 2022f. Water Sector Workforce: New EPA grant program: Innovative water infrastructure workforce development grant program. Available at www.epa.gov/sustainable-water-infrastructure/water-sector-workforce.

EPA (Environmental Protection Administration). 2022g. Climate change indicators: Great Lakes water levels and temperatures.

FEMA (Federal Emergency Management Administration). 2020. National inventory of dams. Available at www.fema.gov/emergency-managers/risk-management/dam-safety/national-inventory-dams.

Lee, Uisung, Jeongwoo Han, Amgad Elgowainy, and Michael Wang. 2018. Regional water consumption for hydro and thermal electricity generation in the United States. *Applied Energy* 210: 661–672.

LSRD (Lower Snake River Dams Options). 2022. Lower Snake River Dams: Benefit replacement draft report. Available at www.lsrdoptions.org/wp-content/uploads/2022/06/LSRD-Benefit-Replacement-Draft-Report_2022-0609.pdf.

NCCOS (National Center for Coastal Ocean Science). 2017. Phytoplankton Monitoring Network (PMN). Available at https://coastalscience.noaa.gov/research/stressor-impacts-mitigation/pmn/.

NOAA Fisheries (National Oceanic and Atmospheric Administration). n/d: How dams affect water and habitat on the West Coast. Available at https://www.fisheries.noaa.gov/west-coast/endangered-species-conservation/how-dams-affect-water-and-habitat-west-coast.

Sakas, M. Elizabeth. 2022. Colorado's snowpack is starting to melt. Warmer temperatures and drought likely mean another year of struggling water supplies. Colorado Public Radio. Available at www.cpr.org/2022/04/14/colorado-water-supply-snowpack-melting-drought-rivers-reservoirs/.

Scott, Michon, and Rebecca Lindsey. 2022. Large declines in the snowpack across the U.S. West. April 7, 2022. NOAA: Climate.gov.

Stevens, Alison. 2021. Which mountain snowpacks are most vulnerable to global warming? NOAA Climate.gov. Available at www.climate.gov/news-features/featured-images/which-mountain-snowpacks-are-most-vulnerable-global-warming.

USACE (U.S. Army Corps of Engineers). 2010. Lower Snake River Dams: A value to the nation. Available at www.nww.usace.army.mil/Missions/Lower-Snake-River-Dams/.

USACE (United States Corp of Engineers). 2021. USACE dam safety program. Available at https://www.usace.army.mil/Missions/Civil-Works/Dam-Safety-Program.

USBR (U. S. Bureau of Reclamation). 2022. Bureau of reclamation: About Us. Available at www.usbr.gov/main/about/fact.html.

USGS (United States Geological Survey). 1965. Eutrophication of the St. Laurence Great Lakes. Available at www.usgs.gov/publications/eutrophication-st-lawrence-great-lackes.

Viñas, Maria-Jose 2015. Historical satellite images reveal snow is melting earlier in Wyoming. NASA's Earth Science News Team, National Aeronautics and Space Administration.

WSGS (Wyoming State Geological Survey). 2022. Water resources in Wyoming. Available at www.wsgs.wyo.gov/water/water.

CHAPTER 8

Repairing Infrastructure and Reusing Water

Clean or dirty, fresh or polluted, water is an essential part of every community. It is a necessary component of producing sectors, a key ingredient to sustaining life for its population, and is indispensable to fire protection and other specialized uses. It makes crops grow and when it comes in torrents, it can take lives and destroy property. Moreover, waterways that import water into the region are vulnerable to natural disasters, terrorist attacks, technological accidents, and regulatory changes. A major disruption of these external water supplies could potentially have devastating effects on the economy and the quality of life of its people. To make it available when it is needed requires investments in infrastructure. Pumps, pipes, dams, reservoirs, treatment facilities, waste water collection and treatment, collecting storm water and used water to avoid harm and damage and put to use again. Water is an essential resource for the vitality of every community. Therefore, governments from the National to the local have invested billions of dollars to attain it, treat it, distribute it, collect it after it has been used, return it to where it was stored, and do all this 24 hours of every day. This chapter is about the role of infrastructure and describes the task water managers play in avoiding water infrastructure crises.

D. E. McNabb and C. R. Swenson, *America's Water Crises*,
https://doi.org/10.1007/978-3-031-27380-3_8

Defining Water Infrastructure

Water systems infrastructure failure is defined as an undesirable or unintended event, occurrence, or situation involving the water infrastructure of a city, a public utility, or an individual or group's private water source the unplanned cessation or significant disruption of water services that could seriously compromise health and public safety.

Along with other federal and state agencies, in the United States the EPA plays a key role in ensuring the cleanliness of the nation's water resources and the safety of the supply. To do so, it must play a large role in assuring the infrastructure needed to collect, cleanse, and return a clean and safe supply of drinking water to the natural environment. The agency's task is complicated by the large numbers of water management organizations and the diverse nature of the sources and the work needed to collect, move, store, treat, and return clean safe water to the streams, lakes, and other sources. The EPA does not do these tasks. Rather, with state and local organizations, the EPA's role is to monitor and protect the water resource, use, and finally return it to its source healthy and unsullied. It must do this with infrastructure that often has spent more than 100 years buried out of site.

The EPA's role was briefly described in September 2021 information release: "The U.S. has invested billions of dollars over the years to build an extensive network of drinking water, wastewater, and stormwater infrastructure to provide the public with safe and clean water. Much of the network of water treatment plants, distribution lines, sewer lines, and storage facilities were installed after World War II. Some of that infrastructure is now over 100 years old. Historically, investment has not been enough to meet the ongoing need to maintain and renew these systems. Over the coming decades, this pattern of underinvestment needs to change and practices put in place to sustain the water services provided by the Nation's public and private water utilities. Doing so is vital to public, economic, and environmental health. While many of the nation's water sector systems have been working hard to move toward greater infrastructure sustainability, the level of renewal and reinvestment in the water sector has not kept pace with the need. This has resulted in a gap between the amount of spending needed and money available to support those needs" (EPA 2021). In a word, what this describes is the potential for a water crisis. This crisis was described in detail in the ASCE's 2021

Infrastructure Report Card on the nation's water resource and the state of the current infrastructure (Box 8.1).

Box 8.1 The state of the nation's drinking water infrastructure

Our nation's drinking water infrastructure system is made up of 2.2 million miles of underground pipes that deliver safe, reliable water to millions of people. Unfortunately, the system is aging and underfunded. There is a water main break every two minutes and an estimated 6 billion gallons of treated water lost each day in the U.S. enough to fill over 9,000 swimming pools. However, there are signs of progress as federal financing programs expand and water utilities raise rates to reinvest in their networks. It is estimated that more than 12,000 miles of water pipes were planned to be replaced by drinking water utilities across the country in the year 2020 alone. In 2019, about a third of all utilities had a robust asset management program in place to help prioritize their capital and operations/maintenance investments, which is an increase from 20% in 2016. Finally, water utilities are improving their resilience by developing and updating risk assessments and emergency response plans, as well as deploying innovative water technologies like sensors and smart water quality monitoring."

Source ASCE (2021) Infrastructure Report Card.

Water Infrastructure Failure

Water system infrastructure failure can be the result of age-related decay or destroyed by an accident that can seriously compromise public health or safety. While infrastructure failure is often considered accidental, well planned and executed investment in repair and replacement of infrastructure can assure that these failures do not occur. Frequently, system failures result from deferred maintenance or infrastructure that has reached the end of its life span and not replaced in a timely way. Like many public services in the U.S., water distribution is the responsibility of local government. The EPA provides expertise, promulgates regulations as to quality, monitors that quality but does not deliver the water. Water is obtained, treated, and distributed by local government, and public and private utilities.

There are more than 50,000 community water systems and approximately 15,000 community wastewater systems in the United States. A water crisis of varying degree can occur at any time in any form at any one of these facilities. Whenever there is a failure, discontinuation, or unintentional disruption of water and wastewater services, or failure to adequately sanitize drinking water brought about by malfeasance by the operators of a public water system, the elected and appointed officials operating these systems hold responsibility. Freshwater and wastewater collection, treatment, and delivery systems include miles of water, and sewerage pipes, as well as a vast network of generators, transformers, and treatment and distribution facilities. A failure of one of these extensive and expensive systems means that a water infrastructure crisis can happen virtually at any time anywhere. Often these failures are not limited to a small or local region but instead affect large areas beyond the failure.

The City of New Orleans included this description of what can be expected from a water system infrastructure failure: "The severity of infrastructure failure can range from very isolated occurrences with limited impacts on the City to large-scale events with catastrophic impacts on lives, property, and operations. Since all citizens and businesses are dependent on the public infrastructure, especially the provision of sewerage and water services, these are essential services, and failure is severe. When one or more of these independent, yet interrelated systems fail due to disaster or other cause, even for a short period, it can have devastating consequences. When the water or wastewater treatment systems in a community are inoperable, serious public health problems arise that must be addressed immediately to prevent outbreaks of disease. When storm drainage systems fail due to damage or an overload of capacity, serious flooding can occur" (City of New Orleans 2021).

An example of the surface damage occurred in Michigan in 2020. The Edenville Dam on the Tittabawassee and Tobacco rivers failed on May 19 when torrential rains sent water over the surface of the dam that resulted in a catastrophic 500-year flood. The flooding reservoir water then caused the subsequent failure of the downriver Sanford Dam and catastrophic flooding in Midland County. Dow Chemical, headquartered in Midland, Michigan, was in the path of the flood. The surface water for all that distance was subject to a variety of human and animal wastes and industrial poisons. Dam failures spread these poisons to long distances and into cities that are mile downstream from the dam failure.

Withdrawals of surface waters have immediate effects on the rate of streamflow at the point of withdrawal. However, the adverse effect of groundwater on surface water may take years or even decades to show up. The effects of groundwater withdrawal include the timing, rates, and locations of streamflow depletions in ways different than that of surface water on groundwater. As shown in the pollution caused by failure of the Edenville Dam in Michigan, decades of discharging industrial waste and chemicals into the Tittabawassee and Saginaw rivers resulted in devasting buildup of pollution. To protect the nation against disasters that follow the breaching of a dam, the Federal government carries out a comprehensive inventor of the number of dams and their potential for failure.

Infrastructure Failure and Public Health

Water system infrastructure failure can result from an action that could seriously compromise public health or safety. This can be an undesirable or unintended event, occurrence, or situation involving the water infrastructure of a public or private water system. It can be the failure, discontinuation, or unintentional disruption of water services, or the catastrophic failure to adequately treat withdrawn freshwater brought about by malfeasance by the managers of a public water system. This infrastructure is public utility infrastructure and, if compromised, would result in a temporary loss of essential function and/or services.

Water System Operations, Management, and Funding

When we turn the faucet, we expect the water to be fit for consumption. Our simple expectation is that the operators of our local water utility have assured that the water is safe. To do this, the leaders of the water utilities organize their service and requisite budgets to assure several key components are in place guaranteeing that the water is safe to drink. First, they must obtain the water resource. This is done through drawing water from wells, tapping into a large regional supply systems, or by managing surface "watersheds" to collect runoff. Often utility operators will use a blending of these sources to provide a water portfolio to assure enough water for the consistent operation of their systems. In each case, they must then transport the water, test and treat the water to assure drinkability, and deliver it to end users in a clean and safe fashion. Each of these highly specialized components require trained and knowledgeable individuals with expertise in their field to assure the water coming out of the tap is always there and always safe to drink. Taken together, this is an expensive undertaking requiring adequate funding, competent management, and considerable technical and scientific expertise.

In the United States, this is done at the local level by city-operated water utilities or by local water utility districts. In the case of a city water utility, the mayor and city council function as the water and wastewater utility commissioners for the purpose of water system oversight. In the case of a water district, commissioners are directly elected from the population of the area severed to provide the oversite for the system. In each case, these elected officials are accountable to the voting public and are responsible for assuring that adequate funds are raised, competent management are hired to oversee the operation, and qualified technical staff are in place to operate the system and assure the water is safe.

Each of these major components: infrastructure repair and replacement; system operations, water quality, and management oversite, require thoughtful financial planning, management, and investment to assure they work well and in a complementary fashion. If one element is underfunded or poorly done, the quality of the water coming from the tap is jeopardized and public health and safety could be put at risk. Where water system failure occurs, it is typically because one or more of these components has been underfunded or suffered lax or insufficient oversight. Utility commissions, be they special district leaders or city council members, are continually under pressure to assure that taxes and fees are as low as possible and their public budget deliberations focus on assuring that the government is functioning in a cost-effective fashion while balancing the competing priorities for funding.

Water Utility Funding

To help assure that funding for water service is not used for other purposes like park maintenance and police protection, cities typically have set up dedicated funds to be used strictly for water services. Best practices in government accounting keep these funds segregated so that the rates charged by the water utility are set aside in a special water fund to be used solely for that purpose. This "fund integrity" is very important to assure that money collected through water rates are not used for other purposes. In this way, rate payers are assured that their payments for water are used only for the acquisition and distribution of clean water.

However, this does not assure that the rate base, or the rate structure is adequate for all the needs of the long-term safe functioning of the water system. Elected officials, always under pressure to keep rates low, may resort to underfunding infrastructure repair and replacement as cuts here are lees apparent than the other areas of utility budgets and the impact for the cuts may not be seen for years. After all, the buried pipes, storage tanks and pumps are "out of site and out of mind" and should district leaders cut funding for chemists, well operators, or utility management, immediate problems in water distribution would likely be immediately apparent. Whereas expensive pipe replacement projects can be postposed, with the problem left for some future leaders to address with the attendant increases in utility rates needed to fund the costly work. This cost avoidance strategy has been employed by cities and utility districts over the years with alarming frequency.

As noted earlier, one of the major vulnerabilities to any water system is the age and condition of the infrastructure used to deliver the water. Much of the nation's water infrastructure has reached the end of its expected life span or is about to, and in many jurisdictions repair and replacement of this vital infrastructure has been deferred to a point where failure is occurring. From large cities such as Jackson Mississippi to small districts like Keystone West Virginia, water systems are failing to provide clean safe water to their populations, as years of underfunding are catching up with them.

Meeting a Crisis with Reused Water

Water for reuse is water that has been used one or more times previously to its proposed reuse. It is collected, cleaned, and returned for reuse, usually in uses not directed for public drinking water but to replace unattainable resources of freshwater for agriculture irrigation or in a community use such as a golf course or park. However, it is also remixed with water in first use in regions where access to freshwater sources are insufficient to meet demand. It can come from many different sources, including previous irrigation use, thermoelectric generation facilities, domestic graywater, municipal wastewater, industrial uses, collected stormwater, and other sources. The process is known as water recycling or water reclamation. The water is reclaimed from a variety of sources, is then cleansed and treated as necessary before its reuse. The treated water is then used for such purposes as irrigation, added to raw potable water supplies for groundwater replenishment, industrial processes, or environmental restoration. Water for reuse can provide alternatives to existing water supplies and be used to enhance water security, sustainability, and resilience.

Incorrectly Defined Wastewater

Water for reuse is incorrectly referred to as wastewater. Wastewater is any water that has been used and then discarded back to its source. Wastewater is used to identify water used once and then may or may not be treated and reused. It is not inherently a problem; it is just the opposite of freshwater. It may have been used for domestic purposes or may be dirty stormwater that is not targeted for sanitation and reuse, or it may the saline water or ocean saltwater that can be treated and used domestically for the first time.

Reused water is clean water that is ready for final treatment and then for use. Although reclaimed and recycled water are often used to refer to the same thing, they are, in fact, different things. Municipal wastewater, for example, that must be cleaned through a series of mechanical, biological, and chemical processes to ensure it is safe and sanitary prior to being released to a natural body of water. Wastewater treatment systems, initially designed to preserve human and environmental health, have advanced to a level where this wastewater can be reclaimed and safely used for other purposes. Recycled or reclaimed water is most often used for such purposes as landscape irrigation, industrial processes, toilet flushing, or even further treated for drinking.

The many sources of water for potential reuse include municipal wastewater, industry process and cooling water, stormwater, agriculture runoff and return flows, and produced water from natural resource extraction activities. These sources of water are treated to meet "fit-for-purpose specifications" for a particular next use. These specifications are the treatment requirements to bring water from a particular source to the quality needed, to ensure public health, environmental protection, or specific user needs. For example, reclaimed water for crop irrigation would need to be of sufficient quality to prevent harm to plants and soils, maintain food safety, and protect the health of farm workers. In uses where there is a greater human exposure, water may require more treatment (EPA 2022b).

Classes of Reused Water Applications

Three general classes of water reuse include the following (excludes agricultural, environmental, and industrial applications):

- Non-potable water reuse: Water that is captured, treated, and used for non-drinking purposes, such as toilet flushing, clothes washing, and irrigation.
- Indirect potable water reuse: Water that will be treated with an environmental buffer and used for drinking water. For example, stormwater or wastewater is first directed to a municipal wastewater treatment plant for treatment. Once treated, it is then directed to an environmental buffer, such as a lake, river, or a groundwater aquifer which is used as a source for drinking water. The water is then treated at a drinking water treatment plant later when withdrawn and directed into the drinking water distribution system.
- Direct potable water reuse: Water will be treated and used for drinking water without an environmental buffer. In this scenario, stormwater or wastewater is directed to a municipal wastewater treatment plant and/or an advanced wastewater treatment facility for treatment. Once treated, it is then directed to a drinking water treatment plant for further treatment or sent directly to a drinking water distribution system.

Reclaimed Water

Reclaimed water is treated municipal wastewater that can be used again. The water that has traditionally been treated enough to be safely discarded is, instead, retained and may have undergone additional, processing for non-potable use. This method of reuse is now common in California, Florida, Arizona, and Texas. Longer periods of drought and more limited surface water, it has also become more common in Nevada. Reclaimed water has been reclaimed and made safe for reuse but is not to recycle water until it has been reused.

In arid states where drought and increasing population are limiting available supplies of fresh potable water, reclaimed water is often used for recharging an aquifer. It can also be water for cooling thermoelectric generation and other industrial uses. This water in, in effect, in a closed-loop system in what was originally withdrawn from surface or groundwater source, heated to become steam and subsequently allowed to cool before either it is returned to the source or used in another purpose in the same facility.

Reclaimed water is not the same as graywater. Graywater is defined as "wastewater from a household or small commercial establishment which specifically excludes water from a toilet, kitchen sink, dishwater or water used for washing diapers" (Nevada Department of Conservation and Natural Resources, n/d). Unlike reclaimed water, graywater does not undergo treatment. In graywater systems, the wastewater from bathroom sinks, tubs, or washing machines is used to water lawns, landscapes, or for other non-potable purposes (Ormerod 2020).

Reclaimed water can provide a source of alternative water to federal and state facilitates, and it is becoming more common for local municipalities to reclaim wastewater and sell it to customers to help lower the community's demand for freshwater. This water is often available at a significantly lower cost than potable water. Using reclaimed water from a local water utility is an excellent way to reduce potable water consumption and support the community's efforts in sustainable water management.

The supply of freshwater is increasingly under stress in many parts of the United States and particularly so in warmer regions of the country. Climate change is changing precipitation patterns, resulting in even less precipitation in already arid locations and more precipitation in others. Semi-arid regions such as the American Southwest are particularly in need of additional supplies of water. For several decades, after tertiary sanitation treatment some recycled water has been accepted as a substitute for drinking water for landscape watering and for irrigating agricultural crops not directly intended for human consumption. The use of this recycled water to augment scarce surface and groundwater supplies has gained more and more acceptance. Moreover, advances in treatment technologies have reduced the cost to the point where recycled water is in many instances cheaper than natural supplies (McNabb 2019).

There are two major categories of wastewater, often divided by its source: domestic (municipal) wastewater, and industrial wastewater. Stormwater is a third type. The two types of domestic wastewater: graywater and blackwater. Graywater is used water from sinks, bathing facilities, and laundry activities. Blackwater is sewage water from domestic and public toilets. Untreated blackwater may contain feces, urine, and used toilet paper. Because body wastes may contain intestinal disease organisms, it must be sanitized and biosolids must be removed before the final cleansed effluent is discharged into a surface water source from which it originated. A less common practice is discharging treated sewage into surface ponds or wetlands where it is allowed to evaporate or be absorbed into the earth. Treatment and purification removes the unwanted components that are necessary before the effluent can be discharged.

Industrial wastewater is water used in industrial and commercial operations, including water that is used for agricultural irrigation. Industrial wastewater that is used in industrial processes can contain such pollutants as residual acids, plating metals, and toxic chemicals. It requires complex and expensive treatment before being discharged. However, industrial wastewater that has only been used for cooling purposes as in thermoelectric power generation, is far less polluted and is often reused for the same purpose. Industrial processes generate many different pollutants in their wastewater. Separate processes are needed to treat each type of contaminant before the water is discharged or reused. After primary and/or secondary treatment, some highly polluted industrial wastewater may be sent for further treatment or entered into the usual municipal wastewater treatment system. Graywater includes wastewater from bathtubs, showers, bathroom sinks, clothes washing machines, and laundry sinks. In most jurisdictions, the official definition of graywater excludes wastewater from kitchen sinks, dishwashers, photo lab sinks, or laundry water from soiled diapers.

Recycled Water

Recycled water is water that was used at least once before, typically in a less than desirable source such as the graywater that is first used as domestic water other than as toilet water, is treated to a required sanitation level and then used for non-potable uses such as for a golf course or a public park, or to be used for some types of agriculture. It can also be water once used in an industrial purpose such as for cooling a thermoelectric generation facility, then cleansed and reused for a similar purpose.

Recycled water generally refers to treated domestic wastewater that is to be passed back into the water cycle. Before it was reclaimed, wastewater targeted for recycling may be mixed with freshwater. Recycled water is water that has been used at least once, is then treated for purity and then may be used again. The reused water is often reffered to as recycled water.

Recycled Water Limits

Recycled water is limited in the number of uses for which it can be reused. The typical source may be domestic sewage, industrial wastewater, and stormwater runoff. Wastewater discharged from buildings and processes, treated at a wastewater treatment facility, is then reused in applications such as irrigation and industrial processes. The degree of treatment needed before reuse and the quality of recycled water depends upon the source and the level of treatment. It often contains more total dissolved salts and nutrients and therefore may be restricted to sustainable landscape irrigation or to recharge groundwater aquifers. The National Water Reuse Database (WRRF) reports on water utilities that supply reclaimed wastewater, including data from water utilities in six states: California, Florida, Texas, Arizona, Colorado, and Nevada. The data mapped include utility name, water reclaimed, facility name, address, city and state, water use type, and the year when the facility provided the latest data to the foundation. General descriptions of reclaimed wastewater uses and types are provided below (Gauron and Coulter 2021; EPA 2022b):

- Irrigation for agriculture
- Irrigation for landscaping such as parks, rights-of-ways, and golf courses
- Municipal water supply
- Process water for power plants, refineries, mills, and factories
- Indoor uses such as toilet flushing
- Dust control or surface cleaning of roads, construction sites, and other trafficked areas
- Concrete mixing and other construction processes
- Supplying artificial lakes and inland or coastal aquifers
- Environmental restoration

- Commercial: landscape irrigation, car washing, commercial operations
- Golf Course: irrigation for golf courses
- Industrial: cooling towers, boiler feeds, processing
- Municipal and Public Areas: irrigation to areas with public access such as parks
- Residential: irrigation to housing areas.

Municipal wastewater is typically cleaned through a series of mechanical, biological, and chemical processes to ensure it is safe and sanitary prior to being released to a natural body of water. Reclaimed water, before it becomes recycled is most often used for such purposes as landscape irrigation, industrial processes, toilet flushing, or drinking. There are no federal standards that directly govern reclaimed water use in the U.S. However, the EPA has established comprehensive guidelines for water reuse but leaves states with the responsibility of developing regulations. Many states, including California, Arizona, Texas, and Nevada, have developed regulations to support and encourage greater water reuse.

In Nevada, for example, reclaimed water is defined as "water that has received at least secondary water treatment and is reused after flowing out of a wastewater treatment facility" (NDCNR, n/d). Secondary wastewater treatment defines the minimum standard for legal discharge of treated wastewater into waterways in the U.S. The quality of reclaimed water depends on the level and type of wastewater treatment. As a result, reclaimed quality can be treated to match predefined water quality standards and disinfected to levels necessary for the chosen method of reuse. In Nevada, reclaimed water use falls into six different categories depending on its quality, ranging from A+ through E. The quality categories and allowable uses for reclaimed water are included in the lowest ratings.

Stormwater Resource

The heavy precipitation that is associated with extreme weather events have long been considered to be a major problem and an even greater disaster with loss of many people and destruction of homes and communities. In addition to being considered a major problem, stormwater and the flood water that follows extreme weather events, when is collected and treated is increasingly being looked upon as a source of freshwater.

Climate is the long-term average of the weather in a given place. While the weather can change in minutes or hours, a change in climate is something that develops over longer periods of decades to centuries.

Climate is defined not only by average temperature and precipitation but also by the type, frequency, duration, and intensity of weather events such as heat waves, cold spells, storms, floods, and droughts. Stormwater is rainwater or melted snow that runs off streets, lawns, and other sites. When stormwater is absorbed into soil, it is filtered and ultimately replenishes aquifers or flows into streams and rivers. In developed areas, impervious surfaces such as pavement and roofs prevent precipitation from naturally soaking into the ground. Instead, water runs rapidly into storm drains, sewer systems, and drainage ditches and can cause downstream flooding, stream bank erosion, pollution of surface water resources, residential and commercial infrastructure and combined storm and sanitary sewer system overflows.

Stormwater harvesting, also called rainwater harvesting, is the collection of rain and storm water collected as runoff from a structure or an area, known as a catchment. The water thus collected can be saved and sanitized for human, animal, or garden crop use, added to groundwater or any other water need. The water can be used immediately or be stored in aboveground ponds or in subsurface reservoirs, such as a cistern or aquifer for later use. Water harvesting is an ancient practice that has enabled some societies to subsist in semiarid or arid areas where other sources of freshwater are neither available nor cannot be consumed without purification.

The 2030 WRG analysis included four estimates of: (1) the impact of economic and social growth on the nation's water resources, (2) costs and requirements of meeting future water demand, (3) comparisons of different growth scenarios, and (4) the implications for stakeholders. The WRG found that combinations of existing low-cost surface watercourse remediation approaches could eliminate up to 100% of the predicted 2030 demand-supply gap in the regions studied. Recommendations included a mix of solutions to increase water supply and to improve water productivity. Greater use of reclaimed stormwater was one of the recommendations of regaining supplies, otherwise wasted water. Also recommended was using stormwater runoff management as a means of contributing to reducing flood damage, support surface water supply infrastructure, protection of the municipal environment, protect public facilities and enhance community.

Stormwater Runoff Pollution

Stormwater runoff picks up pollutants like trash, chemicals, oils, and dirt/sediment that can harm our rivers, streams, lakes, and coastal waters. To protect these resources, communities, construction companies, industries, and others, use stormwater controls, known as Best Management Practices (BMPs). These BMPs filter out pollutants and/or prevent pollution by controlling it at its source. The EPA's National Pollutant Discharge Elimination System (NPDES) stormwater program aids communities in regulating some stormwater discharges from three potential sources: municipal separate storm sewer systems, construction activities, and industrial activities. Operators of these sources might be required to obtain an NPDES permit before they can discharge stormwater. This permitting mechanism is designed to prevent stormwater runoff from washing harmful pollutants into local surface water (EPA).

Most states are authorized to implement the stormwater NPDES permitting program, although the EPA remains the permitting authority in a few states, territories, and on most Tribal lands. Population growth and the development of urban/urbanized areas are major contributors to the amount of pollutants in the runoff as well as the volume and rate of runoff from impervious surfaces. Together, they can cause changes in hydrology and water quality that resulted in habitat modification and loss, increased flooding, decreased aquatic biological diversity, and increased sedimentation and erosion.

Stormwater as a Resource

Weather researchers have determined that extreme weather events such as heat waves and large storms are likely to become more frequent or more intense with continuing climate change. Although average temperatures have risen in the U.S. since records were kept in the early 1900s, the rate of increases has accelerated since the 1990s. Eight of the top ten warmest years on record have occurred since 1998. Within the United States, temperatures in parts of the North, the West, and Alaska have increased the most. Accompanying the temperature increases are changes in weather patterns, one of which is the differences in precipitation and the severity of rainstorm events. The effects of these increases in severe storms are revealing that many if not most of the Nation's water management systems are using outdated design guidelines. A 2019 study published in the AGU journal *Geophysical Research Letters*, reported the "rising number of extreme storms combined with outdated building criteria could overwhelm hydrologic structures like stormwater systems. Current design standards for United States hydrologic infrastructure are unprepared for the increasing frequency and severity of extreme rainstorms, meaning structures like retention ponds and dams will face more frequent and severe flooding" (AGU 2019). If the current water storage systems are unable to collect the stormwater, it cannot be used to augment municipal landscape irrigation.

Produced Water

Produced water is the aqueous liquid (an aqueous liquid is a solution in which the solvent is water) co-produced from a petroleum well which is produced along with the oil and/or gas during normal production operations. This includes the groundwater naturally occurring alongside hydrocarbon deposits, as well as water injected into the ground in the drilling process. Produced water is an inextricable part of the hydrocarbon recovery processes, yet it is by far the largest volume waste stream associated with hydrocarbon recovery. The contaminants present in produced water require deoiling, desalination, degassing, suspended solids removal, organic compounds removal, heavy metal and radionuclides removal, and disinfection. These treatment goals are essentially the same for potable, non-potable reuse, or disposal, although the level of contaminant removal required for potable reuse can be significantly higher, depending on the quality of the produced water. Increasingly stringent environmental regulations require extensive treatment of produced water from oil and gas productions before discharge; hence the treatment and disposal of such volumes costs the industry annually more than US$40 billion. For oil and gas production wells located in water-scarce regions, limited freshwater resources in conjunction with the high treatment cost for produced water discharge makes beneficial reuse of produced water an attractive opportunity. The amount of produced water, and the contaminants and their concentrations present in produced water usually vary significantly over the lifetime of a field. Early on, the water generation rate can be a very small fraction of the oil production rate, but it can increase with time to tens of times the rate of oil produced. In terms of composition, the changes are complex and site-specific because they are a function of the geological formation, the oil and water (both in-situ and injected) chemistry, rock/fluid interactions, the type of production, and required additives for oil-production-related activities.

Conclusion

This chapter discussed two themes in the management of freshwater and wastewater. The first was the shortages in meeting workers in the water and wastewater industries and the critical need to replace aged and rapidly disintegrating infrastructure. Most of these workers are employees of municipal utilities, where salaries do not meet industry levels. The EPA, which is governed by the Clean Water Act to monitor state and local programs for the training of workers and testing of delivered freshwater and sanitizing wastewater before it is discharged into local water courses, describes the growing need for more trained water and wastewater treatment and delivery personnel: "Currently, water utilities face challenges in recruiting, training, and retaining employees. These challenges are exacerbated with roughly one-third of the water sector workforce eligible to retire in the next 10 years. Additionally, as the technologies that are used in the water sector become more advanced (e.g., state of the art water reuse technology), there is a growing need to train and employ water protection specialists with specialized technical skills" (EPA 2022e).

Water infrastructure failure and replacement was a part of the problems facing municipal water utilities and wastewater sanitizing and discharge to natural waterbodies. Water system infrastructure failure can be an unexpected failure or unintended event, occurrence, or situation involving a utility or small private water organization's delivery system's water infrastructure or disruption in the water services that could seriously compromise public safety. This infrastructure can be a reservoir's dam or emergency overflow system, collection water from a stream, or aquifer, the system's incoming water mains, treatment equipment, distribution water pipes, or system laterals from the pipe to a receiving unit. All of these except for the laterals is usually the responsibility of the water utility or supplier. Delivery of impure water to the final recipient as occurred in the Flint, Michigan problem, may not be considered an infrastructure failure, but instead as a breakdown in the organization's management system.

Reclaimed, reused, or recycled water is a solution to a water problem rather than an infrastructure breakdown or management error. It is a readily available means of putting once-used freshwater to a desirable second use while continuing to meet first-use needs in times or locations of water shortage. To be reused, the collected water is commonly restricted to agriculture or recreational uses, as for watering parts and/or golf courses.

References

AGU (American Geophysical Union). 2019. US infrastructure unprepared for increasing frequency of extreme storms. *ScienceDaily*. 1 August 2019. Available at www.sciencedaily.com/releases/2019/08/190801120209.htm.

ASCE (American Society of Engineers). 2021. Report card for America's infrastructure: Our nation's drinking water infrastructure. Available at https://infrastructurereportcard.org/wp-content/uploads/2020/12/National_IRC_2021-report.pdf.

City of New Orleans. 2021. Infrastructure failure: Water systems. Hazard Mitigation Plan. Available at https://ready.nola.gov/hazard-mitigation/hazards/infrastructure-failure-water-systems/.

EPA (Environmental Protection Administration). 2021. Building sustainable water infrastructure. Available at www.epa.gov/sustainable-water-infrastructure/building-sustainable-water-infrastructure.

EPA (Environmental Protection Administration). 2022a. Basic information about water reuse. Available at www.epa.gov/waterreuse/basic-information-about-water-reuse.

EPA (Environmental Protection Administration). 2022b. What is green infrastructure? Available at www.epa.gov/green-infrastructure/what-green-infrastructure.

EPA (Environmental Protection Administration). 2022c. Harmful algal blooms. Available at www.epa.gov/nutrientpollution/harmful-algal-blooms.

EPA (Environmental Protection Administration). 2022d. What is radon? Available at www.epa.gov/radon/what-radon.

EPA (Environmental Protection Administration). 2022e. History of the Clean Water Act. Available at www.epa.gov/laws-regulations/history-clean-water-act.

Gauron, Trevor, and Rebecca Coulter. 2021. Self-audits an efficient way to maintain stormwater management programs and bring communities to a place of preparedness. *Stormwater* 22 (5): 22–25.

McNabb, David E. 2019. *Global pathways to water sustainability*. London: Palgrave-Springer.

Ormerod, Kerri J., Samantha, Redman, and Loretta Singletary. 2020. *Reclaimed water: Uses and definitions*. University of Nevada-Reno College of Agriculture. Available at https://naes.agnt.unr.edu/PMS/Pubs/2020-3813.pdf.

PART III

Regional Problems and Solutions

CHAPTER 9

Water Crises in the West

Large areas of the West are known for their extensive supply of freshwater while other sections are facing water stress that is not far from a water crisis. However, sections of the West are facing a number of environmental crisis along with severe prolonged drought. This section of the nation can be divided into two sections, a coastal region and a mountain region. The three coastal states are considered the wetter states, with the four mountain states the drier states. The 50 states and the District of Columbia are ranked from 1 to 51 according to their total annual precipitation. The wetter states in the West are Washington, with annual participation of 38.4 inches, placing it 30th of the 51 total; Oregon is ranked 20 with 43.62 inches of annual precipitation; and California with 22.2 inches is ranked 41st. The dry states in this region are Idaho, with 16.91 inches of annual precipitation is ranked 45; Montana with 15.3 inches is ranked 47; and Wyoming, ranked 49, receives average precipitation of 13.23 inches. Eastern Wyoming overlies the High Plains aquifer and is, therefore, often included in in the Great Plans section or the northwest region. Nevada, with just 9.46 inches of annual participation, is ranked 51, the driest of all states. Louisiana, not discussed in this chapter, annually receives 56.9 inches of precipitation making it the wettest U.S. state (USA.com. 2022).

All of the seven states depend upon both surface water and groundwater for large proportions of their annual freshwater needs, and all are

D. E. McNabb and C. R. Swenson, *America's Water Crises*,
https://doi.org/10.1007/978-3-031-27380-3_9

subject to periodic droughts. Several of these states are among the highest water users and must often cope with demands for increase in water supply.

Four mountain ranges dominate the West: the Coast Range, the Olympic Range, the Cascade Range, and parts of the Rocky Mountains. Winter storms approaching from the Pacific Ocean collect much of the precipitation that falls on the Western United States. As a result, winter in the region often receive a mix of rain and snow. The mix means sections west of the mountains can expect most of the precipitation they receive to be as rain with many surface water resources, whereas east of the mountain ranges, annual snowpack has been the dominant form, with spring and summer occurring as melting of the snowpack.

The warmer and drier climatic conditions taking place now are resulting in declining surface water resources in the areas east of the mountain ranges, which commensurate with increasing evaporation of surface stored in reservoirs. Melting snowpack had long been the surface water source for agriculture in the region. Earlier melting means less water is available for agriculture over the early growing season, and other sources must make up the loss. Where available, groundwater has been the source used to augment the diminishing surface water resource. Recharging the groundwater aquifers depends on the amount of annual snowpack, precipitation patterns, groundwater aquifer sustainability, and flow controls such as on the internationally regulated Columbia River. Variation in supply makes the region susceptible to drought. Figure 9.1 is a simple map of the region.

The northern five states of this section share a number of climate and precipitation problems. Over the last several decades, much of the region has suffered through periods of extreme drought. The states depend upon surface water that originates outside of the region. Summer water needs have relied upon melting of winter snowpacks to maintain summer flow. However, because of lighter winter snows, lengthy periods of drought and warmer summers, rivers have not met traditional precipitation amounts. Therefore, withdrawals of groundwater have increased significantly since 2000. This, in turn, has resulted in withdrawals that exceed the natural recharging of the regional aquifers.

Access to reliable water resources is thus a growing source of stress on the water supplies of the region. The two sectors with the highest need

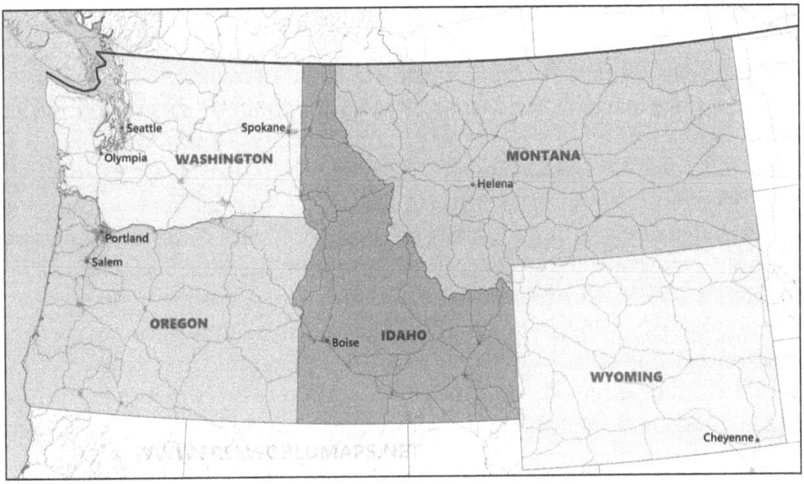

Fig. 9.1 The northern five states of the West Region (*Source* www.freeworld maps.net)

Table 9.1 Shares of state area included in each drought categories July 2022

Portions of state included					
State	*D0:D4 Abnormally dry (%)*	*D1:D4 Moderate drought (%)*	*D2–D4 Severe drought (%)*	*D3:D4 Extreme drought (%)*	*D4 Exceptional drought (%)*
Nevada	100	100	99.5	63.6	29.8
California	100	99.8	97.5	59.8	12.7
Wyoming	89.6	62.9	25.0	7.1	0.0
Montana	39.0	21.2	15.4	3.5	0.0
Idaho	67.9	45.1	17.0	3.7	0.0
Oregon	75.4	66.4	52.8	31.7	1.8
Washington	34.6	8.3	0.0	0.0	0.0

Source NIDIS Drought.gov

for freshwater withdrawals, agriculture and electric power, are important contributors to the nation's economy. Table 9.1 includes how much of the state is impacted at five different levels of drought in July of 2022.

West Coast Drought Data

Although drought can affect every state and territory of the United States, no two states experience the same set of impacts during a drought. Droughts can destroy farm crops, enable wildfires, lower water levels in lakes and reservoirs, and damage ecosystems. In coastal states, drought can also enable saline water intrusion of freshwater aquifers. In an effort to aid states cope with drought's impacts, NOAA's National Integrated Drought Information System (NIDIS) program was reauthorized in 2014 and 2019 to coordinate and integrate drought research, building upon existing federal, tribal, state, and local partnerships in support of creating a national drought early warning information system. An example of the weekly condition reports on drought conditions as of the middle of June 2022 are shown in Table 9.1. More than half of the five states were experiencing moderate drought, and all but Washington was experiencing severe drought. Nearly half of the entire state of Oregon was in extreme drought.

By the end of June 2022, there were some improvements in the share of each state's water supply even as they still were under drought conditions in five of the seven West states: The share of the state in areas of extreme and exceptional drought, both California and Nevada, was either the same as for May or higher. For Nevada, extreme drought increase was observed from 55.4% of the state in May to 63.6% at the end of June. The share of Nevada in exceptional condition increases from 21.3 to 29.8% of the state. For California, extreme drought conditions remained at 59.8% of the state for both periods, although the share in exceptional drought increased from 11.6 to 12.7%.

Table 9.2 Water use in Montana

Annual water use	*Amount acre-feet (m^3)*	*Percent of the annual total (%)*
Hydropower	72,000,000	85.9
Irrigation	10,395,000	12.4
Reservoir evaporation	1,002,000	1.2
Municipal, livestock, industry and domestic	384,000	0.5

Source Cross et al. (2017), MT DNRC (2015)

Oregon was the only state to continue to suffer from large sections under drought conditions. Oregon's May drought conditions were 65.6% of the state in severe drought, 48.1% in extreme drought, and 14.4% in exceptional drought. All three conditions improved to 52.8% in severe conditions, 31.7% in extreme drought and just 1.8% of the state still in exceptional drought conditions. None of the remaining states reported extreme drought, and only three were still in severe drought. Washington had smaller dry conditions and had just 8.3% of the state in moderate drought.

In a drought condition research report, the California Natural Resources Agency collaborated with the state's Environmental Protection Agency, the Governor's Office of Emergency Services, the Department of Food and Agriculture, the Department of Water Resources, the Department of Fish and Wildlife, the State Water Resources Control Board, and the Department of Public Health to produce a weekly drought report on drought conditions in the state. As winter water conditions were coming to an end, the Agency's April 11, 2022 California Drought Conditions weekly report included the following selected points (Drought.ca.gov April 11, 2022):

Columbia Plateau Goundwater Crisis

The aquifer systems encompassing the region include the Columbia Plateau, the Cascade Range, the Modoc Plateau, the High Lava Plains of Oregon, the volcanic Snake River Plain (east to and including parts of the Yellowstone Plateau volcanic field), and the northwest corner of the Basin and Range Physiographic Province (Curtis et al. 2020). The location of these aquifers contribute to the Northwestern Volcanic Aquifer system that provides groundwater for the much of this drought-impacted northwestern region (Fig. 9.2).

The regional Northwest Volcanic Aquifer (also known as the Northwest Volcanic Province) extends across most of the states of Washington, Oregon and Idaho, with smaller segments of northern California, Wyoming and Nevada. Although surface water continues to remain at major source of freshwater withdrawals an average of 75% of all water used in the region, the volcanic terrain supports large groundwater flow systems, as well as geothermal systems that can be tapped for energy. As of a 2017 USGS report, the full size and supply capacity of the overall aquifer had yet to be fully assessed. Both surface water and groundwater

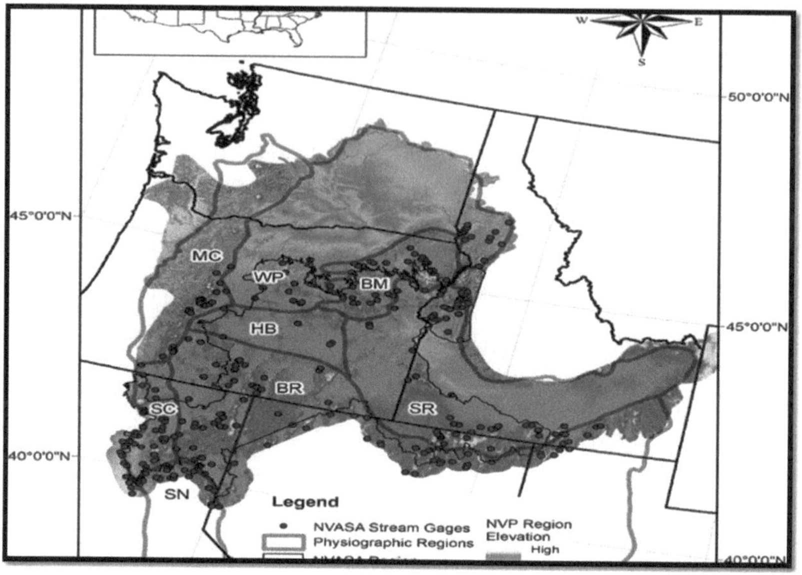

Fig. 9.2 The Northwestern Volcanic Aquifer study area with sub-regions (Key: Dots: research well site. Sub-regions: MC Middle Cascades; SC Southern Cascades; SN Sierra Nevada; WP Walla Walla Plateau; HB Harney Basin; BR Basin and Range; BM Blue Mountain; SR Snake River) (*Source* USGS; Curtis et al. [2020])

accessibility in this region have long been a significant water supply source for all purposes in the region.

The Washington Department of Ecology is responsible for managing water resources throughout the state. This includes implementing projects to meet current and future water needs in the Columbia River basin to support growing communities, the agricultural economy, endangered fish, and the natural environment. The river, which begins in Canada, travels south for most the width of the state to where it joins the Snake River. It then flows westward to the Pacific Ocean. In this journey, the Columbia passes through both Eastern and Western Washington. The area separated by the North-South Cascade Mountain Range, areas with distinct climatology, emphases and uses prerogatives. In addition to statewide issues, the DOE's activities for basins include agricultural areas east of the mountain range, and forests in the areas west of the mountains. The different focus for the primary surface water in the basin is supported by a distinct legislature-approved water budget, as established through the Columbia River Basis Development Account (CRBDA). The focus on this account is to plan, and develop new storage, improve or alter operations of existing storage facilities, implement conservation projects, and develop pump exchanges, or any other actions designed to provide access to new water supplies for both instream and out-of-stream benefits across Eastern Washington. The program focuses on legislature priorities to:

- Deliver replacement water to…subarea farmers, who now rely on a declining aquifer.
- Make water available for current municipal, domestic, industrial, and irrigation needs—especially during drought—and develop water to respond to future needs and climate change, and
- Assure water is available to benefit streamflows and habitat when fish need it most (SWDOE n.d).

The region has experienced multiple droughts since 2000. In 2015 and again in 2018, almost all the region reached historic drought conditions. Oregon and Idaho had been suffering under prolonged drought for multiple years. The Pacific Northwest Drought Early Warning System (DEWS) is a collaborative federal, tribal, state, and local interagency effort to improve early warning capacity and resilience to drought in the region. Selected descriptions of water conditions in individual states of the region are included in the following section.

Drought and Water Crisis in Oregon

In 2021, drought was also covering the Central half of the state of Oregon. NOAA and the U.S. Drought Monitoring reported extreme and severe drought levels which existed in much of the western region of the state. Continual higher than usual higher temperatures and drier days resulted in a lower snowpack and streamflow. With lower soil moisture levels there was less irrigation for farmers and more wildfires. It has also stressed the vegetation's increasing susceptibility to insect damage, and reduced grazing land for livestock, especially in the Coquille, Deschutes, Klamath, Rogue, and Umpqua watersheds.

Surface water accounted for approximately 70% of the state's total withdrawals. Major sources of surface water in Oregon are from the Columbia, Willamette, and Snake Rivers. Despite these resources, in 2021 the state was in the third year of drought. A wider view of Oregon's major water resources was included in a U.S. EPA report: The combination of a declining water supply and increasing population stressed many rivers and brought on a conflict between Oregon and California and between farmers and Native tribes over declining salmon migration. With reduced winter precipitation and much smaller and earlier melting of critical snowpacks, the water levels of rivers and other sources of freshwater in Oregon dropped significantly. The continuing drought drove sections of Oregon and Northern California into what was declared to be an ecological water crisis.

In the third year of a continuing drought, the two-state water crisis began in Oregon, when U.S. Bureau of Reclamation regulators stopped release of water from an important surface water reservoir in Oregon to the Klamath River that crosses into California. The federally operated Klamath Reclamation Project brings Klamath River water from the Upper Klamath Lake just north of the Oregon-California border to more than 130,000 acres of agricultural areas. The Klamath River water is needed for wildlife refuges for migrating birds and to flush and cool river water for passage of large numbers of spawning salmon. Local and national news sources claimed the water problem was turning out to be "shaping up to be the worst water crisis in generations, without the release of water this season into the main canal that feeds the bulk of the massive Klamath Reclamation Project, marking a first for the 114-year-old irrigation system." The Bureau had earlier announced that irrigators would get far less water than usual. However, it later announced because of the continuing drought, no water could be released that season.

The 2022 National Weather Service's Water Supply Outlook was not encouraging: "The water supply forecast for the spring and summer of 2022 is below average for most of Oregon, except for some watersheds draining the North Oregon Cascades and some isolated watersheds in northeast Oregon, where forecasts are near-normal. Most of the areas projected to be below average are already stressed due to drought conditions for much of the past 2 years" (NWS 2022). Groundwater is the major source of year-round dependable water supply in the northeastern section of the state.

Idaho Water Stress

Water stress in 2021 was the fourth consecutive year of drought for Idaho and for most of the northwest, and weather forecasters predicted 2022 would follow the same path. The period after January 8, 2022 was exceptionally dry south of the Clearwater Basin. Snow accumulation was minimal across many basins until widespread early March and April storms brought additional moisture. Southern Snake basins saw larger snow water equivalent (SWE) increases with these early storms compared to the rest of the state. However, a snowy winter would be needed to make up for low 2021 water year (WY). During the annual dry period, reservoir carryover did not occur and the snowpack was well below normal across the entire state. Water users were encouraged to prepare for a short irrigation season and reductions. In general, reservoir storage is below normal across our state. When early April storms did eventually lead to a second peak in snowpack, the outlook was less alarming.

Continuing drought and changing precipitation are restricting access to surface water and groundwater resources for Idaho's two biggest water users, the agricultural and residential sectors. Weather records since the 1950s for the entire northwest region confirm that summers are getting hotter, the wildfire seasons are lasting longer, population growth is increasing the demand for freshwater, and groundwater pumping demand over the Eastern Snake Plain aquifer is increasing. Water withdrawals in general had to be reduced by an average of 12.5% in 2016. After a brief respite, another water crisis began in 2020 when the drier than usual fall season was followed by a below-normal snowpack over the winter and was then followed by a dry spring. During the summer of 2021, Idaho experienced record-high temperatures. Conditions improved somewhat during the winter months of 2022, and sections of Idaho and Oregon were expected to be forced to endure more drought during the 2022 summer.

Surface Water Stress in Montana

Montana depends on surface water for most of its water resource, making the state highly dependent upon snowmelt for maintaining warm season freshwater flow. Groundwater provides 94% of Montana's rural domestic water supply and 39% of the public water supply. Every day approximately

90 million gallons of groundwater are used for irrigation, 16 million gallons to supply water for livestock, and 20 million gallons per day are used to support industry. Amounts used are listed in Table 9.2.

Water Stress in Wyoming

Like its neighbor state of Montana, drought is a continuing threat in Wyoming. The state has to endure periods of moderate to severe drought since the water year of 1999–2000. The severity of the droughts varies from year to year and from sections of the state, with more precipitation declines in eastern Wyoming. Drought conditions eased somewhat in mid-2008 but resumed in 2012 and lasted for two years. Drought returned in 2014 and 2015, but after improving for several years, beginning in 2020 the state was again hit with periods of severe and extreme drought (WRDS 2021). In March of 2022, 100% of the state was under an abnormally dry period with 97.2% of the state in a moderate drought condition and 59.9% in a severe drought condition. However, several winter storms brought the annual snowpack to near the annual averages. In the April snowpack/SWE report for water year 2022 (the 2021–2022 snow season), Wyoming's snowpack/SWE is 89% of median with a basin high of 104% (Laramie) and a basin low of 68% (Cheyenne). At the same time, the state Snow Water Equivalent (SWE) was at 91% of median and at 122% in 2020. Because the SWE measures snowfall rather than snow depth, it measures the amount of water that will be released from the snowpack when it melts.

The 2021–2022 water year was some improvement over the previous two years of drought in Wyoming. In 2018, the USGS estimated that irrigation accounted for about 7790 million gallons per day or about 95.6% of total withdrawals for all uses in Wyoming. Surface water was the source of about 93% of irrigation withdrawals, and groundwater provided about 7% (USGS 2018a, b).

Recent improvements to Wyoming's current 2-year drought have been limited to basins across north Central through northeastern part of the state. Moderate to severe drought conditions continue for almost all basins in Wyoming. Any additional improvements in drought conditions across the state will be dependent on continued wet and cool conditions during the remainder of the spring. The latest outlook for the remainder of May indicates that there will be near average temperatures and precipitation. However, June and July were expected to be drier than average and much warmer than average across most of Wyoming.

Water Stress in Washington State

In an early spring season 2022 report, the EPA announced that 75% of Washington's total water supply is from surface water and 25% is from groundwater. However, over 60% of drinking water in Washington State is supplied by groundwater. Groundwater is therefore an important resource for millions of people in the state. With the declines in the available surface water, the region's agriculture and municipal water user had to turn to groundwater resources.

On July 14, 2021, the Washington State Department of Ecology (DOE) declared a drought emergency for the entire state with the exception of the cities of Seattle, Tacoma, and Everett. The drought was not expected to be as severe as what occurred. The state's mountain regions had received a large snowpack accumulation over the winter, resulting in promising water supply forecasts. However, spring turned out to be the second driest spring in the state since 1895, thereby causing early snowmelt. High temperatures during the summer added to the early snowmelt, creating critical conditions for the state's freshwater resources (Shahpurwala 2021).

Washington's Water Sense program described the importance of water to the economy of Washington in the following brief introduction: "Water is arguably Washington's most valuable resource, supporting a multi-billion-dollar agricultural industry, sustaining one of the nation's most prominent commercial fishing industries, and generating one-third of the nation's hydroelectric power." The report went on to ad this qualifier: "However, the state's water supply—and its economy—are threatened by stress from excessive water withdrawals, increasingly frequent and intense droughts, a growing population, and the potential effects of a changing climate" (EPA 2017a, b, c, d).

Agriculture Water Needs

Washington is a significant contributor to the U.S. agriculture and power sectors, the sectors that are the top users of freshwater. In 2019, Washington state ranked 12th among all 50 states in the value of food production of $9.49 billion (USDA). Washington produces approximately one-third of the nation's hydroelectric power generated in the U.S. Irrigated agriculture in the state requires very large water withdrawals. The entire region suffers from increasingly frequent and intense droughts, an increasing population, and the potential effects of a changing climate. Even recognizing this situation, power generation and irrigated agriculture continue to be the two major users of water in Washington.

In July of 2021, the Washington State Department of Ecology declared a drought emergency for the entire state with the exception of the cities of Seattle, Tacoma, and Everett. Although the state had received a substantial snowpack accumulation over the winter, the year proved to be the second driest spring in Washington since 1895. Very high temperatures that summer accelerated the annual snowmelt, the effect of which was exceptionally low river and lake surface water. The year turned out to be the first time Washington had been given an s "exceptional drought" (D4 classification), the government rating, since 2000, the year the U.S. Drought Monitor began (WWT 2021).

Water Resources in California and Nevada

California and Nevada are the two states forced to cope with attacks on its water resource from more than one primary source. In California, the leading supply in the south is imported from the Colorado River; in the north, it is water from the Sacramento River. In Nevada, the source in the south is also the Colorado River; in the north, it is the Truckee River. Climate change is forcing declines in these and supplemental freshwater sources. The states are often included in the either the Southwest or the West Coast regions. Climate warming in this water-stressed region means both are constantly seeking ways to better ensure they will have enough water to meet all the needs of all the sectors that depend on them. Both are facing declines in sustainable water supplies, with little promise of finding affordable replacement resources of supply.

The weather is getting hotter and drier, droughts get longer, and all research predicts heat waves are likely to become more intense and more prolonged in the near future. Both California's or Nevada's share of the over allocated Colorado River surface water is already declining in quantity and will no longer function as a major surface water supplier. Although Las Vegas, Nevada, and Los Angeles, California, have become more efficient at water conservation, conservation alone cannot solve the region's increasing water scarcity. Both states are enduring a continuing water emergency, with a crisis growing more severe each year.

Emergency in California

In April 2022, California was officially in the third year of a precipitation emergency and had to adopt emergency regulations for drought years. Residents were to prevent the waste, unreasonable methods of water use, and avoid unreasonable methods of diversion of water. The state encouraged greater water recycling and water conservation and curtailed diversions when water was not available under the diverter's priority of water rights. California would also require reporting of stream diversion and its proposed use.

The U.S. Drought Monitor (USDM) classified 100% of California as at least moderately dry, and almost the entire state of California experiencing severe or extreme drought conditions. On April 21, 2021, California's governor declared a drought state of emergency under the provisions of the California Emergency Services Act, to take place in Mendocino and Sonoma counties due to drought conditions in the Russian River waters where most of the Russian River watershed is in severe or extreme drought. On May 10, 2021, the governor expanded the drought proclamation to include counties within the Klamath River, Sacramento-San Joaquin Delta, and Tulare Lake watersheds.

California Land Subsidence

New data indicates that large sections of the California Central Valley land continues to sink. In February 2022, new satellite-based data showed that subsidence due to excessive groundwater pumping continues. The areas experiencing the most subsidence from October 2020 through September 2021 are in the San Joaquin Valley, with a maximum of 1.1 feet of subsidence observed in the region, and the Sacramento Valley, with a maximum of 0.7 feet in the region. Data show that in that period, subsidence of greater than 0.5 feet per year expanded to more areas than observed the previous year. However, fewer areas experienced higher rates of subsidence than at the end of the last drought in 2016. Continued subsidence makes it more difficult for local agencies to bring basins into sustainable conditions by 2040, as state law requires. DWR has intensified statewide subsidence monitoring to help identify impacts and address them collaboratively with local groundwater agencies, counties, and landowners.

California has a diverse climate and geography that ranges from hot, dry deserts to snow covered mountain peaks and a long seacoast. Balancing water supply and demand is a perennial problem for California. The state, no stranger to droughts and water restrictions, is a national leader in water efficiency and conservation initiatives. However, despite its skills in water conservation, extended periods of drought and forest fires have taken their toll on the state's water resources (EPA 2017c). In 2022, it was entering on the third year of the latest in a long history of drought.

The climate of the Central Valley is Mediterranean and Steppe, characterized by hot summers and mild winters, thus allowing for a year-around growing season; at least one crop is under cultivation at all times. About 85% of the precipitation falls from November to April. Most of the precipitation that falls on the valley floor evaporates before it can infiltrate downward to become recharged. The Coast Ranges intercept much of the moisture that moves inland from the Pacific Ocean, so that annual precipitation in the valley is relatively low. Annual precipitation also decreases from north to south, with an average of about 23 inches in the northern part of the Sacramento Valley, to about 6 inches in the southern part of the San Joaquin Valley. During droughts, any benefits of annual precipitation is often lost to evapotranspiration throughout the entire valley, thereby resulting in a net annual moisture deficit.

The mountains that surround the Central Valley intercept moisture from eastward-moving weather systems that have traditionally provided an annual surplus of surface water in the form of rain and snow. Early historical records show the state could exceed 80 inches annually in the Sierra Nevada range. Annual runoff from rainfall and snowmelt has been approximately 32 million acre-feet. Most of the runoff originates in the Cascade Range and the northern Sierra Nevada. This water flows to the valley in perennial streams and provides nearly all the average annual 12 inches of recharge the valley aquifer system receives. Conditions contributing to California's 2022 water crisis in the first two decades of the twenty-first century are highlighted in these points:

- The 2020 water year was the second driest on record. All of California's 58 counties are under a drought emergency proclamation.

- After two dry years, reservoir storage was below 2019 levels, underscoring the need for ongoing water conservation.
- Following substantial rain and snowfall in December, January 2022 was extremely dry. With very little precipitation on the horizon, the statewide Sierra snowpack has gone from 160% of average at the beginning of January to 73% in mid-February. December, January, and February are typically California's three wettest months.
- Californians were asked to reduce their water use by 15% over 2020 levels to protect water reserves and help maintain critical flows for fish and wildlife wherever possible.
- On February 22, 2020, the state's voluntary household dry well reporting system received reports of nine dry wells in the past 30 days.
- The state is providing hauled or bottled water to eight separate small water providers experiencing supply outages in Monterey, Santa Barbara, Santa Clara, Tulare, Shasta, El Dorado, and Los Angeles counties. The providers together serve a population of a little under 3000 people.

California Surface Water

Fertile soil, favorable climate, abundant surface water, and rapid population growth in the Central Valley encouraged the development of agriculture, which soon became one of the major industries of California. Information about what beneficial uses of water resources are recorded for individual Basins. The Basin Plan Portal provides beneficial uses for the resources, services, and qualities of California's aquatic systems. Examples of beneficial uses include swimming, fishing, spawning habitat, protected species habitat, agricultural use, tribal and cultural use, and municipal and domestic supply. Each use represents a different type of benefit provided by California's watersheds, and each Regional Water Board has defined and designated beneficial uses unique to their waters. For example, the 2021 Basin Plan included surface water quality and uses for the Sacramento River and San Joaquin River Basins.

California's Colorado Aqueduct

The Colorado Aqueduct, built in the 1930s, transports water across the desert from the Colorado River to Southern California. Operated by the Metropolitan Water District of Southern California (MWD), it remains the region's primary source of drinking water, stretching 242 miles from the Colorado River at Lake Havasu on the California-Arizona border to its final holding reservoir near Riverside, California. The Colorado River Aqueduct consists of more than 90 miles of tunnels, nearly 55 miles of cut-and-cover conduit, almost 30 miles of siphons, and five pumping stations, supplying approximately 1.2 million acre-feet of water a year, more than a billion gallons a day. Southern California was a classified as semi-desert at the time (Hinds 1932). The range of the California Central Valley is shown in Fig. 9.3.

Surface water satisfied most irrigation needs until the late nineteenth century, when a rapid increase in irrigated acreage produced a demand for water that exceeded the surface water supply, making groundwater supplementation necessary. The drought of 1880 was a major stimulus

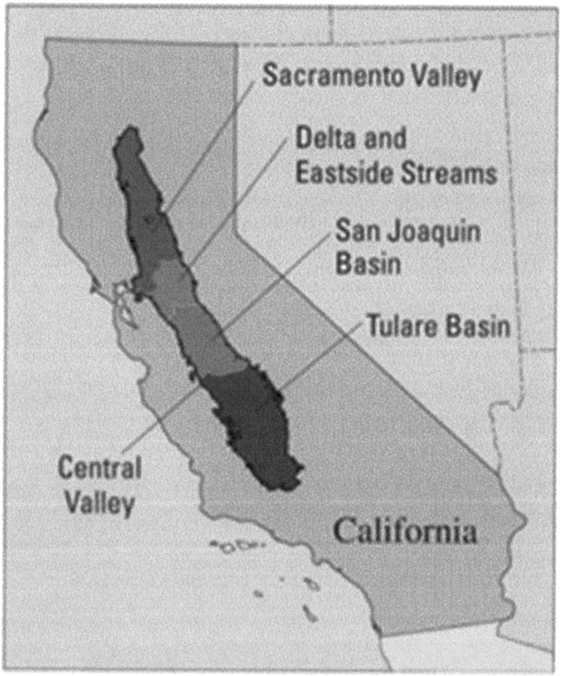

Fig. 9.3 California Central Valley Aquifer system with major surface water basins (*Sources* USGS 2021: Public Domain)

for groundwater development and the first step in what became increasing stress on the groundwater resource. Wells were dependable surface water suppliers, providing water where surface water diversion canals had not yet constructed.

Shallow groundwater was withdrawn easily in 1880, and artesian pressure was sufficient to produce flowing wells in much of the valley. After 1900, groundwater gradually became a more significant part of the total irrigation supply and, eventually, the large number of wells reduced artesian pressure to such an extent that it became necessary to install pumps in order to obtain water.

California has traditionally received as much as 75% of its rain and snow in the watersheds north of Sacramento. However, 80% of California's water demand comes from the southern two-thirds of the state. Until the 1930s, California's mountains and hill barriers made it impossible to move large quantities of that northern water to fields south of San Francisco. The limited surface water sources in Southern California could not meet the drinking water needs of the fast growing population, nor could the limited water supply support the swelling population, despite the mild climate and rich soil. Several water projects were built to import the precious resource to Southern California and the Central Valley

Central Valley Project

Construction of major CVP facilities began in 1938 with ground breaking for building the Shasta Dam on the Sacramento River in Northern California. Over the next five decades, the CVP grew into a system of 20 dams and reservoirs that together can hold nearly 12 million acre-feet. Today, the system extends approximately 400 miles through Central California as a complex, multi-purpose network of dams, reservoirs, canals, hydroelectric power plants, and other facilities. The CVP reduces flood risk for the Central Valley, and supplies valley domestic and industrial water. It also supplies freshwater to major urban centers in the Greater Sacramento and San Francisco Bay areas, as well as producing electrical power and offering various recreational opportunities. In addition, the project provides water to restore and protect fish and wildlife, and to enhance water quality, and supplies water to more than 250 contractors in 29 of California's 58 counties. Deliveries by the CVP include providing an annual average of 5 million acre-feet of water for farms; 600,000 acre-feet of water for municipal and industrial uses (enough water to supply about 2.5 million people for a year); and water for wildlife refuges and maintaining water quality in the Sacramento-San Joaquin Delta. The system as built by the U.S. Bureau of Reclamation also provided aid Depression Era work assistance.

The State Water Project (SWP) was constructed in the 1960s and 1970s to supply water to more than 27 million people and 750,000 acres of farmland in Central and Southern California. Planned, constructed, and operated by California's Department of Water Resources DWR, it is an extensive system of dams, reservoirs, power plants, pumping plants, and aqueducts in Southern California. The SWP is a multi-purpose water storage and delivery system that extends more than 705 miles, approximately two-third of the length of the state. As a system of canals, pipelines, reservoirs, and hydroelectric power facilities, it delivers clean water to 27 million Californians, 750,000 acres of farmland, and businesses throughout our state. Planned, built, operated, and maintained by DWR, the SWP is the nation's largest state-owned water and power generator and user-financed water system. The project is considered an engineering marvel that has helped fuel California's population boom and economic prosperity since its initial construction. For the last 20 years, the system's average water supply use is 34% for agriculture and 66% for residential, municipal, and industrial purposes. Not only does it help California manage its water supply during extremes such as flooding and drought, it is also a major source of hydroelectric power deliveries for the State's power grid.

California Megadrought

California continues to suffer a multi-year drought, resulting in a slew of forest fires and significant declines in available surface water. In April of 2022, 87% of the state was classified as under a severe drought status. The state has been forced to turn to groundwater to maintain the important agriculture industry. The withdrawals are resulting in groundwater depletion far faster than the water can be replenished.

- The California water year that ended September 30, 2021 was the second driest on record. All of California's 58 counties were under a drought emergency proclamation.
- January, February, and March 2022 were the driest on record dating back over 100 years, logging just six inches of precipitation across the Sierra Nevada.

- Statewide precipitation for the water year as of April 13, 2022 was 69% of average. Sierra-Cascades snowpack for the water year was 21% of average, down from 26% the previous week. Statewide reservoir storage was 70% of average for this time of year.
- Californians were being asked to reduce their water use by 15% over 2020 levels to protect water reserves and help maintain critical flows for fish and wildlife.
- As of April 10, 2022 the state's voluntary household dry well reporting system received reports of 24 dry wells in the last 30 days.
- $3.3 million in funding to the San Luis and Delta-Mendota Water Authority to repair segments of the Delta-Mendota Canal in the San Joaquin Valley were damaged by land subsidence tied to over-pumping of groundwater.

California Groundwater Use

The invention of the deep-well turbine pump around 1930 allowed withdrawals from greater depths, which encouraged further development of groundwater resources for irrigation. Withdrawals increased sharply from 1940 to 1960, averaging about 11.5 million acre-feet per year by the 1960s and 1970s, which was approximately 20% of the total irrigation withdrawals for the United States at that time. Withdrawals reached a maximum of 15 million acre-feet per year during 1977, a drought year. During the 1960s and 1970s, withdrawals greatly exceeded recharge, and water levels declined precipitously, as much as 400 feet in places. The declines caused a major reduction in the amount of groundwater in storage and resulted in widespread land subsidence, mainly in the western and southern parts of the San Joaquin Valley.

Increased rainfall and construction of additional surface water delivery systems halted most of the serious water-level declines after 1977, and water levels recovered to pre-1960 levels. The network of aqueducts in the Central Valley is usually sufficient to provide one-half or more of the water needed for irrigation in years of average or above-average precipitation. In dry years, however, reliance on groundwater supplies is greater, and aquifers are again being subjected to withdrawals in excess of recharge during a severe drought.

The Central Valley has become one of the most important agricultural areas in the world. No single region of comparable size in the United States produces more fruits, vegetables, and nuts. More than 7 million acres are currently under irrigation. During 1985, crop irrigation accounted for 96% of the surface water and 89% of the groundwater withdrawn in the Central Valley.

The California Aquifer System

The California Central Valley System is shaped into five sub-basins that are, from north to south: the Sacramento Valley, the Delta, the Eastside Streams, the San Joaquin Basin, and the Tulare Basin. The Sacramento Valley is in a trough about 400 miles long and from 20 to 70 miles wide, including more than 20,000 square miles of exceptional agricultural land. The trough is filled to great depths by marine and continental sediments, which are the result of millions of years of inundation by the ocean and erosion of the rocks that form the surrounding mountains. Sand and gravel beds in this great thickness of basin-fill material form an important aquifer system. From north to south, the aquifer system is divided into the Sacramento Valley, the Sacramento, San Joaquin Delta, and the San Joaquin Valley sub-regions, based on different characteristics of surface water basins. The Central Valley is considered one of the most important agricultural areas in the world. No single region of comparable size in the United States produces more fruits, vegetables, and nuts. In 1995, more than 7 million acres were under irrigation. During 1985, crop irrigation accounted for 96% of the surface water and 89% of the ground water withdrawn in the Central Valley.

Water Scarcity in Nevada

All water in Nevada, including groundwater, belongs to the public and is appropriable for beneficial use. Nevada began regulating its water in 1866 and made major reforms to its water laws in 1913. In 1939, the legislature enacted the Nevada Underground Water Act, making all groundwater subject to appropriation and regulation by the State Engineer.

Southern Nevada gets nearly 90% of its surface water from the Colorado River, which begins as snowmelt in the Rocky Mountains. A series of research studies from 2003, 2005, 2006, 2009, and 2019 agree that from 1930 the average annual snowpack has declined in large sections of the West. Snowpack in the North Rocky mountains is experiencing greater declines than other locations and will continue to do so for at least the rest of century. A recent study found that the most drastic declines were occurring in Western Montana and the Pacific Northwest. Since the 1950s, more of regional precipitation was falling as rain than as snow (Cross et al. 2017).

Groundwater in the Las Vegas Valley is the source for the remaining 10% of the total water needs for the state. Most of the groundwater withdrawn supplies agriculture and almost all of the community needs. The groundwater is withdrawn from three major aquifer zones that are from 300 to 1500 feet below land surface. With more than 71% of all groundwater withdrawn for agriculture irrigation, 26.7% of all groundwater is used for public supply. This drinking water supply is protected from surface contamination by a layer of clay and fine-grained sediments throughout most of the valley. Prior to the 1960s, Nevada used very little groundwater sources. However, as pumping and overdraft increased, Nevada enacted a system for more effective groundwater management, including designation of critical groundwater areas where withdrawals may be restricted to preserve supplies. Groundwater accounts for 45.6% of the state's total freshwater supply, about half is used for agriculture, and 30% is used for mining. Groundwater accounts for 100% of individual household water use and 22.9% of public supplies (Stanford University Woods Institute, n.d).

Both of these sources are under stress that was brought on by climate change and prolonged drought. On August 16, 2021, the Federal Government issued a water shortage declaration on the Colorado River. In January 2022, Southern Nevada's Colorado River water allocation was reduced by 7 billion gallons, enough water to serve 45,000 homes. The new allocation was 279,000 acre-feet of water, although the state's conservation efforts were able to use only 242,000 acre-feet in 2021.

Conclusion

Water shortage throughout the West is one of the consequences of the continuing climate warming and the resulting water scarcity occurring in the Western United States. Changing weather patterns, rapid population growth, and the increases in water demand occurring as a result of inefficient usage are all contributing to water scarcity. According to the U.S. Government Accountability Office, an estimate of 80% of the CONUS will experience some kind of water shortage in the next five to ten years. Climate change is already resulting in western states receiving less precipitation compared to the eastern half of the U.S. More precipitation is falling as rain and less by snow. And, the snow that does fall is melting faster, changing both the amount and timing of streamflows. Droughts are having serious impacts on fish and wildlife, recreation-based economies, agriculture, and some municipalities. The timing of these changes is not only occurring more rapidly, the effects are growing in severity as the climate continues to change. Droughts are occurring more often and lasting longer in the West, even as precipitation in eastern states is increasing.

To survive during protracted droughts in the important agriculture industry in the West, farmers have had to increase their dependence upon groundwater. As a result, excessive groundwater withdrawals are diminishing the existing aquifers faster than they can be replenished. Aquifers are considered essential resources for human survival and to meet those needs, they cannot be over-withdrawn. Yet, far more groundwater is being pumped out than can be naturally replenished. Many if not most dry-area aquifers such as those found in the Southwest are being significantly depleted. These include the two primary groundwater systems in the Western United States: California's Central Valley Aquifer and the Ogallala Aquifer.

References

Cross, Wyatt F., John LaFave, Alex Leone, Whitney Lonsdale, Alisa Royem, Tom Patton, and Stephanie McGinnis. 2017. Water and climate change in Montana. Chapter 3: 2017. *Montana Climate Assessment*. Available at https://leg.mt.gov/content/Committees/Interim/2019-2020/Water-Policy/Meetings/Sept-2019/MontanaClimateAssessment-Water.pdf.

Curtis, Jennifer A., Erick R. Burns, and Roy Sando. 2020. Regional patterns in hydrologic response, a new three-component metric for hydrologic analysis and implications for ecohydrology, Northwest Volcanic Aquifer Study Area, USA. *Journal of Hydrology: Regional Studies*. Available at https://doi.org/10.1016/j.ejrh.2020.100698.

Drought.ca.gov. 2022. California drought upgrade, April 11. Available at https://drought.ca.gov/media/2022b/02/CA-Drought-Update-4-1-22.pdf

EPA (Environmental Protection Administration). 2017a. What climate change means for Montana. Available at https://19january2017snapshot.epa.gov/sites/production/files/2016-9/documents/climate-change-mt.pdf.

EPA (Environmental Protection Administration). 2017b. Climate impacts in the Northwest. Available at https://19january2017snapshot.epa.gov/climate-impacts/climate-impacts-northwest_.html.

EPA (Environmental Protection Administration). 2017c. State water facts. Available at https://19january2017snapshot.epa.gov/www3/watersense/our_water/state_facts.html#content.

EPA (Environmental Protection Administration). 2017d. Saving water in Washington. Available at www.epa.gov/sites/default/files/2017c-02/documents/ws-ourwater-washington-state-fact-sheet.pdf.

Hinds, Julian. 1932. The Colorado River Aqueduct. *The Military Engineer* 24 (134): 15–19.

NWS (National Weather Service). 2022. Oregon water supply summary as of February 8, 2022. Available at www.weather.gov/media/pqr/WaterSupplyOutlook.pdf.

Shahpurwala, Aiman. 2021. *Washington's freshwater under stress from drought*. Olympia, WA: Washington Water Trust, August 11. Available at www.washingtonwatertrust.org/2021/08/11/washingtons-freshwater-under-stress-from-drought/

SWDOE (State of Washington Department of Ecology). n.d. Columbia River water supply projects. Available at https://ecology.wa.gov/Water-Shorelines/Water-supply/Water-supply-projects-EW/Columbia-River-Basin-projects

USA.com. 2022. U.S. average precipitation state rank. Available at www.usa.com/rank/us--average-precipitation--state-rank.htm

USGS (United States Geological Survey). 2018a. Groundwater decline and depletion. Available at www.usgs.gov/special-topic/water-science-school/science/groundwater-decline-and-depletion?qt-science_center_objects=0#qt-science_center_objects

USGS (United States Geological Survey). 2018b. Wyoming water use in the U.S., 2015. Available at www.usgs.gov/media/images/wyoming-water-use-us-2015

USGS (United States Geological Survey). 2021. National water census: Regional groundwater availability studies. Available at www.usgs.gov/mission-areas/water-resources/science/national-water-census-regional-groundwater-availability?qt-science_center_objects=0#qt-science_center_objects

WRDS (Water Resources Data System). 2021. Wyoming drought. Available at www.wrds.uwyo.edu/drought/drought.html

CHAPTER 10

Water Crisis in the Southwest

The four states in the Southwest region included in this region are Arizona, New Mexico, Utah, and Colorado—the *Four Corners* states. The Four Corners refers to the point where the four states—Colorado, New Mexico, Arizona, and Utah—meet on the high desert of the Colorado Plateau in the Southwest. These four states meet at the northeast corner of Arizona, the Southwest corner of Colorado, the southeast corner of Utah, and the northwest corner of New Mexico. Both New Mexico and Colorado are among the western regions with some access to groundwater from the Ogallala aquifer system. In 2022, drought was severe in the entire region.

Water in the Dry Southwest

Drought is a major contributor to water scarcity in many locations; it is particularly a problem in the American Southwest. In September of 2022, all four of states had been abnormally dry for most of the year and portions of all four states were under extreme drought. Utah was suffering most from drought, with more than 60% in extreme drought, and nearly forty percent of the New Mexico were under severe drought (Table 10.1).

D. E. McNabb and C. R. Swenson, *America's Water Crises*, https://doi.org/10.1007/978-3-031-27380-3_10

Table 10.1 Drought in the Southwest states first week of September 2022

Sections of Southwest states in drought category, August 2022					
State	*D0:D4 Abnormally dry (%)*	*D1: D4 Moderate drought (%)*	*D2:D4 Severe drought (%)*	*D3:D4 Extreme drought (%)*	*D4 Exceptional drought (%)*
Utah	100.0	100.0	95.8	61.3	3.6
New Mexico	99.1	87.8	58.6	10.4	0.2
Arizona	100.0	66.9	23.5	2.9	0.0
Colorado	89.9	46.4	17.0	4.2	0.6

Source NIDIS Drought.gov

For much of the first two decades of the twenty-first century, coping with drought and water stress has been exceptionally hard, pressing problems for all four Southwest states. In May of 2022, more than half of all the four were under severe drought conditions, led by 99.5% of Utah and 95.8% of New Mexico, which also had 79% of the state in extreme drought and 24.7% of the state under exceptional drought. The two points that distinguish the exceptional drought condition are: (1) exceptional and widespread crop/pasture loses; (2), shortages of water in reservoirs, streams, and wells creating waster emergencies.

Drought in Arizona

As shown in Table 10.1, in the first week of September 2022, more than 60% of Utah and 10% of New Mexico were under extreme drought. This was the 22nd year of drought in the Southwest. It is now considered a megadrought.

Climate change's effects of higher temperatures and declines in precipitation have been exceptionally hard on the Southwest and West Coast states during the first two decades of the twenty-first century. Population growth in these same states has contributed to increases in water stress. Megadrought and accompanying increasing temperatures have been hard on agriculture in this arid Southwest. Reductions in winter snowpack have reduced this important contribution to spring and summer stream flows, further contributing to the region's growing water crisis.

Both surface water and groundwater are under stress in parts of this region. Water stress from drought and extensive groundwater withdrawals in these states are exacerbated by extensive population growth. Arizona has always been hot and dry, with living conditions continuing to make the state a desired place to live. For more than 50 years, temperatures in Arizona have been steadily rising with average summer temperatures, now 1.8 °F (1°C) higher compared to 1970 levels. Over the same period, the state has recorded 6.2 more days that are above 110 °F. Droughts have been common events in the American Southwest, in the past, commonly reoccurring approximately every 20 years or so. In 2022, the region was well affected by drought that has lasted since 2000. Megadroughts are intense drought events that last for at least 20 years. NOAA and UCLA-sponsored researchers found that megadroughts are not uncommon in the Southwest. Megadroughts are known to have occurred repeatedly in this region from 800 to 1600. Relatively milder droughts have been common occurring intermittently along with these earlier extreme droughts over hundreds of years. However, researchers believe that the current drought began as a turn of the twenty-first-century drought and will very likely persist through 2022 and possibly longer, matching the duration of earlier megadroughts. Current records show that this is the driest this region has been in at least 1200 years. Moreover, these long droughts are predicted to continue to occur beyond the end of this century (NIDIS 2022).

Arizona Water Stress

Arizona depends on surface water for much of its water. The Colorado River provides the largest share of this supply. The Colorado and other surface water sources provide up to 54% of the state's total water supply. However, seven states share water from the Colorado, contributing to the stress. Arizona receives the largest share of the Colorado River, which is 19% of the total allotments. The Department of Agriculture's Bureau of Reclamation determines amounts in allotments. In 2021, water is stored in both Lake Mead and Lake Powell, which together is a primary water supply for Arizona's 7 million residents. The stored water has dropped so low that withdrawals must be curtailed to avoid dropping below entry to electric generating plants. The State of Arizona has a complex set of requirements designed to assure that new development has a 100-year supply of water identified before approval. This system has worked for the area in the past, but as with any such prediction model, it is based on assumptions that must be tested and updated regularly to assure they are relevant as climate change impacts future water supply. Consequently, some Arizona communities are now restricting population growth, with the concern that there will not be enough water for any more residents. Most Arizona cities, however, have developed complex multi-source water portfolios that include surface water from various sources, and groundwater from wells which is augmented by recharge of municipal wastewater. The City of Peoria Arizona, for example, includes in their portfolio an allotment of water from the Colorado River, the Salt/Verde Rivers, groundwater from numerous wells, and recharges a large percentage of their municipal waste water. Taken together, this portfolio creates their multi-source 100-year supply.

In 2021, Arizona, California, and Nevada were forced to reduce what they could take from the Colorado River. A second reduction occurred in 2022. From the 2021 reduction, Arizona saw an 18-percent reduction of what the state could take from the river. According to the Arizona Department of Water Resources (ADWR), 41% of the state's water supply comes from groundwater, 36% from Colorado River, 18% from in-state rivers, and 5% from reclaimed water. The 2023 reduction is expected to cut another 21% from Arizona's share of Colorado River water.

The results of the Lake Powell and Lake Mead in the August 2022, 24-month study, provide operating guidelines for the Lower Basin shortages and the operations of Lake Powell and Lake Mead from 2022 to 2024. Shortages are determined by the system storage and elevations of the reservoir water surface. These measurements set the operations for the coordinated operation of Lake Powell and Lake Mead. Actions occur when the elevation of Lake Powell is less than 3575 feet or above 3525 feet; the Lake Mead elevation basis point is at or above 1025 feet (USBR 2022c). On August 16, 2022, the Interior Department announced reduction actions to protect the long-term sustainability of the Colorado River system, which includes commitments for continued engagement with impacted states and tribes. The 2022 actions particularly focused on a reduction of the water being accessible to Mexico, Nevada and Arizona. Arizona will lose 592,000 acre-feet of Lake Powell water in 2023—21% of its annual allotment, Nevada will lose 89% of its 2023 allotment, and Mexico will lose 7% of its 2023 allotment. No additional reductions were to be required for California. The reductions come after the preliminary restrictions on the Colorado River Basin water surface determined in the August 2022 24-month study. Although the reductions were set the earlier annual operations based on predictions for for Lake Mead and Lake Powell in 2023, most of the 2023 reductions will apply to withdrawals from Lake Powell.

Final release of the study in August found that Lake Mead's elevation had fallen below 1075 feet, which means that Arizona, Nevada and Mexico will receive less water than they hoped. Much of Arizona's reduced amount of water will primarily effect Central Arizona agriculture.

The additions to earlier reductions in withdrawals from Lake mead for Arizona, Nevada and Mexico mean that Lake Mead will be forced to operate in its first-ever Level 2a Shortage Condition in calendar year 2023 (Jan. 1, 2023, through Dec. 31, 2023). The August 24-Month Study projects Lake Mead's January 1, 2023, operating determination elevation to be 1,047.61. elevation (1,040.78 feet) and adding the 480,000 acre-feet of water held back in Lake Powell to Lake Mead's capacity to maintain operational neutrality.

Drought and Lake Evaporation

The megadroughts during hotter and drier days are also increasing the rate of evaporation in water bodies stored behind the Colorado River dams, as a result leading to water bodies and sources becoming more stressed, and threatening further reductions. Arizona has a large agriculture industry, the majority of which requires irrigation. The increasing heat is resulting in higher rates of evaporation from surface water and farm and landscape vegetation. The typical plant, including any found in a landscape, absorbs water from the soil through its roots. This moisture is released to the atmosphere when transported through plants, a process known as transpiration. Hotter days and climates will result in greater water loss due to higher evaporation and transpiration rates. This higher water loss increases stress on Arizona's water resources. This is happening throughout the Colorado River system as well as other surface water river systems, including the Salt, Verde, and Gila rivers and includes the reservoirs of Lake Mead, Lake Powell, and Lake Pleasant. As this occurs and accelerates, greater stress in the water resources of the entire American Southwest is expected.

Water Management in Arizona

Not surprisingly given the hot desert conditions prevalent in Arizona, the state has given considerable attention to the acquisition and management of their water resources. Each state is free to create their own water management systems within the context of Federal Government oversite and within the structures created by Congress to manage and protect water resources for the nation. For states in the arid west, the 1922 Colorado River Compact created the foundation for states that ultimately depend on the Colorado River for significant portions of their water. This Compact created the cornerstone for Arizona's future water supply and the management constructs in place to assure water delivery and storage for the state.

Along with California and Nevada, Arizona is one of the three states designated in the Compact as members of the Southern Basin, which guarantees 2.8 million acre-feet of Colorado River water to the state. In 1968 President Lynden Johnson signed the Colorado River Basin Project Act which created the Central Arizona Project system to construct a series of canals to transport Colorado River water to the Arizona Counties of Maricopa, Pima and Pinal where population growth was anticipated to grow. This 336-mile system provides transport and storage of this important water resource to these arid counties in central and southern Arizona. In 1971 the Arizona State Legislature created the Central Arizona Water Conservation District (CAWCD) and set up the Central Arizona Project (CAP) organization as the management structure for this district. The CAP is governed by a 15-member directly elected Board of Directors who oversee the operation of all the assets of the district and assure that the water is allocated in accordance with applicable laws and agreements over the use of water in the area. The Board also determines the price of the water to be charged to users. This rate structure includes the acquisition and transportation costs, system operations, and infrastructure repair and replacement funds, as well as a transfer of funds to the Federal Government for the initial cost of system construction.

In 2022, 80% of Arizona's population use CAP water which is supplied to local municipal water customers, and Indian Tribes who in turn treat and deliver water to end users. While the CAP is primarily focused on surface water from the Colorado River, it also oversees and manages the Central Arizona Groundwater Replenishment District (CAGRD). CAGRD is an outgrowth of Arizona's 1980 Groundwater Management

Act which strictly manages groundwater pumping in the most populous areas of the state. Established in the mid-1990's CAGRD provides a mechanism and resources to members to assist in the replenishment of the aquafers in accordance with provisions of the Groundwater Act. In this way the requirement of 100-year assured supply is enforced before new construction is permitted. For its member jurisdictions and entities, CAGRD supports and enables the replenishment of the critically important groundwater resources within the area. Together the CAP and CAGRD create a balance to manage and conserve surface and groundwater in the Arizona.

Hot and Dry in New Mexico

New Mexico relies on both groundwater and surface water sources, but about 87% of the state's public water supply is from groundwater. No other Southwestern state gets as large a percentage of its domestic water from groundwater sources. Clearly, such a heavy dependence upon groundwater has its downsides; for the water to continue to be available, the aquifers must be recharged. Severe declines in groundwater levels have occurred in some parts of the state. Some municipalities such as Santa Fe have seen that water levels in their groundwater wells decline as much as 300 feet in a recent 10-year period. Megadrought has resulted in even greater reliance upon groundwater as the declines in available surface water result in an even greater reliance upon groundwater. Santa Fe is fortunate, however, because it has both groundwater and surface water sources; many other New Mexico communities are not located near a surface water body such as the Rio Grande River. However, water withdrawals from New Mexico, Colorado, Texas, and Mexico have greatly reduced the volume of the Rio Grande over the past 50 years. New Mexico and several other states also draw water from the Colorado River Basin, and demand for water from that river has exceeded supply.

The Rio Grande Basin provides water for irrigation, households, environmental, and recreational uses in Colorado, New Mexico, and Texas, as well as Mexico. Reclamation projects within the basin include the San Juan-Chama trans-mountain diversion project between Colorado and New Mexico; the Middle Rio Grande Project in New Mexico; and the Rio Grande Project in New Mexico and Texas. These projects support approximately 200,000 acres of irrigated agriculture, which produces alfalfa, cotton, vegetables, pecans and grain, for municipalities, tribes, and industry. Reclamation's facilities provide critical water and power for industry and communities, including Albuquerque and Las Cruces, New Mexico. A 2016 Reclamation Bureau Water Act report identified climate challenges the Rio Grande River Basin could likely face:

- Climate projections suggest that temperatures throughout the Rio Grande are projected to increase by roughly 5–6 °F during the twenty-first century.
- Climate projections suggest that annual precipitation in the Rio Grande Basin will remain variable over the next century.
- Warmer conditions are projected to transition snowfall to rainfall, decreasing the annual April 1 snowpack and April–July runoff from this snowpack in the Rio Grande Basin (USBRd n.d).

New Mexico also has long periods of drought and inconsistent precipitation, so relying on surface water can lead to shortages as well. Water withdrawals from New Mexico, Colorado, Texas, and Mexico have greatly reduced the volume of the Rio Grande over the past 50 years. New Mexico and several other states also draw water from the Colorado River Basin, and demand for water from that river exceeded supply more than a decade ago.

The Upper Colorado River Basin, which extends into western New Mexico, has been experiencing a protracted, multi-year drought that began in 1999. Prolonged droughts, in conjunction with climate change, have the potential to reduce snowpack, increase temperatures, and create earlier mountain snow thaws, which all reduce water supplies. Climate change could also create a longer, hotter warm season in New Mexico. The state is concerned whether this will result in prolonged periods of extremely low water flow, as warmer temperatures increase water loss from surface sources by evaporation.

Drought and Wildfires

The Upper Colorado River Basin, which extends into western New Mexico, has been experiencing a protracted, multi-year drought that began in 1999. What is now a megadrought, combined with the precipitation declines and other climate change impacts, has reduced snowpack, increased temperatures, and created earlier mountain snow thaws, all of which reduce water supplies. Climate change could also create a longer, hotter warm season in New Mexico. The state is concerned that this will result in prolonged periods of extremely low water flow, as warmer temperatures increase water loss from surface sources by evaporation.

Windy, dry conditions have exacerbated fire-weather conditions from several large early season wildfires in 2021 which impacted New Mexico and the other Four Corners states. In early May 2022, wildfires had burned 147,909 acres in the Sangre de Cristo Mountains in northern New Mexico and there was only 20% of the remaining that contained, according to the National Interagency Coordination Center. By the middle of the month, there were 72 separate wildfires burning in 888,945 acres in New Mexico.

New Mexico residents gets 85 to 87% of its drinking water from groundwater. In the east- central portion of the state, the major source of most of this water is the groundwater withdrawn from the already stressed High Plains aquifer. This groundwater continues to be pumped at an unsustainable rate, just as is happening in much of this multi-state—but distinctly—limited aquifer. The water withdrawn in huge quantities in the counties that are located above the aquifer is used primarily for irrigated agriculture. The aquifer in eastern New Mexico and North Texas is being mined and not replaced. In New Mexico's Curry and Roosevelt Counties, which are completely dependent on groundwater for all water uses, 93% of the total groundwater withdrawn is for agriculture. Both of these counties share a border with Northern Texas, where similar pumping from the aquifer for agriculture is also dominant. Annual agricultural products from the two New Mexico counties alone have been valued at more than $710 million in cattle, calves, and milk products account for most of the water withdrawn here.

The rate of change in the level of groundwater in the New Mexico's section of the aquifer can vary significantly. In much of southeast Curry and northeast Roosevelt Counties, water levels in 2012 were declining more than five feet per year, and locally much more. This is the area with the greatest concentration of irrigated agriculture with many wells. In the western portions of the area, there is less than ten inches precipitation, but still without much irrigation and, therefore, few wells. Much of the western edges of Ogallala aquifer in New Mexico has never very actively used for irrigated agriculture; the groundwater supply there is too thin and varies in dependable quantity.

Groundwater withdrawn for livestock uses includes supplies for animal watering, feedlot operations, dairy operations, and other on-farm needs.

Table 10.2 New Mexico water withdrawals in 2015 (billion gallons/day) and source

Use category	*Total withdrawals*	*Surface water share (%)*	*Groundwater share (%)*
Livestock	2.0 BGD	62	38
Aquaculture	7.5 BGD	79	21
Mining	4.0 BGD	72	28

Source USGS (2018a, b)

Livestock includes dairy cows and heifers, beef cattle and calves, sheep and lambs, goats, hogs and pigs, horses, and poultry. Other livestock water uses include cooling of facilities for the animals and products, dairy sanitation, and daily washing of facilities. Other uses contributing to the estimated water withdrawals include animal waste disposal systems, and various incidental water losses. The livestock category excludes on-farm domestic use, lawn and garden watering, and irrigation water use (USGS 2018a, b).

The USGS collects data from and estimating total water withdrawals of a five-year average of freshwater withdrawals for eight categories of uses. The smallest water user categories were for livestock, aquaculture, and mining. In 2015, estimated total freshwater withdrawals of groundwater for aquiculture alone was 1234 million gallons per day (MGD), while total surface water withdrawals for all freshwater aquaculture uses were 5950 MGD. Total water withdrawals for miming in 2015 were close to 4.00 BGD, or about 1% of total withdrawals from all uses and 2% of total withdrawals from all uses. Nearly 2870 million gallons per day are for mining purposes. Table 10.2 includes total annual average withdrawals of surface and groundwater portions for the three categories.

Water Stress in Utah

Prolonged drought is a life-threatening condition to Utah, and in 2022, the state was subjected to yet another prolonged drought. Very large areas of Utah were in high levels of drought for eight of the previous 10 years, and the lack of the snowpack melt needed to provide spring and summer streamflow was not enough to end it. The drought-caused water stress crisis in Utah is apparent in the Great Salt Lake, where water levels in the lake have dropped to the lowest levels ever recorded.

Utah is one of the states in this region that suffered from prolonged drought for ten years between 2012 and 2022. The damage from drought in Utah is an example of the damage drought can have on a state or a region and its water resources. The problems are is endemic and an example of the damage both can have on a region. The negative impact that drought can have on a region makes it among the nation's most costly event ever to take place. However, the long-term cost of a drought is among the hardest to disaster pin down. Unlike impacts from floods, hurricanes, tornados, or other weather-related disasters, the damage a drought can do is not always visible. For example, the loss of forage during a drought in one growing season can show up again years later when for years after the end of the drought.

Loss of Snowpack

In addition to the declines in the annual snowfall, Utah snowpacks are more than 25% or more below what had been considered to be normal. Annual mountain snowpacks are also melting almost two weeks earlier than what they have been. Snowpacks provide the basis for spring and summer streamflow and for groundwater recharges. The changes occurring to the size of the annual snowfall and the earlier snowmelt is what supports the spring and summer surface water streamflow the water for recharging groundwater aquifers. Climate change reductions in winter snowfall and the earlier timing of the snowpack melt are affecting the ability of a mountain states to meet agriculture water needs through the dry years. Extended drought's hot, dry conditions will take years to refill the drained reservoirs (UDWR 2012).

Like many western U.S. states, Utah's spring and summer use of surface water depends on the melt of the annual snowpack. During the recent drought conditions, the reduced annual snowfall greatly reduces the annual snowmelt, resulting in lower groundwater streamflow and reservoirs that resupply the depleted summer-season surface water resources. Early melt due to warmer weather means earlier snowmelt and less water available later in the year. Where available, surface water has long been the preferred water source. During a drought, however, more water users must withdraw groundwater. In good snowpack years, when available, the agriculture and recreational industries rely on surface water to meet their regular needs. In lower precipitation years, these water users must increase their dependence on groundwater. The following points were included in an early 2022 drought conditions report:

- 99.45% of the state was in severe drought or worse, with 43.46% of Utah in extreme drought.
- Statewide snow water equivalent (SWE), or how much water would be in the snow if melted, peaked at 12 inches. This is 75% of the typical median peak of 16 inches for the water year.
- Twenty-two of Utah's largest 45 reservoirs are below 55% of available capacity. Overall, statewide storage is 60% of capacity. This time in 2021 reservoirs were about 67% of capacity.
- Of the 96 measured streams, 56 are flowing below normal despite spring runoff. Two streams were flowing at record low conditions.
- The district has received very little new water reservoir storage since 2020 and is expected to receive very little again in 2022. To mitigate the effects of the drought on their storage reservoirs, one water district reduced the amount of water it intended to deliver to contract holders in 2022.

Utah's Water Needs

Utah is the second driest state in the United States. Situated in the rain shadow region of the Sierra Nevada Mountain range that blocks storms from the Pacific, the average annual rainfall is 13.6 inches. With an average annual precipitation of 9.5 inches, Nevada is the driest state. Both of these two dry states have been arguing over who will get what share of the freshwater provided by the Snake River.

Approximately two-thirds of all freshwater used in Utah is used for agriculture irrigation, accounting for as much as 72% of freshwater withdrawn. Of the total, groundwater use accounts for close to 25% of the state's total freshwater use. Approximately half of the groundwater withdrawn is by agriculture. Groundwater is important for domestic consumption, on average accounting for 100% of individual household water use and 54% of municipal supplies.

Groundwater from the Upper Colorado River Basin (UCRB) makes as much as half of the total baseload contribution of the Colorado River and is withdrawn from the Lake Powell reservoir (USGS 2021a, b, c). However, current weather changes and predicted more and longer droughts means the Colorado River can no longer able be expected to meet the water needs of Nevada. To remedy this discrepancy, Nevada had to find other sources. One of these was the under-used groundwater beneath a farm valley on the border between Nevada and Utah (Fig. 10.1) By the end of this century, the Colorado River water resource is predicted to decline by as much as a third, affecting water supplies all the states that depend on Colorado River water for their base supply of freshwater. The UCRB has a drainage area of about 114,000 square miles, covering portions of Utah, Colorado, Arizona, and New Mexico, plus Wyoming.

Although the Snake River flows through neither Utah nor Nevada, the Snake Valley aquifer extends across and far approximately 100 miles along the borders of both states (Philpot et al. 2016). Groundwater from this aquifer in the land on both sides of the Utah-Nevada border is used primarily for agriculture. Most of the population of approximately 1000 in the valley are located on the Utah side of the border. Both Utah and Nevada are arguing over access to the aquifer. Nevada wants to ship

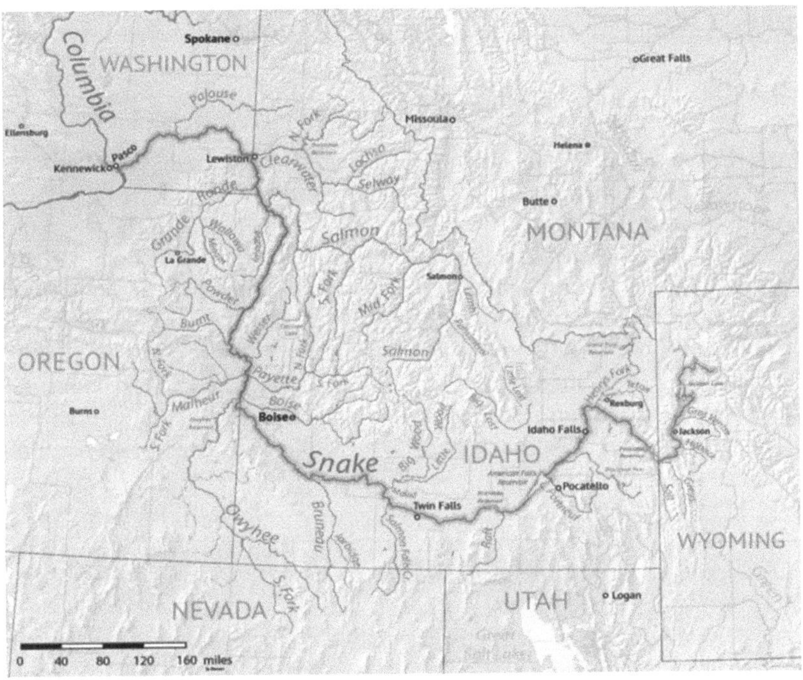

Fig. 10.1 Map of the Snake River in Idaho and other states (*Source* Wikimedia Commons; Map of the Snake River watershed, USA. Intended to replace older File:SnakeRiverNicerMap.jpg. Created using public domain USGS National Map data. *Source* = Own work Author = [User:Shannon1|Sh])

the water to large urban areas in the south of the state (SNWA 2011). Utah wants to keep access to the water for local agriculture, believing that Nevada's withdrawals of the aquifer would be more than could be recharged. They fear the decline of the annual snowpack, with a strong likelihood of even greater declines in future, will permanently destroy access for agriculture for Utah farmers. They cite as an example what happened to farms in California when farmers lost water rights to the Los Angeles area.

Nevada merged the groundwater access licenses of a number of the smaller water organizations to form the Southern Nevada Water Authority (SNWA) with plans to secure unallocated rural groundwater for use by urban area in the southern section of the state, primarily for Las Vegas. The SNWB plans to construct a 300-plus mile pipeline from a point near the Utah border to ship the water to the Las Vegas and other locations.

Water Stress in Colorado

From 1984 to 2012, total streamflow deliveries from the Upper Colorado River basin's outlet at Lees Ferry, Arizona, to the Lower Colorado River Basin averaged 10.3 million acre-feet/year. Baseflow accounted for nearly a third of this. Baseflow is the movement of groundwater into streams and, on average, accounts for more than 50% of annual streamflow in this Upper Basin. It is vital for sustaining flows in the Colorado River during dry periods. Over the 1916–2015 period, streamflow in the Upper Basin declined by 16.5% (Xiao et al. 2018). A U.S. Geological Survey and the Bureau of Reclamation study projects that a hot and dry future climate may lead to a 29-percent decline in Upper Colorado River Basin baseflow at the basin outlet by the 2050s, affecting both people and ecosystems (Fig. 10.2).

Colorado is the birthplace of one of, if not, the most important surface water resource in the western United States, the Colorado River. It is also the longest river in the west, providing drinking water for more than 40 million people and irrigation water for parts of seven states: Wyoming, Nevada, Arizona, California and Colorado and for the Mexican states of Sonora and Baja California (Xiao et al. 2018). Measured at Lees Ferry (the start of the Grand Canyon), between 1916 and 2014, streamflow on the Colorado declined by more than 16%, although annual precipitation over the Colorado River Basin increased by approximately 1.4% over the 100 plus years. High mountain snowpack in headwater

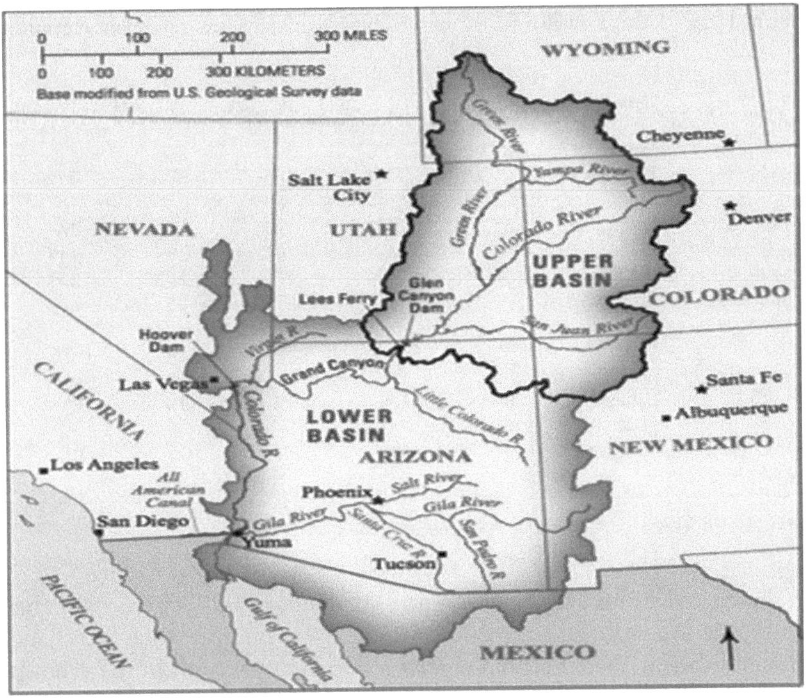

Fig. 10.2 The two basins of the Colorado River (*Source* USGS Public Domain)

basins of the Colorado's streamflow contributed about 70% to the annual streamflow (Christensen et al. 2014). Analysis of the streamflow of the Colorado at a number of points indicate that 53% of the decreasing runoff from Colorado snowpacks is associated with warming over the entire basin. Most of the flow of the Colorado River originates in the Rocky Mountains.

As of September 2021, the April-to-July runoff into Lake Powell totaled just 26% of average, despite near-average snowfall in the upper basins. Total Colorado River system storage, when measured for the September Bureau of Reclamation release, was 39% of capacity, down from 49% in 2020. Five-year shows predictions for the lower Colorado River region as of February 2022 in Table 10.2 for Lake Powell and 10.3 for Lake Mead. The projections are the percent chance of reaching critically low elevations for water at Lake Powell and Lake Mead. The changes

Table 10.3 Lake Powell chance or reaching critically low reservoir elevations in 2022

Height	*Run*	*WY 2022*	*WY 2023*	*WY 2024*	*WY 2025*	*WY 2026*
Less than 3,524 ft	January 2021	87%	42%	39%	36%	37%
	February 2022	**90%**	**77%**	**50%**	**50%**	**37%**
	Difference:	3%	35%	11%	14%	0%
Less than 3,490 ft (minimum power pool)	January 2021	N	8%	20%	23%	26%
	February 2022	**N**	**23%**	**27%**	**27%**	**23%**
	Difference:	–	15%	7%	4%	-3%
Less than 3,375 ft	January 2021	0%	0%	0%	0%	0%
	February 2022	**0%**	**0%**	**0%**	**0%**	**0%**
Dead pool = 3,370 feet	Difference:	0%	0%	0%	0%	0%

Source USDR 2022

in these predictions are the differences for January and February for their water levels in years 2022 to 2028. Results were interpreted as the chance of falling below the threshold surface value in any month in the calendar water year (USBR 2022a, b, c). Included are the minimum level for power production and for what is considered the "dead pool" which is the point where water levels in the river are so low that they are below where it would be impossible to release water from dam to the river (Table 10.3 and Fig. 10.3).

Colorado Water Crises

Commonly, with information about surface water and groundwater in Colorado, most of the data are typically led where word of the region-wide crisis focuses on the Colorado River. For example, in June of 2021, the Colorado River water stored behind Hoover Dam had dropped to its lowest level ever since the dam was built, underscoring the gravity of the extreme drought then affecting most of the American Southwest. "Lake Mead, formed in the 1930s from the damming of the Colorado River at the Nevada-Arizona border about 30 miles east of Las Vegas, is the largest reservoir in the United States. It is crucial to the water supply of 25 million people including in the metropolitan areas of Los Angeles, San Diego, Phoenix, Tucson and Las Vegas. By midnight, the lake surface fell

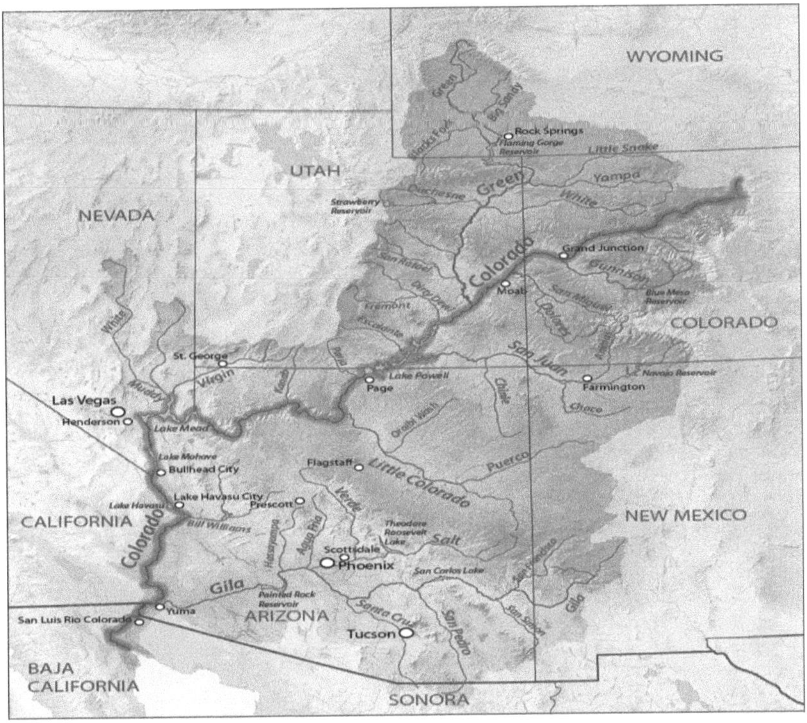

Fig. 10.3 USGS map of the Colorado River Southwest to Mexico (USGS: Colorado Water Science Center 2012)

to 1,071.56 feet above sea level, dipping below the previous record low set on July 1, 2016. It has fallen 140 feet since 2000. The drought that has brought Lake Mead low has gripped California, the Pacific Northwest, the Great Basin spanning Nevada, Oregon and Utah, plus the Southwestern states of Arizona and New Mexico and even part of the Northern High Plains" (Januta and Trotta 2021 online) (Table 10.4).

However, there also problems with the state's groundwater. The USGS identifies seven principal aquifers or aquifer systems in Colorado: South Platte Aquifer, Arkansas Aquifer, High Plains Aquifer, San Luis Valley Aquifer System, Denver Basin Aquifer System, Piceance Creek Basin Aquifer, and the Leadville Limestone Aquifer of west-central Colorado. Nineteen of Colorado's 63 counties rely solely on these important

Table 10.4 Lake Mead chance of critical reservoir elevations in 5 Future Water Years

Height	*Run*	*WY 2022*	*WY 2023*	*WY 2024*	*WY 2025*	*WY 2026*
Less than 1,020 feet	January 20,212	0.0%	<1%	36%	40%	47%
	February 2022	**0%**	**0%**	**0%**	**33%**	**37%**
	Difference:	0%	0%	−16%	−7%	−10%
Less than 1,000 feet	January 2022	0%	0%	3%	17%	20%
	February 2022	**0%**	**0%**	7%	**13%**	**17%**
	Difference:	0%	0%	4%	−4%	−3%
Less than 950 feet (Minimum power pool)	January 2022	0%	0%	0%	<1%	3%
	February 2022	**0%**	**0%**	**0%**	**0%**	**0%**
	Difference	0%	0%	0%	0%	−3%
Less than 900 feet	January 2022	0%	0%	0%	0%	<1
	February 2022	**0%**	**0%**	**0%**	**0%**	**0%**
	Difference:	0%	0%	0%	0%	0%

Source SDR (2022)

groundwater resources for drinking water and other domestic uses. The Colorado Geological Survey (CGS) has provided a thorough description of groundwater in the state and its issues in the Colorado Groundwater Atlas (0N-010).

Private wells and public water supply systems serve an estimated 20% of the state's population. Agriculture is the largest user of groundwater for irrigation to meet livestock and rural domestic water needs. However, the status of the groundwater supply has become a "critical natural resource issue in Colorado" (CGS 2022).

Colorado Use of Groundwater

The High Plains aquifer that reaches Colorado is present beneath all or parts of 13 counties. Land use in eastern Colorado that depends on this aquifer system is almost exclusively agricultural with most groundwater withdrawn used for irrigation and crop-related purposes. According to the Colorado Geological Survey (CGS), eastern Colorado's topography is largely flat to gently rolling terrain that is bisected by mostly eastward-flowing rivers and streams. Eastern Colorado's population has been low, with densities rarely exceeding four to nine residents per square mile. Most of the aquifer is a part of Colorado's Northern or Southern High Plains' Designated Ground-Water Basins and their associated groundwater management districts.

Land use in eastern Colorado is almost exclusively agricultural with most groundwater withdrawal used for irrigation and crop-related purposes. The High Plains aquifer is present beneath all or parts of 13 Colorado counties. According to the Colorado Geological Survey (CGS), eastern Colorado's topography is characterized by flat to gently rolling terrain that is bisected by mostly eastward-flowing rivers and streams. Eastern Colorado's population remains low, with population densities rarely exceeding four to nine residents per square mile. Most of the aquifer is a part of Colorado's Northern or Southern High Plains Designated Ground-Water Basins and their associated groundwater management districts.

The USGS identifies seven principal aquifer systems in Colorado: The High Plains Aquifer, South Platte Aquifer, Arkansas Aquifer, San Luis Valley Aquifer System, Denver Basin Aquifer System, Piceance Creek Basin Aquifer and the Leadville Limestone Aquifer (CGS 2003). The High Plains aquifer is present beneath all or parts of eastern Colorado's Weld, Logan, Sedgwick, Phillips, Washington, Yuma, Lincoln, Kit Carson, Cheyenne, Kiowa, Prowers, Las Animas, and Baca counties. Land use in this portion of the state is almost exclusively agricultural with most groundwater withdrawal used for irrigation and crop-related purposes. Most of this land is a part of Colorado's Northern or Southern High Plains Designated Ground-Water Basins and their associated groundwater management districts. An estimated 15,600 wells were recorded in the state by the end of the twentieth century.

Conclusion

The Four Corner region of the Southwest is among the country' fastest growing states—as well as being the CONUS's hottest and driest states. The largest is Arizona, with a 2022 population 7.67 million; the smallest of the four, New Mexico, has a population of 2.1 million. The other two are Colorado, population 5.9 million and Utah, population 3.3

million. Some of the largest cities in the country are located in the Southwest, where water scarcity is fast becoming a common problem. Phoenix, Arizona is the largest city in the Four Corners, with over 1.7 million people. Phoenix is also has the highest average temperature of all major U.S. cities, averaging 111 days of temperatures over 111 degrees heat annually.

Drought is only one of the weather factors contributing to water scarcity in the Southwest. For instance, the snowpack across much of the mountains in the northwest and Southwest has been below average. What snow does accumulate is melting sooner than traditionally, leaving less surface water available in spring and summer when it is needed for agriculture and other uses. The annual snowmelt that is now happening earlier is due to climate warming. Much of the snowmelt that occurs in Utah and Colorado feeds the Colorado River which is now absorbed by the dry mountain soils or evaporates into the atmosphere. This long drought has decreased Colorado River levels to lowest levels since construction of the river's major dams and forming of Lake Powell and Lake Mead, the two largest reservoirs in the country by multiple states for power supply, irrigation, and water supplies.

References

CGS (Colorado Geological Survey). 2022. Colorado groundwater atlas. Available at https://coloradogeologicalsurvey.org/water/colorado-groundwater-atlas/#gwa00

Christensen, Niklas S., Andrew Wood, Nathalie Voisin, Dennis P. Lettenmaier, and Richard N. Palmer. 2014. The effects of climate change on the hydrology and water resources of the Colorado River basin. *Climatic Change* 62: 337–363. https://doi.org/10.1023/B:CLIM.0000013684.13621.1f.

Januta, Andrea, and Daniel Trotta. 2021. Hoover Dam reservoir hits record low, in sign of extreme western U.S. drought. *U.S. News*, June 10. Available at www.reuters.com/world/us/hoover-dam-reservoir-hits-record-low-sign-extreme-western-us-drought-2021-06-10/.

NIDIS (National Integrated Drought Information System). 2022. Research spotlight: Climate-driven megadrought. Available at www.drought.gov/research-spotlight-climate-driven-megadrought.

Philpot, Simone, Keith Hipel, and Peter Johnson. 2016. Strategic analysis of a water rights conflict in the south western United States. *Journal of Environmental Management* 180: 247–256.

SNWA (Southern Nevada Water Authority). 2011. Southern Nevada Water authority Clark, Lincoln and White Pine Counties Groundwater Development Project: Conceptual Plan of Development. Available at http://water.nv.gov/hearings/past/Spring%20-%20Cave%20-%20Dry%20Lake%20and%20Delamar%20Valleys%202011/Exhibits/SNWA%20Exhibits/SNWA_Exh_191_2011%20Plan%20of%20Development.pdf

UDWR (Utah Division of Water Resources). 2012. As winter ends and runoff begins, Utah remains in drought. Available at https://water.utah.gov/waterdata/drought/

USBR (U. S. Bureau of Reclamation). 2022a. Colorado River 5-year probabilistic projections. Available at www.usbr.gov/lc/region/g4000/riverops/crss-5year-projections.html

USBR (U. S. Bureau of Reclamation). 2022b. Bureau of reclamation: About us. Available at www.usbr.gov/main/about/fact.html

USBR (U. S. Bureau of Reclamation). 2022c. Bureau of reclamation release of 2022 August operating guidelines for Tier 1. Available at www.usbr.gov/lc/region/g4000/24mo.pdf

USGS (United States Geological Survey). 2018a. Groundwater decline and depletion. Available at www.usgs.gov/special-topic/water-science-school/science/groundwater-decline-and-depletion?qt-science_center_objects=0#qt-science_center_objects

USGS (United States Geological Survey). 2018b. Wyoming water use in the U.S., 2015. Available at www.usgs.gov/media/images/wyoming-water-use-us-2015

USGS (United States Geological Survey). 2021a. National water census: Regional groundwater availability studies. Available at www.usgs.gov/mission-areas/water-resources/science/national-water-census-regional-groundwater-availability?qt-science_center_objects=0#qt-science_center_objects

USGS (United States Geological Survey). 2021b. USGS water resources. Pennsylvanian and Mississippian Aquifers of the Appalachian Plateaus. Available at www.usgs.gov/mission-areas/water-resources/science/pennsylvanian-and-mississippian-aquifers-appalachian-plateaus#publications

USGS (United States Geological Survey). 2021c. Mississippi embayment aquifer system. Available at www.usgs.gov/mission-areas/water-resources/science/mississippi-embayment-aquifer-system

Xiao, Mu, Bradley Udall and Dennis Lettenmaier. 2018. On the causes of declining Colorado River streamflows. *Water Resources Research* 54, https://doi.org/10.1029/2018WR023153

CHAPTER 11

Water Crises in the High Plains

The broad Midwestern section of the contiguous United States can be divided into two sections: the Northern High Plains, and the Prairie Planes. The six states that overlay most of the High Plains aquifer as shown in the USGS Fig. 11.1. The Northern High Plains include North Dakota, South Dakota, Nebraska, and the eastern third of Wyoming. States in the Southern High Plains section are Kansas, Oklahoma, Texas, and sections of eastern Colorado. The Southern High Plains states section form the eastern boundary of the High Plains. This chapter concentrates on the six core Plains aquifer states, addressing Colorado, New Mexico, and Wyoming in more than one section.

The High Plains Aquifer

The High Plains regional area has a dry middle-latitude continental climate with abundant sunshine, moderate precipitation, frequent winds, low humidity, and a high rate of evaporation. The weather for most of the summer of 2022 had been abnormally dry for most of the region, with most in high levels of drought. Mean annual precipitation ranges from 12 inches in the west to 33 inches in the east. During most years in much of the area, irrigation is required for economic yields of typical crops—alfalfa, corn, cotton, sorghum, soybeans, peanuts, and wheat. Groundwater for

D. E. McNabb and C. R. Swenson, *America's Water Crises*,
https://doi.org/10.1007/978-3-031-27380-3_11

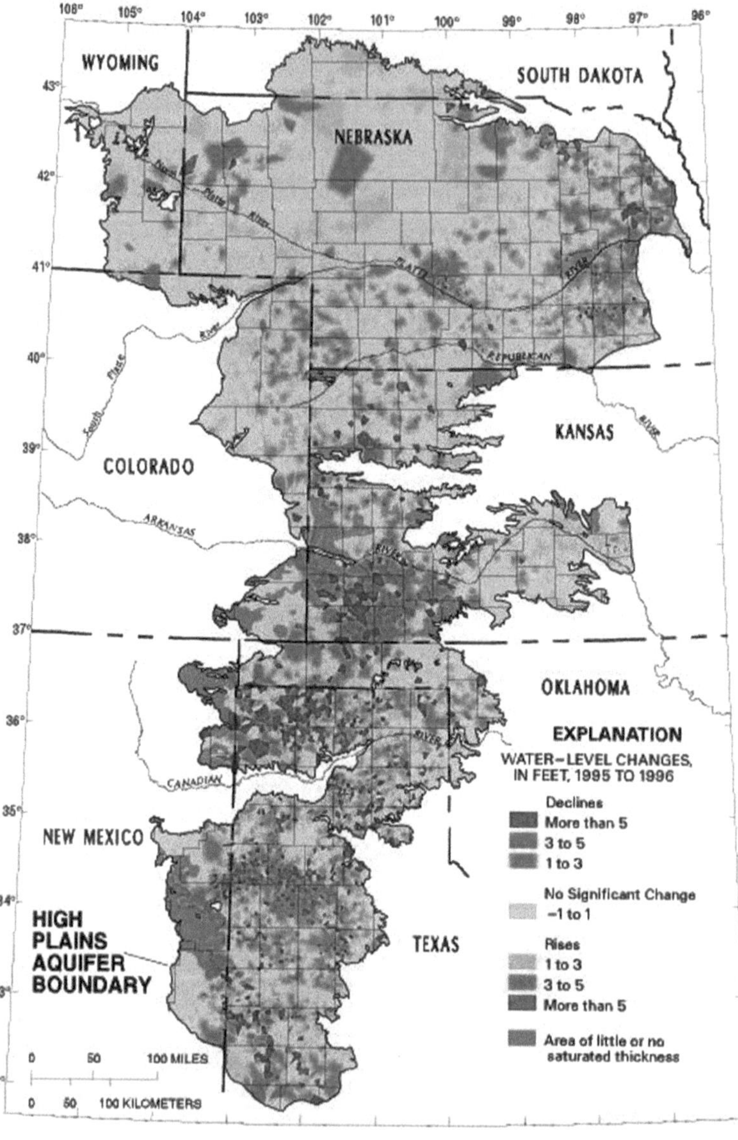

Fig. 11.1 Extent of the High Plains aquifer (*Source* USGS Nebraska Water Science Center [2014])

irrigated agriculture is common in this section, where precipitation cannot always be depended on.

In the area that overlies the High Plains aquifer, estimated amounts of acreage irrigated with groundwater increased rapidly from the 1940s to 1980 but did not change greatly from 1980 to 2002: 1949—2.1 million acres, 1980—13.7 million acres, 1997—13.9 million acres, and 2002—12.7 million acres. However, the proportion of irrigated acreage in each state relative to total irrigated acres did change substantially over time in some states.

All the states in the High Plains have concerns with some aspect of their water resources and the effects of prolonged drought. Sections of the national drought condition weekly reports on the then-current drought history indicate that 2022 was apparently going to be an extension of the drought that began in 2020. Although heavy precipitation events during early months of 2022 had benefited many locations, by September, most states in section of the country were still experiencing some degree of drought. Active weather events were recorded wide areas of the Pacific Northwest, Northern Rockies, the Great Plains, Lower Midwest, and isolated areas of the South and Southeast, with the most severe weather occurring in the Central Plains and areas of the Midwest. The more central sections, known as Tornado Alley, are where numerous tornadoes touch down in eastern Kansas, southeastern Nebraska, and parts of Illinois, to name a few of the states where they regularly occur.

Widespread heavy rainfall accumulations are common in this region, with the heaviest accumulations common in eastern Nebraska. These rainfall events provide much-needed moisture to the region-boosting soil moisture levels across parched areas from Kansas to South Dakota. However, during the hot summer months most of the soil water evaporates quickly during a drought. The May 7, 2022 NOAA drought report for the Northern High Plains included the following points:

- Widespread moderate to extreme drought (D1–D3) continues to persist across much of the Northern Great Plains. The hardest hit state, North Dakota, had Extreme Drought (D3) impacting 85% of the state. Extreme drought also covered roughly 20% of South Dakota. Earlier in the month, more than 40% of both Kansas and Nebraska were in extreme drought.
- The most prominent drought impacts include significant and record-breaking wildfires (over 800 fires in North Dakota alone), lack of

water and poor water quality for livestock, poor pasture and rangelands, and reduced ability to grow some commercial and specialty crops.

- Widespread burn bans were in effect, and some states are offering water assistance programs and coordinating hay hotlines/feed lists. Some producers reducing their herd size, and some farmers were not able to plant spring wheat.
- Despite the potential for some short-term rain relief, the summer climate outlook does not look promising to alleviate drought conditions. There is a greater chance for above-normal temperatures across the entire Northern Great Plains, and a greater potential for below-normal precipitation in the western parts of the region.
- It is likely that drought impacts will continue throughout summer, with the potential for more wildfires and decreased water supply. A secondary peak of wildfire is expected in late summer/early fall if conditions do not improve before then.

The Northern High Plains region holds one of the nation's most important source of the groundwater, the Ogallala Aquifer. This deep groundwater resource makes it possible for the middle states of the contiguous United States to supply much of the nation's grains and livestock. This complex aquifer system underlies sections of the Midwest from South Dakota to northern Texas. Depletion of the groundwater supply in the Central section of the High Plains Aquifer remains a serious issue in many locations. Natural weather patterns make precipitation and surface water supplies less dependable in the Central and southern sections of the High Plains. As a result, agriculture is increasingly dependent upon groundwater, resulting in a considerable imbalance between water-level changes (KGS 2021).

The Northern High Plans

The three Northern High Plains states are subject to similar climate and weather conditions. They endure unreliable precipitation and long hot, dry summers that often lead into prolonged bouts of severe drought. Therefore, they have become increasingly dependent upon groundwater for irrigation. The wettest time of the year is late spring and early summer, while the driest season is winter. Warmer winters in all of the western mountain ranges have resulted in lower levels of the mountain snowpack. Year-to-year differences in the amount of the snowpack or the timing of the melting and subsequent runoff have implications for areas downstream across the plains and beyond.

The Northern Plains portion consists of the High Plains includes North Dakota, South Dakota, and Nebraska, with a portion of the eastern section of Colorado that can be categorized as being located in the Northern High Plains, includes some of the driest portions of the central United States. Table 11.1 shows the end of summer 2022 drought status for these states.

The U.S. Department of Agriculture's Climate Resilience Tool Kit (CRTK) identified the climate of the Northern High Plains as a variable climate with icy winters and hot summers. This makes the region prone to extreme events such as droughts, floods, heat waves, cold waves, blizzards, and severe weather. A feature of the region is its east-west precipitation variation. Winds and storms moving from west mean areas in the eastern states can receive two to three times the precipitation of areas in the mountainous western states. This precipitation gradient greatly shapes the ground cover. Traditionally tall-grass prairie is typical in the east, transitioning to short-grass prairie and sagebrush in the west, and rain fed row crops in the east transitioning to irrigated crops and rangeland in the west.

The Rocky Mountains play a critical role in the Northern High Plains as they feed several of the rivers that flow across the region,

Table 11.1 Northern High Plains drought by area under drought level, September 2022

Sections of state included					
State	*D0:D4 abnormally dry*	*D1:D4 moderate drought*	*D2:D4 severe drought*	*D3:D4 extreme drought*	*D4 exceptional drought*
Inclusions	*D0–D4 (%)*	*D1–D4 (%)*	*D2–D4 (%)*	*D3–D4 (%)*	*D4 only (%)*
Nebraska	100.0	87.5	65.0	29.1	10.5
South Dakota	81.8	41.0	13.1	4.6	0.0
North Dakota	84.3	5.3	1.4	0.0	0.0
Colorado	84.3	46.3	17.0	3.6	0.3

Source NIDIS Drought.gov

providing surface water for communities, ecosystems, and agriculture. The USDA Climate Hubs are organizations established in February of 2014 to collect and deliver science-based knowledge, practical information to farmers, ranchers, forest landowners, and resource managers to aid in climate-informed decision-making in light of the increased risks and vulnerabilities associated with a changing climate. The function in partnership with the National Oceanic and Atmospheric Administration (NOAA) and Department of Interior (DOI), are supporting regional climate resilience through the Climate Resilience Toolkit. The Toolkit is a website that offers information and climate tools from across the U.S. Federal Government in one location (CRTK, n.d.).

Since the 1980s, temperatures in the Northern High Plains have increased more than any other region in the U.S., with an average warming of 1.7 degrees F. Generally, nighttime low temperatures have increased somewhat more than daytime high temperatures and increases more noticeable in winter. In the future, average temperatures are predicted to continue to increase more rapidly. Those changes were also expected to see decreases in the occurrence of cold waves and increases in more severe heat waves. The region is suffering from both severe drought and extreme flooding. Flooding in 2011 was followed by widespread drought in 2012. The 2017 flash drought across northern portions of the region was preceded by a near-record wet period. Large and severe wildfires have occurred in the west. The extreme wetness of 2019 resulted in widespread. This was followed by the beginning of another drought in 2020 (CRTK 2021). The five levels of drought underway in the five Northern five are shown in Table 11.2.

Prolonged drought is a major concern of all the states in the Central and Southern Plains region. In 2022, both Texas, with 40% of the state in extreme drought, and Oklahoma with 31% in extreme drought, were under drought conditions. The water managers in these states were forced to take actions to limit access to freshwater to many communities. Table 11.2 lists the various drought levels in the states in this section as of March and early May or 2022.

High levels of drought were common in nearly all sections of the Plains states that overlie large portions of the Great Plains aquifer. In areas where possible, the dry heat has resulted in large amounts of groundwater withdrawn from the underground rervoirs. The tables also include data of the two states with large sections over the major aquifers.

Table 11.2 Central/Southern High Plains levels of drought in last week of summer 2022

Sections of state included					
State	*D0:D4 abnormally dry*	*D1:D4 moderate drought*	*D2:D4 severe drought*	*D3:D4 extreme drought*	*D4 exceptional drought*
Inclusions	*D0–D4 (%)*	*D1–D4 (%)*	*D2–D4 (%)*	*D3–D4 (%)*	*D4 only (%)*
Texas	78.4	59.4	31.9	8.3	0.6
Oklahoma	100.0	96.5	85.4	45.8	3.1
New Mexico	99.1	84.2	46.8	7.1	0.0
Kansas	96.4	75.3	64.0	42.3	12.1

Source NIDIS Drought.gov

Northern Plains Water Issues

The surface water resource in the High Plains has been severely stressed by the prolong drought. An example is how the Colorado River Basin water situation continues to deteriorate due to the long-term impacts of drought with water levels at Lake Powell and Lake Mead currently at 24% full and 31% full, respectively. With Lake Powell's water surface elevation currently at 3,522 feet, it is quickly approaching the 3,490-foot threshold level at which Glen Canyon Dam can continue to generate hydropower. The fortuitous heavy but short-term precipitation in spring of 2022 led to targeted improvements in droughts in the Pacific Northwest, the Great Plains, the South and the Southeast, while decreases were registered in the Southwest, Texas, Southeast, and the Mid-Atlantic.

Until the early 1940s, depth to water in the aquifer and the slope of the land surface were major factors controlling the distribution of irrigated acres. However, by the early 1960s, technological advances in pumps allowed irrigation in areas with deeper (greater than 100 feet below land surface) groundwater wells. The development of center-pivot irrigation systems then allowed irrigation on rolling terrain.

Irrigated agriculture is essential in meeting much of the world's food needs. It is also the world's largest water user, accounting for more than 70% of annual global water consumption. For most of its use, irrigation water is freshwater that is not stored, delivered, scheduled, or applied efficiently. Unrestricted groundwater or surface water withdrawals threaten sustained availability of both ground and surface water. In some areas of the Central United States, the over-withdrawal of water from sections of the Ogallala (High Plains) Aquifer has resulted in the water table dropping more than 150 feet in many locations. The Ogallala Aquifer is one of the world's largest aquifers consisting of low water table aquifer surrounded by sand, silt, clay, and gravel located beneath the lands east of the Rocky Mountains and extending east to nearly the Mississippi River. It provides water for necessary for agriculture from South Dakota to Texas.

The main aquifer systems in Northern High Plains underlie parts of South Dakota and Wyoming, Kansas, and Nebraska. Groundwater production from the aquifers of the High Plains system is used primarily for irrigation, with smaller quantities used for rural domestic, livestock, municipal, mining, manufacturing, and power purposes. Nebraska and Kansas are the states using the largest amount of High Plains groundwater.

Water Use in North and South Dakota

North Dakota has little access to the High Plains or Ogallala Aquifer systems. The state is situated over a series of rock layers that thicken to the west and can reach thousands of feet in thickness. Bedrock aquifers are among these layers. These aquifers in this state are more continuous than those that are farther to the south. The water in deeper parts of these aquifers is salty and does not provide quality drinking water. The Fort Union aquifer, Fox Hills-Hell Creek aquifer, and Dakota aquifer are major bedrock aquifers in North Dakota. The Dakota aquifer is the deepest branch at more than a mile below the surface of western North Dakota.

In an early April 2022, NOAA's Drought Monitor showed two-thirds of South Dakota were in moderate or severe drought. Although the climate outlook was for warmer and drier than average climate through June, South Dakota was one of the Northern High Plains states expected to improve due to heavy precipitation early in the period. The current drought in southwestern South Dakota began in mid-2020. In May of 2022, many areas in the state had soil moisture shortages from previous dry years and reservoirs and surface water resources were lower than average.

Nebraska Water Use

There are at least seven major aquifer systems underlying Nebraska. The Ogallala/High Plains Aquifer is the largest and most important of the major aquifer systems in Nebraska in terms of volume of water in storage and withdrawals for irrigation. Several secondary aquifers exist in areas outside of the High Plains aquifer, providing water for irrigation, municipal supplies, and domestic use in parts of far western and far eastern Nebraska. About 80% of the state's public drinking water and nearly all of its private water supply are from groundwater sources.

Nebraska is the fourth largest user of groundwater in the United States, just behind California, Texas, and Arkansas. Nebraska has more than 100,000 registered irrigation wells and an additional 16,000 registered water wells. Nebraska experienced the largest expansion of irrigation of the top-ten irrigated states in both acres and percentage change. While Nebraska was adding about 934,000 acres, the remaining nine states experienced a net decrease in irrigated acres of 186,000 acres. By county the state's irrigated acres correspond to those areas having both productive soils and water availability (primarily from High Plains Aquifer). In 1969, Nebraska had roughly 3.8 million irrigated acres, increasing every decade to reach 9.3 million acres in 2019. Center-pivot irrigation systems constitute nearly 98% of the acres under sprinkler irrigation in the state. Despite the increase in irrigation, there is little or no increase in the decline in the Ogallala water level from 2021 to 2015 (McGuire 2017).

The South/Central High Plains

The states included in the south/CENTRAL High Plains are Kansas, Texas, and Oklahoma. This combination is because all these states overlie portions of the High Plains and Ogallala Aquifer system. Although New Mexico is often included in descriptions of the American Southwest, we have included a discussion of its water condition in this chapter as well as in the Four-Corners region.

Dependence on groundwater from the High Plains aquifers also justifies their position on the Central High Plains. Withdrawals from the aquifer in Texas are significantly greater than that occur in New Mexico and Oklahoma. Combined, however, they add significantly to the total withdrawals from this important resource. Traditionally, the High Plains underground groundwater system is often referred to as the Ogallala

Aquifer, named after the Ogallala tribe of the Sioux Indians. Together, this large sector constitutes the most important and productive agriculture production. The states benefit from an estimated 134,000 square miles of the Central Plains aquifer system. Sections of this asset are also among the most stressed groundwater resources.

Texas Water Stress

Drought in Texas improved somewhat over the summer of 2022, largely due to several severe rainstorms. At the beginning of summer, 71.5% of the state was in severe drought conditions, with 40.6% in extreme drought. By the second week of September, Texas severe drought had dropped to 31.9% of the state and just 8.3% was still under extreme drought.

More than 60% of water used in Texas comes from groundwater in nine major and 21 minor aquifers. These aquifers are a critical source of water for Texas, supplying 13.7 million acre-feet of water used in the state in 2014. About 76% of this water is for irrigation, with irrigators withdrawing most of this water from the Ogallala Aquifer alone. About 36% of water used to meet municipal demands is from groundwater.

The Ogallala Aquifer is the most important aquifer in Texas, where it is separated into northern and southern parts by the Canadian River valley (TWDB 2016). Although some wells had been used as early as 1919, groundwater in North Texas was not exploited extensively until the mid-twentieth century (Hernández et al. 2013; Musick et al. 1990). Irrigation development in the Texas High Plains began when farmers started drilling irrigation wells in the Ogallala Aquifer during the major drought of the 1930s. Rapid irrigation development began in the late 1950s and peaked in the late 1970s, but then reached a higher peak in 2000. When center-pivot systems allowing previously non-irrigated land are to be irrigated.

In the Texas Ogallala Aquifer, post-development irrigation pumping has lowered the Ogallala Aquifer water table and locally affected groundwater flow direction. Pumping is now the largest discharge mechanism in the Ogallala Aquifer. Most of the pumping discharge is offset by a decrease in aquifer storage. The largest amount of pumping and corresponding drawdown occur where the initial saturated thickness was largest. Some of the pumping discharge is offset by capture, with decreased discharge to springs, streams, and other surface discharge features. Many springs have either experienced reduced flow or dried up completely. Streams and draws that were originally fed by springs or aquifer discharge also have reduced or no flow (Deeds and Hamlin 2015, ES2).

Additional studies indicate that recharge represents a small fraction of current water usage, the result being existing groundwater supplies are declining. Deeds and Hamlin divided the Ogallala into two recharge regions. Recharge in the southern region has been affected by agricultural development and ranges from miniscule amounts (0.007 inch) to more than 3 inches per year, with the most recharge in areas where irrigated crops are raised on relatively permeable soils. In the northern region, relatively unproductive clay soils limit agriculture and less water is withdrawn from the aquifer to influence recharge.

Oklahoma Water Stress

Drought conditions occurred in Oklahoma in 2022 in all categories all drought except the highest, exceptional drought, which dropped from 9% of the state to 3.1%. The areas under drought at the end of May were 76.5% for moderate drought, 60.5% for severe drought, and 31.1% of the state in extreme drought, moderate drought had increased to 96.5% of the state, severe drought increased to 85.4% and extreme drought increased to 48.8%.

There are 22 major groundwater basins containing approximately 390 million acre-feet of water in storage, although only one-half of this amount may be currently recoverable. Groundwater is the major source of water in the western half of the state, while most of the surface water is in the eastern half of Oklahoma. The state's largest groundwater basin is the Ogallala Aquifer in western Oklahoma. It contains 90 million acre-feet of water, nearly one-fourth of all groundwater in the state. Groundwater depletion is currently a problem in western Oklahoma, and this loss is primarily due to crop irrigation.

Groundwater Stress in Kansas

In Kansas, the High Plains aquifer consists of three distinct groundwater resources: the Ogallala Aquifer, the Great Bend Prairie aquifer, and the Equus Beds. The Ogallala Aquifer underlies most of western Kansas and consists mainly of the Ogallala Formation, a geologic unit that formed from sediment eroded off the uplifting Rocky Mountains. The southwest

Kansas portion of the Ogallala has suffered the greatest losses since levels started dropping in the 1950s. Initially, water levels fell less than a foot each year until 2020. Over the drought years between 2010 and 2014, declines averaged about three feet per year. Since 2015 and the renewal of drought, the water level declined about 1.2 feet.

Since the late 1880s, farming has been the dominant industry in the High Plains area. In some sections farmers have irrigated using shallow wells to access groundwater to supplement surface water. For example, in the late 1800s, some farmers were irrigating crops using surface water diverted from the Platte and Arkansas Rivers or some were irrigating using shallow groundwater pumped windmills. Through the early 1940s, these shallow wells and the slope of the farmland surface were major factors limiting the extent of irrigated acres. It was not until the early 1960s that advances in pump technology allowed irrigation in areas with the extraction of groundwater from depths of greater than 100 feet below the surface. Prior to this, groundwater irrigation from the aquifers had received only a modest boost in the 1930s and 1940s, when pumps powered by automobile engines replaced windmills, allowing access to water from greater depths. Once federally funded rural electrification became possible to power irrigation pumps with electricity-powered pumps, irrigators laid pipes. The development of center-pivot irrigation systems allowed irrigation on rolling terrain. Increases in annual groundwater withdrawals for irrigation from the High Plains aquifer occurred about every five years by the U.S. Geological Survey and state agencies.

Kansas Water Issues

Kansas state law requires the owner of a water right or permit to appropriate water for beneficial use, except for domestic use, to file a complete and accurate water use report on or before March 1, following the end of the previous calendar year. Kansas water users generally divert about 5 million to 6 million acre-feet per year of surface water and groundwater for beneficial uses—irrigation, power generation, public water supplies, industrial processes, stock watering, and other purposes. Ten to 20% of the diverted water is used for non-consumptive uses such as hydroelectric power generation or once-through cooling at thermoelectric power plants. Eighty to 90% of the diverted water is used for consumptive uses—that is, uses in which all or most of the diverted water is evaporated, ingested, or otherwise permanently removed from the local source of supply (KDOA 2016).

Prior to the COVID-19 pandemic, 84% of the consumptive use of water in Kansas was for agriculture. The remaining 10 to 16 percent was for municipal use (public water supply) of total consumptive use of water in the state. The remaining five percent of consumptive water use is for industrial, recreation, stock watering, hydraulic dredging, and other uses. Water use varies from year to year along with climate, the time of day, and the percent of annual precipitation. For example, at the beginning of 2017, southwest Kansas was categorized to be in a severe drought (U.S. Drought Monitor). By May, no portion of the state was categorized as in a drought. Kansas was in moderate drought from August through the end of 2017 (KDOA 2017). After three years of drought, the end of the 2021 water year (March 2022), 100% of the state was still moderately dry, 55.9% of the state in a severe drought, 9.7% in extreme, and 1.7% in an exceptional drought stage (drought.gov 2017).

Water Stress in New Mexico

The Great Plains extends into eastern New Mexico and Colorado, where flat to rolling prairie with scattered hills and bluffs gradually rises westward to 5,000 and 7,000 feet above sea level and the frontal ranges of the Rocky Mountains and the Southern Rocky Mountain Basin. West of the frontal ranges in northern New Mexico are additional and higher mountain ranges, with many spurs and extensions. Most of the high peaks are located near the Continental Divide, which extends approximately north and south through western New Mexico and central Colorado. The altitude of the divide decreases in southern New Mexico to less than 4,500 feet in a few areas (Robson and Banta 1985).

In the first week of May 2022, the U.S. Drought Monitor reported the entire state was under some level of drought: 98.9% of the state was under a moderate drought, 95.8% under a severe drought, 68% extreme drought, and 15.7% of the state was under exceptional drought conditions. Conditions had changed somewhat by the end of the summer, when moderate drought had dropped to 84.2%, severe drought to 46.8%, and extreme drought to 7.1%. No parts of the state remained on exceptional drought conditions.

The drought that does remain is still enough to put the state on a critical condition when it comes to water resources. New Mexico relies on both groundwater and surface water sources, but about 87% of New

Mexico's public water supply comes from groundwater. No other southwestern state gets such a large percentage of its domestic water from groundwater sources. Such a heavy dependence upon groundwater has its downsides. Severe declines in groundwater levels have occurred in some parts of the state. Some municipalities such as Santa Fe have seen water levels in their groundwater wells decline as much as 300 feet in the past 10 years.

New Mexico also has long periods of drought and inconsistent precipitation. Water withdrawals from New Mexico, Colorado, Texas, and Mexico have greatly reduced the volume of the Rio Grande over the past 50 years. New Mexico and several other states also draw water from the Colorado River Basin, and demand for water from that river has long exceeded supply.

Drought and Climate Change

The Upper Colorado River Basin, which extends into western New Mexico, has been experiencing a protracted, multi-year drought that began in 1999. Prolonged droughts, in conjunction with climate change, have the potential to reduce snowpack, increase temperatures, and create earlier mountain snow thaws, which all reduce water supplies (EPA 2013).

Climate change is already creating a longer, hotter summer season in New Mexico. The state is experiencing extremely low surface water flow in the Colorado and Rio Grande rivers; warmer temperatures are increasing water loss from other streams and lake surfaces by evaporation. The Upper Colorado River Basin, which extends into western New Mexico, has been experiencing protracted, multi-year droughts that began in 1999. Prolonged droughts, in conjunction with climate change, are reducing snowpack and creating earlier mountain snow thaws, which all reduce snowpack and snowmelt contribution to surface and groundwater resources.

Conclusion

The High Plains aquifer, also known as the Ogallala Aquifer, underlies about 112 million acres, or 175,000 square miles, in parts of 8 states, including Colorado, Kansas, Nebraska, New Mexico, Oklahoma, South Dakota, Texas, and Wyoming. The USGS has published reports on water-level changes in the High Plains aquifer since 1988. Substantial water-level declines continue to occur in large areas of the aquifer. In 2015, total recoverable water in storage in the aquifer was about 2.91 billion acre-feet, which is an overall decline of about 273.2 million acre-feet, or 9%, since predevelopment. Average area-weighted water-level change in the aquifer was a decline of 15.8 feet from predevelopment to 2015 and a decline of 0.6 feet from 2013 to 2015. Recoverable water in storage from 2013 to 2015 for the aquifer declined 10.7 million acre-feet, which is about 30% of the recoverable water in storage change from 2011 to 2013. This difference is likely related to reduce groundwater pumpage during the 2013 and 2014 irrigation seasons as compared to the 2011 and 2012 irrigation seasons (USGS public domain).

On the last week of the 2022 summer, prolonged drought conditions continued in areas of Kansas, Nebraska, South Dakota, and North Dakota in response to continued dryness, decreased soil moisture, poor pasture and rangeland conditions, and impacts to dryland crops. In Kansas and Nebraska, streamflow levels were below-normal flows. Average temperatures were above-normal in the Dakotas, while Nebraska, eastern Wyoming, and Kansas were near-normal to slightly below-normal. According to NOAA, the Great Plains Region saw its 5th warmest and 19th driest June–August on record. Statewide, Nebraska experienced its 3rd driest June–August as well as its 2nd driest August on record.

References

CRT (Climate Resilience Toolkit). 2021. The Northern Great Plains. Available at https://toolkit.climate.gov/regions/northern-great-plains.

Deeds, Neil E., and Scott Hamline. 2015. Final conceptual model report for the High Plains Aquifer System groundwater availability model. TWDB Report. Available at www.twdb.texas.gov/groundwater/models/gam/hpas/HPAS_GAM_Conceptual_Report.pdf.

EPA (Environmental Protection Administration). 2013. Saving Water in New Mexico. Available at www.epa.gov/sites/default/files/2017-02/documents/ws-ourwater-new-mexico-state-fact-sheet.pdf.

Hernandez, Jarro H., Prasanna H. Marek, Terry A. Howell, and Wonsook Ha. 2013. Groundwater levels in Northern Texas High Plains: Baseline for existing agricultural management practices. *Texas Water Journal* 4 (1): 22–34.

KDOA (Kansas Department of Agriculture). 2016. Water use reporting. Available at https://agriculture.ks.gov/divisions-programs/dwr/water-appropriation/water-use-reporting.

KDOA (Kansas Department of Agriculture). 2017. Irrigation Water use in Kansas. Available at https://agriculture.ks.gov/docs/default-source/dwr-water-appropriation-documents/ks_irr_2017.pdf?sfvrsn=6b8d88c1_0.

KGS (Kansas Geological Survey). 2021. Groundwater levels fall in central and western Kansas. Available at www.kgs.ku.edu/General/News/2021/groundwater_levels_fall.html.

McGuire, Virginia L. 2017. Water-level and recoverable water in storage changes, High Plains aquifer, predevelopment to 2015 and 2013–15: U.S. Geological Survey Scientific Investigations Report 2017–5040. Available at https://doi.org/10.3133/sir20175040.

Musick, J.T., Pringle, F.B., Harman, W.L. and Stewart, B.A. 1990. Long-Term Irrigation Trends: Texas High Plains. *Applied Engineering in Agriculture*, 6: 717–724. https://doi.org/10.13031/2013.26454

Robson, Stanley G., and Edward R. Banta. 1985. *Ground-Water Atlas of the United States: Segment 2, New Mexico.* USGS. Available at https://doi.org/10.3133/ha730C.

TWDB (Texas Water Development Board). 2016. Texas aquifers study groundwater quantity: Quality, flow, and contributions to surface water. Available at www.twdb.texas.gov/groundwater/docs/studies/TexasAquifersStudy_2016.pdf#page125.

USGS (United States Geological Survey). 2014. *High plains aquifer resources and information water-level changes in the high plains aquifer, 1980 to 1996.* Nebraska Water Science Center. Available at https://ne.water.usgs.gov/ogw/hpwlms/fs96/hp96fig3.html.

CHAPTER 12

Water Use in the Great Lakes Region

The Great Lakes region includes Iowa, Michigan, Minnesota, Wisconsin Ohio, Indiana, and Illinois. Climate in these states is typical for the interior Middle-west coterminous United States. In addition to rain and hail precipitation, winters in the region can include snow and ice. These states and Ontario, Canada almost form a complete ring around these five lakes: Lake Superior, Lake Michigan, Lake Huron, Lake Erie, and Lake Ontario. The coldest winter weather occurs in Minnesota and the mildest winters are in Missouri. While spring and fall months tend to have mild, summers these seasons are often hot and humid, with temperatures more than 100 degrees Fahrenheit in the summer months. Combined, these five Great Lakes hold the largest source of freshwater on the globe. However, the states that surround the lakes have helped to introduce significant pollution into them all.

The Midwest weather in general is subject to extremely cold air masses from the far north, and warm, humid air masses from the Gulf of Mexico, resulting in a wide range of both temperature and precipitation extremes. From spring through early fall, rain will typically fall a few days a week in many parts of the region. Snow is also standard in winter months, although it varies depending on the area. In the southwestern Minnesota counties, there is an average of 36 inches of snowfall each year, while an average of 70 inches of snow is typical of the northeastern part of the state that borders Lake Superior.

D. E. McNabb and C. R. Swenson, *America's Water Crises*,
https://doi.org/10.1007/978-3-031-27380-3_12

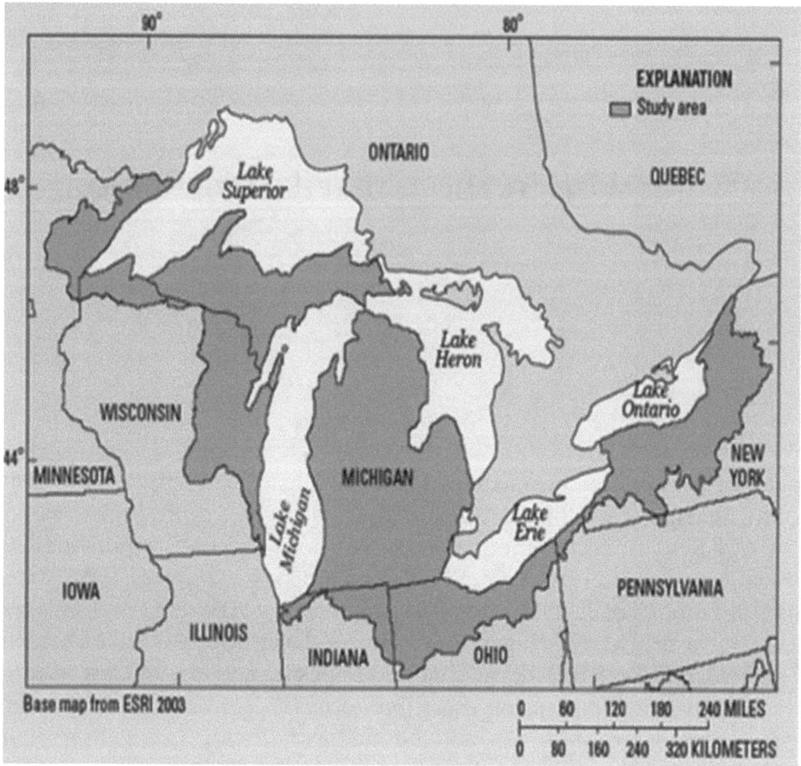

Fig. 12.1 The location of the Great Lakes states (*Source* ESRI Base Map from ESR)

The development pattern throughout this area saw major immigration as the country grew westward beginning in the 1700s and accelerated in the industrial area. Consequently, most water infrastructure has aged and the need to restore the rapid crumbling water and wastewater systems is a serious problem that is occurring in all states of this region (Fig. 12.1).

Since 2000, the longest duration of drought in Indiana lasted 44 weeks beginning on August 25, 2020 and ending on June 22, 2021. In September 2022, the state was still free of drought. Ohio was the only other one of the seven states we include in this region to be completely free of drought in September. Five of the seven states were reported to

Table 12.1 Great Lakes Region levels of drought by category, June 2022

Sections of state included					
State	*D0:D4 abnormally dry*	*D1·D4 moderate drought*	*D2:D4 severe drought*	*D3.D4 extreme drought*	*D4 exceptional drought*
Inclusions	*D0–D4 (%)*	*D1–D4 (%)*	*D2–D4 (%)*	*D3–D4 (%)*	*D4 only (%)*
Iowa	62.4	40.2	20.2	2.2	0.0
Michigan	20.7	6.0.8	0.0	0.0	0.0
Minnesota	26.0	8.9	1.0	0.0	0.0
Wisconsin	18.9	6.4	0.0	0.0	0.0
Ohio	4.5	0.0	0.0	0.0	0.0
Indiana	31.3	0.0	0.0	0.0	0.0
Illinois	29.6	7.1	0.2	0.0	0.0

Source NIDIS Drought.gov

be under moderate drought and three were in severe drought conditions. Iowa was the only state to report any area in extreme drought (2.2%). All of the region's states were reporting some portions of the state was abnormally dry. More than 60% of Iowa was abnormally dry while Ohio was only 4.5% dry (Table 12.1).

The pipes, structures, and buildings that make up water infrastructure systems are aging and rusting away. In the Great Lakes region, and around the country, funds for water infrastructure investments are inadequate to keep up with needs. The effects of drought apparent in Table 12.1 reveal that Iowa, without any direct connection to any of the lakes was the state most hit by drought, whereas for the 2022 summer season, Ohio was least effected by drought.

The report includes condensed and modified information prepared by technical staff from Iowa DNR, the Iowa Department of Agriculture and Land Stewardship, IIHR—Hydroscience and Engineering, and the U.S. Geological Survey, in collaboration with Iowa.

In June of 2022, Iowa was the Great Lakes region's driest state and the hardest hit by drought; 40.2% of the state was in moderate drought, 20.2% was in severe drought, and just 2.2% was in extreme drought. By the second week of September, increases had occurred in all these conditions: 67.1% was abnormally dry, 41.4% was in moderate drought, 23% in severe drought, and 5.6 in extreme drought. August 2022 was the fifth month in a row of below normal statewide average precipitation. Only northeastern Iowa reported wetter than average conditions with pockets of three to four inches of above-normal totals. For August, the statewide average precipitation totaled 3.53 inches, or 0.60 inches below normal. For the entire state, monthly precipitation totals ranged from 0.25 inches at Keokuk Lock and Dam to 8.88 inches at a gauge near Center Junction.

Some improvement in drought conditions had occurred in southwest and central Iowa, but drought conditions worsened in southeast Iowa over the summer. For the 2022 three-month summer season, temperatures averaged 72.6 F degrees, 1.2 degrees above normal. During the same period, precipitation totaled 10.26 inches or 3.30 inches below normal. During 2021, warmer summer prevailed, while the summer of 2022 was drier (Iowa DNR 2022).

August temperatures were from one to two degrees above average across Iowa with near-normal temperatures reported in northern and northeast. The statewide preliminary average temperature was 72.3 degrees, 1.3 degrees warmer than normal, with a warmer August occurring last year. Sioux City Airport reported the month's high temperature of 102 degrees on the August 2nd. This was 18 degrees above normal for that date and location. Anamosa reported the month's low temperature of 47 degrees on the 10th, 14 degrees below normal.

The 28-day average stream baseflow trends are used as an indicator of long-term water-level changes in shallow aquifers, and during August 2022 below-normal baseflow was observed from northwest to southeast Iowa. Streamflow, soil moisture, and shallow groundwater levels are all lower than normal in northwest and southeast Iowa. During August, streamflow conditions were below-normal flow in two-thirds of the state, while in the northeast corner of the state streamflow remained above-normal. Many rivers were in much below-normal conditions, although others moved into the above-normal condition (Fig. 12.2).

Missouri River Basin

In the Missouri River Basin, drought is not an uncommon climate event. Significant drought events occurred in the 1930s and 1950s that substantially affected water supplies, crops and livestock, energy, transportation of goods, and the ecosystem. A large-scale drought event occurred in 2012 that followed a devastating flood in 2011. The drought of 2012 followed a devastating drought event. The upper basin was in drought again in

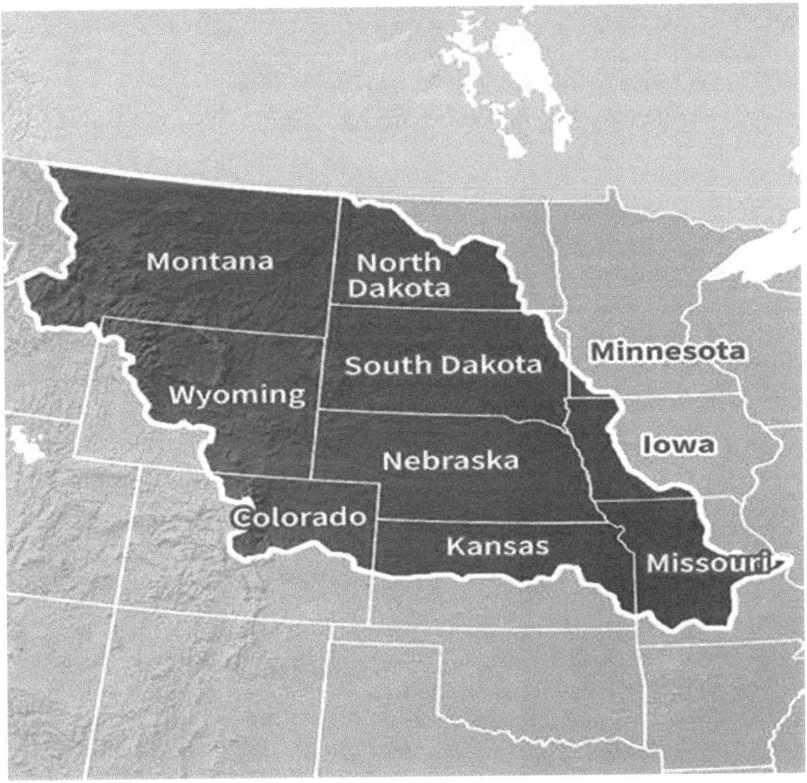

Fig. 12.2 Extent of the Missouri River Basin (*Source* NIDIS [2022])

2017 with a flash drought that was characterized by a rapid decline in soil moisture, low spring rainfall, high temperatures, and above-average wind speeds. The current drought began in 2020. In September 2022, drought was affecting over 46% of the Basin, with 74% of the Basin abnormally dry (D0), according to the U.S. Drought Monitor (Drought.gov).

Although only Iowa and Missouri and a small southeast corner of Minnesota are included in the Missouri River Basin, (Fig. 12.2) the bulk of the Basin in the Great Plains region is drier than the Great Lakes region. Across much of the upper Missouri River Basin, precipitation is projected to increase in winter, spring, and fall and decrease in the summer. Much of that increase will also occur in the upper Great Lakes region. In the southern part of the basin, the largest increases are expected to occur during the spring. In the central and southern parts of the basin, the largest decreases will occur during the summer months. Extended summer droughts will likely be common.

August shallow groundwater levels were statistically normal across much of the north and northeast parts of Iowa. However, declining groundwater levels leading to areas of increased water stress and vulnerability occurred across much of the rest of the state. The September 2022 a National Agricultural Statistics Service (NASS) report indicated that over August topsoil and subsoil moisture levels improved in north central, northeast, and east central Iowa, but had worsened in the west central, central, and southeast parts of the state. More than 80% of topsoil and subsoil moisture is rated as short of very short in the southeast area. At the same time, in northeast Iowa, more than 85% of topsoil and subsoil moisture was rated as adequate or surplus.

Drought Future

The Great Lakes region has become warmer over the last several decades, despite the heat absorbing lakes. Between 1900 and 2010, the average air temperature increased by more than 1.5 degrees Fahrenheit. Moreover, the rate of increase has accelerated in recent decades, particularly temperatures during periods accompanied with drought conditions. Summers in Illinois and Michigan might feel like current summers in Texas or Oklahoma by the end of the century.

Temperatures are projected to continue increasing throughout the Midwest, with the greatest increases in average temperature expected in northern areas. Extreme weather events are predicted to occur more often, particularly in the eastern section and less toward the west. Heavy rains, followed by devastating floods, are already common, but climate change is expected to intensify storms and lead to greater precipitation across the entire region during this century. Annual precipitation has already risen by as much as 20% in some areas. Projections of future precipitation indicate that heavy downpours are likely to occur primarily in winter and spring months while summers will become drier, especially in the southern portion of the region (Climate.gov 2021). Increases in temperatures are just one of the factors contributing to water stress in this and other U.S. watersheds. Land-use changes, and pollution increases are additional contributors. Tavernia et al. analyzed data from six different scenarios in 551 watersheds in the Midwest and Northeastern U.S. to identify long-term potential increases in water stress (Tavernia et al. 2013).

The 2019 U.S. Climate Reliance Toolkit (USCRTK) monograph on what water managers in all regions can do to be prepared to better cope with the significant climate change-influence expected to occur over the remainder of this century included the following selected water-related points as they relate to this region in dealing with climate change.

Box 12.1 Failing infrastructure in the Great Lakes water region

"The Great Lakes are a national treasure. They contain 20% of the earth's fresh surface water and are the drinking water source for more than 40 million people. However, dangerously outdated infrastructure remains a huge threat to [many] lakes. From combined sewer overflows to old lead service pipelines, water infrastructure needs to be updated to protect water resources for future generations. Failing wastewater infrastructure continues to release sewage and polluted storm water into local waterways every year, contaminating Great Lakes beaches, threatening public health and damaging local economies. Similarly, aging drinking water infrastructure is a costly challenge for many communities, as illustrated by threats to drinking water in Toledo, Ohio, from toxic algae in Lake Erie, and in Flint, Michigan, from lead contamination from improper water treatment and aging pipes."

"Water and wastewater infrastructure challenges are exacerbated by climate change impacts, such as increased runoff caused by more severe and frequent storm events that lead to increased flooding, sanitary and storm sewer overflows, and risks to public health and the health of the huge snow and ice-fed lakes."

"According the U.S. Environmental Protection Agency (EPA), over the next 20 years almost $180 billion is needed to repair and upgrade…drinking water and wastewater infrastructure in the Great Lakes region. In Minnesota alone, the figure tops $11 billion. Nationally, the cost is even higher."

Source *Clean Water Action* 2022.

Climate Change in the Great Lakes Region

In addition to the extreme rainfall and flood events now taking place in the region, they have been increasing over the last century. More are expected over the rest of the century. Increased precipitation has and likely will continue to lead to flooding, erosion, all of which reduce water quality. Other changes occurring or that are likely to change are:

- Longer growing seasons and rising carbon dioxide levels will increase yields of some crops. However, these benefits will be offset by more extreme weather events. In the long term, more decreases in agricultural productivity are expected.
- Increased heat wave intensity and frequency, increased humidity, degraded air quality, reduced water quality, and change in disease patterns will increase public health risks.
- The Great Lakes region has a highly energy-intensive economy, with per capita emissions of greenhouse gases more than 20% higher than the national average. Climate change restrictions may force these organizations to invest in new production methods or go out of business.
- Climate change will exacerbate a range of risks to the Great Lakes, including changes in harmful blooms of algae, and declines in beach health.
- In a positive change, ice cover declines will lengthen the commercial navigation season.
- Finally, at-risk communities in the Great Lakes region are becoming more vulnerable to climate change impacts such as flooding, drought, and increases in urban heat islands.

Box 12.2 Factors influencing groundwater in the states of the Great Lakes region

"Ground water is important to ecosystems in the Great Lakes Region because it is, in effect, a large, subsurface reservoir from which water is released slowly to provide a reliable minimum level of water flow to streams, lakes, and wetlands. Ground-water discharge to streams generally provides good quality water that, in turn, promotes habitat for aquatic animals and sustains aquatic plants during periods of low precipitation. Because of the slow movement of ground water, the effects of surface activities on ground-water flow and quality can take years to manifest themselves. As a result, issues relative to ground water are often seemingly less dire than issues related to surface water alone."

"Ground water is a major natural resource in the Great Lakes Region that helps link the Great Lakes and their watershed. This linkage needs to be more fully understood and quantified before society can address some of the important water-resources issues in the region.

The major ground-water resources issues in the Great Lakes Region revolve around (1) the quantity of ground water, (2) ground-water and surface-water interaction, (3) changes in ground-water quality as development expands, and (4) ecosystem health in relation to quantity and quality of water."

"A major attraction of the Great Lakes Region is the abundant water supply on which manufacturing, power generation, transportation, agricultural, and recreational sectors have historically relied. Most large public water supplies are obtained from the lakes themselves, but ground water is the source of drinking water for about 8.2 million people within the watershed."

Source Alley et al. (1999) with USGS update (2016).

Regional Groundwater Report

The latest report on the condition of the Great Lakes reported the 2020 total withdrawal amount for the Great Lakes-St. Lawrence River basin, excluding instream hydroelectric water use, was 37,856 million gallons per day (mgd) or 143,300 million liters per day. In-stream hydroelectric power water use accounted for 93% of the water use in the region but is not considered a withdrawal in the traditional sense because it includes "run of the river" use, where the water remains in the water body and has negligible water consumption. Therefore, despite being an important water use for the Great Lakes-St. Lawrence River region.

The 2020 water withdrawal total represents a decrease of approximately 3% from the 2019 reported total withdrawal amount of 38,859 mgd (147,098 mld). "It is normal to see some fluctuation in water use from year to year, but some sectors, like self-supply irrigation, may exhibit greater variability due to the influence of weather patterns that increase or decrease seasonal use. Approximately 5% of the total reported amount withdrawn (1,997 mgd or 7,559 mld) was consumed or otherwise lost from the Basin".

The GLC report indicated public water supplies for cities and other municipal activities, and industrial water consumption continued to be the largest users and consumers of water from the basin. In 2020, public water users accounted for 5,165 mgd with 575 mgd consumed daily. Industrial users withdrew 4,943 mgd, consuming 532 mgd (water user return the water to the source after its use; water consumers do not) 2022 Great Lakes Commission.

Prior to the COVID-19 pandemic, the USGS published results of a study of the importance of groundwater resources in the states within the Great Lakes region. The western boundary of the Great Lakes region is roughly adjacent to the states with access to the over-withdrawn High Plains aquifer systems. The states in this region periodically suffer through severe drought conditions. The study concluded that four factors can be expected to affect the health and sustainability of the major groundwater resource in the states of the Great Lakes region that will continue to be influenced by these four issues:

1. Maintaining the quantity of groundwater during and after withdrawals for all uses.
2. Ensuring withdrawal rates occur at rates sustaining interaction of groundwater and surface water.
3. Real-time monitoring of changes in groundwater quality as development and population increase.
4. Sustained maintenance of the health of the ecosystem in relation to quantity and quality or water.

Water Crisis Throught the Region

The lead in the Flint Michigan public water system is perhaps the most well-documented drinking crisis ever to occur in the Great Lakes region.

For more than a century, the Flint River served as an unofficial waste disposal site for treated and untreated refuse from the many local industries that have sprouted along its shores, from carriage and car factories to meatpacking plants and lumber and paper mills. The waterway has also received raw sewage from the city's waste treatment plant, agricultural and urban runoff, and toxics from leaching landfills. As the industries along the river's shores evolved, so too did the city's economy. In the mid-twentieth century, Flint was the home to nearly 200,000 people, many employed by the booming automobile industry. Beginning in the 1980s, the auto industry changed; the plants in Flint closed; autoworkers had to find work elsewhere; and utilities income plummeted. The water supply lost more than half of its customers as people left for jobs elsewhere. Flint's population plummeted to just 100,000 people, about 45% of whom now live below the poverty line. Nearly one in six of the city's homes were abandoned. System administrators overseeing the city water utility chose to address the problem by finding a less expensive source of raw water. They turned to the polluted Flint River, having decided to temporarily distribute poorly treated water until a new water pipeline from Lake Huron was built. Although the Flint River water was highly corrosive, Flint officials failed to treat it properly, and lead leached out from aging lead supply pipes into thousands of homes. Later many other harmful chemicals from the decades of dumping pollutants into the river surfaced.

The Flint water crisis was not the only water crisis to occur in Michigan. The Benton Harbor water system exceeded the federal drinking water standards for the level of lead for three consecutive years until a required six-month compliance sample found decreased levels of lead, putting the city just at the federal action level after three years above that level. In addition to other actions, the utility provided free certified filters for lead and free bottled water. "After analyzing water samples from about 200 homes, results show that when used properly, filters are effective at reducing lead in drinking water. Free bottled water continues to be provided ...for City of Benton Harbor residents for cooking, drinking, brushing teeth, rinsing foods, and mixing powdered infant formula" (EPA 2021g Filter Study Results). These stop-gap remedial actions came after the public was exposed to water with dangerously high levels of lead and do not address the underlying problem of aging infrastructure leaching contaminates.

Safe Drinking Water: A Right or a Privilege?

The Flint and Benton Harbor crises highlight the challenge that decision-makers face when prioritizing the allocation of scarce financial resources and setting utility rates to address aging infrastructure. Replacing outdated and failing water supply and delivery systems are very expensive. In many cases, utilities across the country have not set water rates high enough over the years to create capital replacement funds for their systems and now face repair and replacement costs that they are financially unprepared to undertake. Faced with this scenario, decision-makers in these two communities failed to protect the public from the negative impacts of unsafe drinking water. They sought to delay needed repairs further or reached for low-cost solutions that ultimately created the unsafe conditions.

These two examples also highlight the disparity that can exist in the distribution of water between financially well-off communities and low-income areas. This crisis disproportionally impacts predominantly low-income areas. Here, in response to public pressure over the years to keep rates low, past utility commissioners all too often failed to adequately fund capital repair and replacement accounts resulting in inadequate funds being available to keep the systems functioning properly when the systems began to fail due to age. This creates a scenario where utility decision-makers in these areas are now faced with failing systems and lack adequate funds to repair and replace them. They then seek low-cost solutions to their infrastructure failures that can lead to major public health problems if water becomes unsafe to drink. This also exacerbates the social problem of racial disparity in the United States as low-income areas are more likely to have predominately minority populations. This translates into a disproportionately high risk of poor drinking water quality in predominately minority areas. The 2020 census showed Flint's population of 81,381 was 54% African American and 4.6% Latino. Benton Harbor's 2020 population of 9,087 was 85.1% African American and 3.9% Latino.

The public health emergencies in these two communities were preventable. Had adequate funds been budgeted over the years, today's commissioner and mangers would have been able to repair and replace the systems in a timely way to assure the water remained safe. This is a looming crisis in many of the country's aging water distribution systems, particularly in low-income areas, where those systems have reached the end of their expected life cycles and funds have not been budgeted for infrastructure replacement.

The citizens in these two communities were exposed to unhealthy drinking water as a result of cost-cutting or cost avoidance by those overseeing water distribution. Future decision-makers would be wise to keep these two public health failures front of mind as they make decisions regarding water rates and capital repair and replacement funding. Public health and safety depend on clean drinking water, and everyone should be confident that the water that comes from their taps is safe to drink. Fundamentally, these examples highlight a critical public health question facing the country: Is clean safe drinking water a right or a privilege?

Preventing a Water Crisis in Minnesota

Minnesota is not showing signs of a coming water crisis that is and will hit much of the United States, particularly in the west and southwest (drought or unsafe domestic water supplies): but the state is not without potential water problems, as the following water protection organization report suggests: "Minnesota is the land of 10,000 lakes, but thousands of them are not drinkable, fishable, and swimmable. 40% of our lakes, rivers and streams are impaired, and this number is growing despite millions of dollars of investments. The groundwater that three of four Minnesotans depend on for drinking water is increasingly polluted with nitrates. Agricultural drainage projects and urban and suburban development are speeding water off the land, creating erosion and reshaping our waterways. In short, Minnesota is squandering our clean water heritage, and unless we act, we may lose it" (MCTA 2022).

Minnesota is recognized as having an abundance of clean and readily accessible water. However, it is neither unlimited nor is it evenly distributed. "Despite this disparity, the residents of Minnesota tend to take water for granted. The problem that Minnesota faces is some locations within Minnesota may experience shortages of clean water by 2030" (Teiken 2012).

Water Problems in Illinois

According to the Illinois Environmental Council (IEC), Illinois has abundant water resources: there are over 119,000 miles of streams; 900 miles of large rivers; more than 91,000 freshwater lakes and ponds; and, has jurisdiction over more than 1,500 square miles of open water in Lake Michigan. This does not mean it is without water problems: "Unfortunately, surface and groundwater resources in Illinois are being impaired or degraded. Many Illinois rivers, streams, and lakes are contaminated with E. coli bacteria; contain fish that have been exposed to mercury and PCBs; exhibit low oxygen levels; and contain excessive phosphorus, manganese, siltation, suspended solids, and algae, all of which negatively impact aquatic life" (EIC 2021).

On August 3, 2021, the U.S. EPA ordered the city of Cahokia Heights, Illinois, and the several drinking water utilities in the city to take immediate actions to prevent drinking-water contamination and to protect the health of residents. Cahokia Heights, located in Southern Illinois, faces numerous environmental challenges. The City of Cahokia Heights has had overflows from the City's sewers that could impact human health. City residents have reported drinking water issues. The United States EPA and the Illinois EPA are working to ensure that sewage and drinking water-related issues within the City of Cahokia Heights are addressed. Tackling this potential human health risk is a top priority for both agencies. Community advocates and residents raised their concerns to EPA, including their concerns about environmental justice. The water systems serve approximately 21,000 people.

Conclusion

The Great Lakes system's five large lakes, the one small lake, four connecting channels, and the St. Lawrence Seaway together make this one large freshwater system that is currently undergoing significant changes to human activity and the effects of global warming. The states that surround the lakes have contributed to these problems. Lakes Superior, Michigan, Huron, Erie, and Ontario hold about 90% of the freshwater in the United States and approximately 20% of the world's freshwater supply. In 2019, forty million residents of the United States and Canada depend on this system for clean drinking water. However, all the water is not always as clean as it once was. Threats to the Great Lakes' ecosystems, include invasive species, industrial and agricultural chemical and animal waste pollution, and habitat destruction are among the major threats to the system. Climate change is raising surface water temperatures, shifting weather patterns with changes to traditional precipitation forms and periods, and lake levels. Pollutants from residential, agricultural, and industrial areas reduce water quality. Land development decreases the amount of wildlife habitat. Fish populations have been declining in recent years as a result of these threats and increased fishing pressure (NOAA 2019).

Water withdrawals for human activity and warmer surface water temperatures in the Great Lakes have contributed to lower water levels by increasing rates of evaporation and causing lake ice to form later than usual. This extends the season for evaporation. Lower water levels in the Great Lakes forced ships to reduce their cargo tonnage by 5 to 8% between 1997 and 2000, which increased shipping costs.

Another effect of warmer water, reduced ice cover, and increased evaporation is a corresponding increase in precipitation over nearby land, especially "lake effect" snowfall. Rising water temperatures are also expected to expand the ranges of and give new advantages to some invasive species such as the zebra mussel, and to encourage the growth of toxic waterborne bacteria.

References

Alley, William M., Thomas E. Reilly, and O. Lehn Franke. 1999 (and 2013 update). Sustainability of Groundwater Resources. USGS Survey Circular 1186. Available at https://pubs.usgs.gov/circ/circ1186/pdf/p1-5.pdf.

Climate.gov. 2021. Climate change to increase water stress in many parts of U.S. Available at https:\\www.climate.gov/news-features/featured-images/climate-change-increase-water-stress-many-parts-us

EIC (Illinois Environmental Council). 2021. Protecting clean water in Illinois is essential to community, economic and ecosystem health. Available at https://ilenviro.org/clean-water/.

EPA (Environmental Protection Administration). 2021a. Water reuse research. Available at www.epa.gov/water-research/water-reuse-research.

EPA (Environmental Protection Administration). 2021b. Building sustainable water infrastructure. Available at www.epa.gov/sustainable-water-infrastructure/building-sustainable-water-infrastructure.

EPA (Environmental Protection Administration). 2021c. A closer look: Temperature and drought in the Southwest. Available at www.epa.gov/climate-indicators/southwest.

EPA (Environmental Protection Administration). 2021d. Climate Change Indicators: Weather and Climate. Available at www.epa.gov/climate-indicators/weather-climate.

EPA (Environmental Protection Administration). 2021e. Climate Change Indicators: Snowpack. Available at www.epa.gov/climate-indicators/climate-change-indicators-snowpack.

EPA (Environmental Protection Administration). 2021f. A closer look: Land loss along the Atlantic Coast. Available at www.epa.gov/climate-indicators/atlantic-coast.

EPA (Environmental Protection Administration). 2021g. Grantee Research Results. Available at https://cfpub.epa.gov/ncer_abstracts/index.cfm/fuseaction/display.abstractDetail/abstract_id/10996/report/2021

Iowa DNR (Iowa Department of Natural Resources). 2022. Water Summary Update: A Snapshot of water resource trends for August 2022. Issue 135. Available at www.iowadnr.gov/Portals/idnr/uploads/Water%20Summary%20Update/20220908wsu.pdf.

MCTA (Minnesota Center for Environmental Advocacy). 2022. Clean Water. Available at https://www.mncenter.org/clean-water.

NIDIS (National Integrated Drought Information System). 2022d. Missouri River Basin. Available at NIDIS www.drought.gov/dews/missouri-river-basin.

NOAA (National Oceanic and Atmospheric Administration). 2019. Drought: August 2019. National Centers for Environmental Information, September 11. Available at www.ncdc.noaa.gov/sotc/drought/201908.

Tavernia, Brian G., Mark D. Nelson, Peter Caldwell, and Ge Sun. 2013. Water stress projections for the northeastern and midwestern United States in 2060: Anthropogenic and ecological consequences. *Journal of the American Water Resources Association* 49 (4): 938–952.

Teiken, Katherine. 2012. *Water scarcity in Minnesota a policy analysis and recommendation*. University thesis, Twin Cities: University of Minnesota.

USGS (United States Geological Survey). 2016. The role of snowpack, rainfall, and reservoirs in buffering California against drought impacts. Available at www.usgs.gov/news/role-snowpack-rainfall-and-reservoirs-buffering-california-against-drought-impacts.

CHAPTER 13

Water Crises in the South and Southcentral Regions

States in the Southcentral region included here are Arkansas, Louisiana, Mississippi, Tennessee, and Alabama. The Mississippi embayment aquifer is the dominant groundwater resource supplying this region. Groundwater-storage related (hydrogeologic) units of the Mississippi embayment aquifer system include the Mississippi River Valley alluvial aquifer, Vicksburg-Jackson confining unit, the upper Claiborne aquifer, the middle Claiborne confining unit, the middle Claiborne aquifer, the lower Claiborne confining unit, the lower Claiborne aquifer, the middle Wilcox aquifer, the lower Wilcox aquifer, and the Midway confining unit. This extended aquifer system is more than 5,000 feet thick in a wide area of south-central Louisiana and southwestern Mississippi. The Mississippi embayment aquifer system is linked to the Mississippi River Valley alluvial aquifer. Figure 13.1 outlines the extent of the aquifer system as provided by the USGS. The central wavy line in the center is the Mississippi River.

Mississippi embayment groundwater is an important resource for agricultural and municipal uses in the region. Production of important rice and cotton crops in the region depends on groundwater for irrigation. Many municipalities rely on the groundwater resource for industrial and public use. The demand for the groundwater resource has resulted in groundwater availability issues in the Mississippi embayment include (1) declining groundwater levels of 50 feet or more in the Mississippi

D. E. McNabb and C. R. Swenson, *America's Water Crises*,
https://doi.org/10.1007/978-3-031-27380-3_13

Fig. 13.1 Range of the Mississippi Embayment aquifer system (*Source* USGS)

River Valley alluvial aquifer in parts of eastern Arkansas from agricultural pumping, (2) declining groundwater levels of over 360 feet over the last 90 years in the confined middle Claiborne aquifer in southern Arkansas and northern Louisiana from municipal pumping, and (3) litigation between the State of Mississippi and a Memphis water utility over water rights in the middle Claiborne aquifer (Clark et al. 2011).

The Mississippi embayment that spreads across parts of seven states near the foot of the river was formed from thick layers of the river's silt deposits. This soil and the abundance of surface and groundwater make it one of the most fertile regions of the coterminous United States. Because of the substantial growth of cities and the larger riverboats and barges that replaced steamboats, the first decades of the twentieth century saw the construction of massive engineering works such as levees, locks, and dams, often built in combination. A major focus of this work has been to prevent the lower Mississippi from shifting into the channel of the Atchafalaya River and bypassing New Orleans. When storms such as Hurricane Katrina overflow the levies, miles of surface lands become inundated.

Throughout the twentieth century, the Mississippi River has also experienced major pollution and environmental degradation, primarily from high agricultural nutrients, including nitrogen, phosphorus, potassium, calcium, magnesium, sulfur, and silicon and chemicals from agricultural and industrial runoff. Algal growth and biomass accumulation, harmful to aquatic life, result from excess nutrients. Algal growth can interfere with drinking water treatment and use of water by industrial facilities.

Drought in the South

This region includes four of the ten wettest states in the 48 contiguous states. Louisiana with average annual rainfall of 56.9 inches of rain makes it the wettest state in the nation. The second wettest is Mississippi with average rainfall of 55.5 inches, and Alabama is a close third with 55.3 inches. Arkansas is the seventh wettest state with annual rainfall of 59.6 inches. Yet, despite these heavy rainfall averages, the region can also endure period of drought (Engström et al. 2021). For example, in early June of 2022, roughly three quarters of Louisiana was continuing to be abnormally dry; nearly 45% of the state was undergoing moderate drought, and nearly a quarter of the state was in severe drought. In the southwest corner of the state, 8% of the state was in extreme drought conditions. The region had experienced dry weather conditions for most of the spring and warming temperatures were expected to continue with less than normal precipitation. From November 2021 to February 2022, Louisiana experienced its second driest period for those four months since 1895. Almost 25% of Mississippi, Louisiana's neighboring state, was also abnormally dry (Table 13.1).

Drought in Louisiana

In 2022, Louisiana, which is considered the wettest state in the continental U.S., was under a prolonged period of abnormally dry conditions over most of the state. In early May, 65.8% of the states was abnormally dry, with close to half of the state under moderate drought and nearly 25% under severe drought conditions; it was also the only state in this region with extreme drought, with 8.4% in this category. As August was beginning, abnormally dry conditions had expanded to reach

Table 13.1 Drought conditions in the South, Spring 2022

Sections of state included					
State	*D0:D4 abnormally dry (%)*	*D1:D4 moderate drought (%)*	*D2:D4 severe drought (%)*	*D3:D4 extreme drought (%)*	*D4 exceptional drought (%)*
Louisiana	65.8	44.6	24.7	8.4	0.0
Mississippi	24.7	0.4	0.0	0.0	0.0
Alabama	3.9	0.0	0.0	0.0	0.0
Arkansas	3.8	0.0	0.0	0.0	0.0
Tennessee	0.5	0.0	0.0	0.0	0.0

Source NIDIS Drought.gov

83.5% of the state, moderate drought had increased to nearly 50% of the state. However, declines for both the severe drought, 3.8%, and extreme drought had dropped to just half of one percent.

Groundwater Use in Louisiana

Louisiana aquifers are being overdrawn by about 350 million gallons a day, losing water faster than it can be replenished. This is creating saltwater intrusion—another threat to the region's groundwater resources. As more groundwater is removed from existing aquifers for irrigation and for community use, saltwater intrusion is occurring to some degree in many of the coastal aquifers of the United States. Louisiana, California, and Florida are states where this is occurring now. Saltwater cannot be used to irrigate crops or be consumed by people. USGS studies how excessive groundwater pumping, sea level rise, and other factors contribute to the encroachment of seawater into fresh groundwater supplies.

In Louisiana, the freshwater aquifers under Baton Rouge are being contaminated by saltwater intrusion due to excessive groundwater withdrawals. The Baton Rouge area is located in the southeast portion of the state along the Mississippi River. In 2021 the area was under extreme drought, further increasing groundwater withdrawals. However, by the end of May 2022, Baton Rouge's 15.17 inches of rain was still far below the 25.94 inches of rain that is usually received by June. The USGS monitors saltwater intrusion; Box 13.1 describes the intrusion process and dangers that can ensue from this process.

Arkansas Water Issues

Although drought may not be the most worrisome water concern in the Southcentral region, it occurs often enough to be a concern. One of the reasons for this is that the lack of surface water leads to increased dependence upon groundwater. When this happens, the likelihood of excessive withdrawal of water from the region's extensive aquifer system becomes a very big water concern to communities in south and east parts of the state as it contributes to the continued decline in water levels in wells that supply those communities. "The Sparta aquifer [water table] has had large drops in Jefferson, Union and Colombia Counties because of heavy use by cities and industry. If not corrected the aquifer would become depleted and wells would then go dry. During the last few years the recognition of this problem has led to the communities and local industries in the region to switch to surface sources in order to allow the Sparta aquifer to recover" (AGS 2020).

The U.S. Geological Survey estimated that, according to 2010 data, Arkansas was second in the nation in groundwater usage, behind California and just ahead of Texas. Arkansas accounted for 10% of all U.S. groundwater withdrawals that year, drawing 7.8 billion gallons of water per day. The state was 26th in surface water withdrawals that year, drawing 3.5 billion gallons of water per day. It was seventh in total of surface and groundwater withdrawals.

Much of Arkansas' groundwater comes from deposits of sand and gravel in the Mississippi River Embayment. Irrigation wells, with depths ranging from 100 to 200 feet, commonly produce 1,000 to 2,000 gallons per minute (GPM). Although usable for irrigation and some domestic uses, the high iron content of these aquifers makes the water generally unsuitable for human consumption in many areas. The deeper aquifers which underlie the majority of the Gulf Coastal Plain supply most domestic and municipal needs in this part of the State.

An ongoing water use issue is the withdrawals from of the Mississippi Alluvial aquifer in eastern Arkansas. This one aquifer is the primary source of irrigation water and as a consequence is experiencing large areas of declining water levels. Water conservation, using different farming methods or growing different crops and finding surface sources of water, will be needed in the future or this most used of the states aquifers could be lost. This alluvial aquifer is the most important aquifer in Arkansas and is used primarily for agriculture irrigation. Farm withdrawals from this aquifer account for approximately 94% of all groundwater used in the State. That supply is currently over withdrawn.

Box 13.1 Saltwater intrusion process and impact

"Under natural conditions, the seaward movement of freshwater prevents saltwater from encroaching on freshwater coastal aquifers. This interface between freshwater and saltwater is maintained near the coast or far below the land surface. The interface actually is a diffuse zone where freshwater and saltwater mix. This zone is referred to as the zone of dispersion or the zone of transition."

"Groundwater pumping can reduce freshwater flow toward coastal areas and cause saltwater to be drawn toward the freshwater zones of the aquifer. Saltwater intrusion decreases freshwater storage in the aquifers, and, in extreme cases, can result in the abandonment of wells. Saltwater intrusion occurs by many ways, including lateral encroachment from coastal waters and vertical movement of saltwater near discharging wells. The intrusion of saltwater caused by withdrawals of freshwater from the groundwater system can make the resource unsuitable for use. Thus, groundwater management plans should take into account potential changes in water quality that might occur because of saltwater intrusion."

Source USGS (2016).

The Sparta aquifer is the second most important aquifer in terms of use, and the aquifer was used in the past dominantly as a source of public and industrial supply, although increasing irrigation use is occurring because of critically declining water levels in the Mississippi River Valley alluvial aquifer. Other aquifers of the Coastal Plain generally are used as important local sources of domestic, industrial, and public supply, in addition to other minor uses (USGS, n.d.).

Arkansas Surface Water Resources

Surface water sources supply 34% for the state's water needs. Public-supply systems provide drinking water to 93% of the people of Arkansas and 66% of this water is from surface sources. As the population of the state continues to increase the most likely source for more water will be the lakes and rivers of the state. These are the greatest single source of water in Arkansas and are predominately used for hydroelectric power generation at dams. Some 62 billion gallons a day produce this power and then return the water back into the rivers. The rivers in Arkansas's five major drainage basins are important sources of water and in places with major dams they form, they include the state's largest lakes.

Composition of the Aquifer

Arkansas is divided into a highland area in the northwest and a lowland region in the south and east. Abridged sections of a 2014 USGS report of the aquifers of Arkansas identified 16 major aquifers in Arkansas, divided into two major regions: They are the Coastal Plain Province of eastern and southern Arkansas, with 11 aquifers, and the Interior Highlands Division of western Arkansas, with the remaining five aquifers (Kresse et al. 2014). Groundwater in the Coastal Plain is the most valuable natural resources in the State, with its importance to agriculture, while also supplying abundant water for commercial, industrial, and public-supply use.

The quality of surface water varies across the state, due primarily to the composition of the underground material. In eastern and southern Arkansas, the "soft" rock exposed at the surface easily erodes, causing the water to be cloudy or turbid when compared to streams in western and northern parts of the state that are underlain by "hard" or consolidated rock (AGS 2020).

Interior Highlands

The interior highlands of western Arkansas have less reported groundwater use than other areas of the State, reflecting a combination of factors. These factors include prevalent and increasing use of surface water, less intensive agricultural uses, lower population and industry densities, lesser potential yield of the resource, and lack of detailed reporting. The overall low yields of aquifers of the Interior Highlands result in domestic supply as the dominant use, with minor industrial, public, and commercial-supply use. Where greater volumes are required for growth of population and industry, surface water is the greatest supplier of water needs in the Interior Highlands. The various aquifers of the Interior Highlands generally occur in shallow, fractured, well-indurated, structurally modified bedrock of this mountainous region of the State, as compared to the relatively flat-lying, unconsolidated sediments of the Coastal Plain.

According to the USGS, the states in this region extend from the Mississippi Embayment Aquifer System (MEAS) that underlies most of the South Coastal Plains and the Mississippi Alluvial Plain sections of the Coastal Plain Province east to the Atlantic Coast. The westernmost aquifer systems extend eastward from Arkansas to northwestern Mississippi and include six identified separate aquifers in a band of bedded sand, silt, and clay. Units of the Mississippi embayment aquifer system include the Mississippi River Valley alluvial aquifer. Of the six major aquifers included in the system, the middle Claiborne aquifer is the most heavily used.

Mississippi Water Crisis

Mississippi uses about 3½ billion gallons of water every day. Only one-fourth of that comes from surface water like lakes or reservoirs. The rest is pumped from the underground aquifer that supplies the groundwater that meets 75% of all Mississippi's water needs, with 66% of the total of combined surface and groundwater is used for irrigated agriculture. Of the 322,000 millions of gallons per day (mgd) withdrawn nationally, in 2010... Irrigation accounts for 98% of water use in from the Mississippi alluvial aquifer. Mississippians use an average of 2,690 mgd or 0.83%. In addition to irrigation, other chief users are public supplies and self-supplied industrial use.

Jackson is the largest city in Mississippi. The City's water system includes more than 71,000 water connections. The city's drinking water treatment system had been having problems for years. In 2010, a winter storm caused several water main breaks and a widespread outage. City hospitals increased privately owned hospital well patient capacity as a response to that emergency. As a result, in the August 2022 crisis core medical services were able to continue operating with running water. In 2012, Jackson failed a U.S. Environmental Protection Agency (EPA) inspection enforcing Safe Drinking Water Act standards, resulting in a November 2012 settlement requiring that the city improve maintenance.

In the middle of September 2022, Jackson, the state's capital city, was in the 40th day of an ongoing water crisis (Wikipedia, 15 September 2022). The trigger of the current crisis began in late August 2022 after the Pearl River flooded due to severe storms in the state. The flooding caused the Curtis Water Treatment Plant, the city's largest water treatment facility, which was already running on backup pumps due to failures the month prior, to stop the treatment of drinking water indefinitely.

An approximately 150,000 residents of the city were without access to safe drinking water. United States President Joe Biden declared a federal disaster to trigger federal aid.

Box 13.2 2020 Notices of noncompliance to the Safe Drinking Water Act

The City of Jackson, Mississippi Public Water System failed to comply with the lead and copper rule (LCR) treatment technique requirements for pH and/or alkalinity Water Quality Parameters (WQPs) for the following monitoring periods:

- January–June 2016 (186 days of excursions of WQPs);
- July–December 2016 (221 days of excursions of WQPs);
- January–June 2017 (200 days of excursions of WQPs);
- July–December 2017 (258 days of excursions of WQPs);
- January–June 2018 (91 days of excursions of WQPs);
- July–December 2018 (166 days of excursions of WQPs);
- January–June 2019 (211 days of excursions of WQPs);
- July–December 2019 (113 days of excursions of WQPs); and
- January–June 2020 (62 days of excursions of WQPs based on data through April 2020).

The City failed to report the WQP violations to SDWIS and did not provide public notification for the following monitoring periods: July–December 2016; January–June 2017; and July–December 2017. Therefore, the City is in noncompliance with 40 C.F.R. §§ 141.82(g) and 141.203 and Mississippi Code Ann. §§ 15-20-72.1.4.3 and 72.1.5.2 for failure to maintain optimal WQPs and provide the appropriate public notification.

Source EPA (May 11, 2020).

In February 2021, a winter storm shut Curtis Water Treatment Plant, leaving residents without water for a month. City leaders asked the state for $47 million for sewer repairs but received only $3 million. The city and the EPA agreed on a repair plan in July 2021, but the city's water infrastructure continued to deteriorate. An example of failure to meet federal water quality standards in the section of the 2020 EPA's noncompliance notice of the Jackson's water and wastewater system is included in Box 13.2.

Mississippi 2022 Water Crises
Residents complained of low water pressure and sewage floating in the streets. The city issued many boil water orders after the February storm, including one in July 2022 when the treatment plant damage forced a move to backup pumps; that order was still in effect when the August 2022 flooding began.

The water crisis in Jackson shows similar characteristics to water system crises in Flint and Benton Harbor, Michigan. While the Mississippi and the Michigan water systems are far apart and do not share the same water sources, they are very similar in that their water crisis was triggered by the same causes: system failure resulting from inadequate funding to operate, maintain, and upgrade aging systems. Years of underfunding, deferring needed rate increases, resulted in the ultimate catastrophic failures of these systems leaving residents with no reliable source of safe drinking water. Fitting the pattern seen in Flint and Benton Harbor, Jackson is also a relatively low-income city with a large minority population. In 2020 Jackson had a population of 153,701 of which 82.5% were African American, and a median household income of $40,064 and 24.5% living in poverty.

Water and Power in Alabama

Insufficient water is not a regular problem in Alabama. Rather, for large numbers of residents in these southern states, the problem is the inability to access clean freshwater and to get rid of it in a sanitary method. In 2019, a combination of national health and sanitation organizations considered the lack of access to sanitation to be the most serious water access concern in the rural South. For example, in rural Lowndes County, Alabama, only 20% of homes are connected to sewer systems; the rest must install and finance their own $30,000 septic system. This is in an area where the median value for a mobile home was only $23,900 Instead, some residents use surface PVC pipes to remove wastewater away from homes but is then discharged untreated onto the backyards—a practice known as "straight-piping." In Alabama, homeowners with septic system violations can be fined, and for many years they were subject to arrest for unpaid fines. Low-income residents worry about fines, jail time, or eviction as a result of seeking help with wastewater issues, as well as distrust in public authorities (McGraw and Fox 2019).

In USGS data published in 2009, total water use in Alabama was about 9,957 million gallons per day (mgd). Estimates of withdrawals by source indicate that total surface water withdrawals were about 9,467 mgd (95% of the total withdrawals) and the remaining 491 mgd (5%) were from groundwater. More surface water than groundwater was withdrawn for all categories except aquaculture, mining, and self-supplied residential. During 2005, estimated withdrawals by thermoelectric power, 8,274 mgd, public supply, was the overwhelming leader in water use (USGS 2009b). Almost all of the thermoelectric power water use (about 8,274 mgd) was from surface water; nearly all of the water (98%) was used for once-through cooling, and most of that water was returned to a surface water source. Water withdrawals by sector for 2005 are listed in Table 13.2.

Thermoelectric power accounts for 87% of all the freshwater withdrawals that take place in Alabama, just 2 percentage points away from copying the similar distribution of the 89% of all the electricity produced with thermally driven, water-cooled electricity produced in the United States. Almost all of this water is surface water. The thermoelectric power process withdraws large amount of water, but only a small percentage

Table 13.2 Water withdrawals in Alabama by water use category in million gallons/day

Rank	*Water use category*	*Average amount in million gallons/day*
1	Thermoelectric generation	8,274
2	Public supply	802
3	Self-supplied industrial	550
4	Agriculture (irrigation)	161
5	Aquiculture	75
6	Self-supplied residential	39
7	Livestock	28
7	Mining	28
	Total	9,957 Mgal/day

Source USGS (2009, 2022)

is lost to evaporation (Torcellini et al. 2003). It is used for cooling the power generating equipment, allowed to cool, and then returned to the source. In the cooling process, some of the water evaporates along with portions of any water stored in reservoirs. Because evaporated water cannot be returned to where it was withdrawn, it is counted as water that was consumed. Thermoelectric plants consume slightly less than half a gallon for of freshwater per kilowatt hour (kWh) of electricity. A kilowatt hour is the energy needed to keep a 1,000-watt appliance running for one hour, whereas hydroelectric power plants evaporate an average of 18 gallons of water used.

Power Generation Water in Alabama

Alabama is the second largest thermoelectric producer in the eastern U.S. (after New York), and its Browns Ferry Nuclear Plant is the nation's second largest nuclear generating facility. Plants with other or a combination of fuels in 2121 list included six plants that were some or all coal fueled, 22 by natural gas, one 52-year-old petroleum-fired operation, and 13 biomass-fired plants that were associated with timber mill operations. Although none were reported as contributing significantly to the amount of freshwater lost to evaporation, some loss from evaporation from open reservoirs was noted. Plants with other fuels listed in the state included 18 that were hydroelectric and two solar photovoltaic plants.

The Lower Mississippi Water Science Center (LMWSC) collects hydrologic data and conducts research on Alabama, Arkansas, Louisiana, Mississippi, and Tennessee water resources. In 2016, the Center's Water Center researchers reported water withdrawals have more than doubled in Alabama from 1960 to 2005. The growth in withdrawals actually occurred from 1960 to 1980, from 4,220 million gallons per day to more than 10,350 million gallons per day. Water withdrawals in following years have remained nearly constant from 2000 to 2005 with 8.6 million gallons per day in 1985 to 9.96 million in 2005.

Most of the remaining increases in-sector water withdrawals were mostly the result of slow but steady population growth. Population increased about 19% from 3.3 million in 1960 to 3.9 million in 1980 and increased another 17% from 1980 to 4,569,805 in 2005, followed by continued slow growth to 5,039,877 in 2021. As a result of the leveling off of withdrawals as population has increased, gross per capita use has declined. The data indicate that gross per capita for all uses of freshwater use increased from about 1,292 gallons per day for 1960 to a high of about 2,661 gallons for 1980, and then decreased to about 2,185 gallons for 2005. The thermoelectric power production withdrawals influence the trends for surface water withdrawals in Alabama and accounted for more than 80% of total withdrawals in 2015. Surface water was the source for 94% of all withdrawals and 70% of the water withdrawals excluding the thermoelectric sector.

More water continues to be withdrawn for thermoelectric power generation than for any other use. Thermoelectric power withdrawals are nearly all from surface water, and, therefore, dominate the Alabama surface water resources. From 1960 to 1980, total water withdrawals by thermoelectric plants in Alabama steadily increased from about 4,220 million gallons per day to 10,350 million gallons per day. From 1980 to 2000, changes in thermoelectric power production processes reduced total water withdrawals. From 2000 to 2010, total water withdrawals were relatively constant before declining in 2015 partly due to changes in thermoelectric power production processes. Thermoelectric power plants that require cooling water account for 85% of electricity generation. Once-through cooling systems, representing 37% of thermoelectric generation, draw cooling water from nearby rivers and streams, but high stream temperatures can reduce cooling efficiency while environmental regulations limit effluent temperatures. Elevated stream temperatures in the future threaten to reduce thermoelectric power generation capacity from these once-through systems. Water lost in this process is largely due to evaporation.

Groundwater withdrawals have steadily increased over the same 55-year period, but only represent 6% of the total 2015 water withdrawals. These water withdrawal changes have occurred while Alabama's population has increased from 3.3 million to 4.9 million. Groundwater withdrawals slowly increased from 1960 to 2005, primarily because of increased groundwater withdrawals for public supply. Since 1985, public-supply withdrawals have accounted for more than 50% of groundwater withdrawals in the State.

Water Issues in Tennessee

In Tennessee, approximately 4.6 million people rely on public water systems that use surface water as a source for their drinking water. Approximately 1.6 million people rely on public water systems that use groundwater as a source for their drinking water, and approximately

Table 13.3 Tennessee surface and groundwater use in MGD by use from 2015

Use sector	*Surface water*	*Groundwater*
Public	594	256
Domestic	0.0	42.8
Irrigation	27.4	36.4
Livestock	1.5	12.0
Aquaculture	45.2	11.7
Industrial	682	51.6
Mining	142	17.1
Thermoelectric	4620	2.18
Totals	5,990	430

260,000 people that receive their drinking water from a public water system that use combination of groundwater and surface water. A 2014 USGS report estimated that an additional 538,000 people get their drinking water from private wells and springs. Most west Tennessee citizens rely on groundwater for their drinking water. Table 13.3 displays the distribution of surface and groundwater used by nine primary use sectors.

Products that typically pollute groundwater, particularly in urban areas, are typically chlorinated solvents or degreasers and gasoline. These all evaporate rapidly and most often are not a problem in surface water. However, they are a serious problem in groundwater where they do not biodegrade and can be in the groundwater for decades. In agricultural areas, surface water contamination sources are also typically nitrate from fertilizer and animal waste, bacteria, protozoa, and urban runoff that includes runoff from yards, asphalt, etc. that has heavy metals and pesticides/herbicides.

Groundwater in karst areas that are impacted by surface water is also subject to these same contaminants. Karst water problems are characterized by sinkholes, springs, disappearing streams, and caves. Karst systems have rapid, highly directional groundwater flow in discrete channels or conduits. Karst aquifers have very high flow and pollutant transport rates, particularly under rapid recharge events such as severe rainstorms. This is a particular concern for public or private water supplies using wells or springs in karst areas where pathogenic organisms that would not be present in true groundwater can survive under the influence of large amounts of surface water. Karst systems are quite easily contaminated since the waters can travel long distances through conduits with no chance for natural filtering processes of soil or bacterial action to diminish the contamination. Tennessee geology is such that the naturally occurring arsenic that plagues a number of the western states is not present in this state. Neither does there appear to be a problem with uranium. However, studies of public groundwater supplies across the state have determined that there are locations with elevated levels of radon, a radioactive gas that forms naturally when uranium, thorium, or radium, which are radioactive metals break down in rocks, soil, and groundwater. People can be exposed to radon primarily from breathing radon in air that comes through cracks and gaps in buildings and homes. Because radon comes naturally from the earth, people are always exposed to it (EPA 2022e).

According to the EPA, every two years, states are required to publish a report about the state's water quality for federal regulators. The report lists waterways that are "impaired," meaning they are too polluted to support at least one of their uses. Uses can include recreation, fishing, conservation, and drinking, among other things. In 2010, Tennessee researchers said 32.4% of the river miles they tested were impaired. A decade later, in the state's 2020 report, that number jumps to 55.4%, meaning more than half of the waterways that they sampled were too polluted to support their basic functions.

Florida Water Crises

Among the water problems that Florida must deal with are the effects of ocean warming, rising sea level, and weather changes that included increasing precipitation. The flooding that results from these and the increasing number and severity of weather extremes are a result of severe storms and storm surges from hurricanes and typhoons and other heavy weather events. These are all associated with climate warming. Hurricanes are large, swirling storms that form over warm ocean waters and end after significant damage on the surface regions of the east and southeast states. Florida's mean elevation above sea level is just 100 feet above sea level, with 18.5% of the state covered by water (USGS 2018a).

Sea level rise is an understandable concern for the state. Two climate warming-related factors are behind the rise: (1) the water from snow and ice melt in glaciers that is adding water to the oceans and (2) the expansion of seawater that occurs as the oceans get warmer (Erickson et al. 2006). The melting of glaciers and ice sheets are the two main factors contributing to rising sea levels around the globe. From roughly 2010, the warming-related expansion of water and melting land ice contributed roughly equally to rising sea levels. Since then, however, melting land ice is now responsible for twice the increase of thermal expansion (explorati on.edu 2022).

The Saffir-Simpson Hurricane Scale rates different storm categories by wind speeds that range from Category 1 with winds from 119 kilometers per hour (74 mph) to Category 5, with winds more than 252 km/hour (157 mph). Hurricane Ian was a Category 4 Atlantic hurricane that grew to Category 4 as it hit southern Florida on September 28 after it reached a measured windspeed of 155 mph. Ian caused at least 130 fatalities in Florida, five in Cuba, another five in North Carolina, and one in Virginia, making Ian the deadliest hurricane to strike Florida since 1935. Damage to properties was estimated to be at least $67 billion, most of which was due to flooding (Wikipedia 2022).

Tri-State Water Wars

Georgia, Alabama, and Florida have been fighting since 1990 over the future allocation of water in two major river basins that cross their borders (APR 2021). Georgia and Alabama have been fighting over the Alabama-Coosa-Tallapoosa (ACT) basin, and all three states are in conflict over the Apalachicola-Chattahoochee-Flint (ACF) basin. Each state has its own concerns about the proper allocation of water. The following revised description of reasons for the three-state "water war" was aired by Alabama Public Radio (APR) on July 23, 2021 (APR 2021):

- **Georgia**: Georgia wants to have enough water to continue growing, particularly in booming metro Atlanta and to supply cities such as Columbus and the agricultural usage in the state's southwest corner. The problem is that Atlanta is not located in a water-rich area of the state, multiple river systems drain into either the Atlantic Ocean or the Gulf of Mexico. Moreover, Atlanta is limited in access to supplemental groundwater resources. The granite that underlies the city restricts its access to the groundwater alternative supplies to cover for its reduced summer supplies of surface water.
- **Alabama**: Alabama is worried that Atlanta's growing need for water will limit its own use of water for power generation, which already accounts for as much as 80% of all water used in Alabama, municipal supply, fisheries, and other current and future needs. Although Alabama is not suffering from rapid population growth such as

Atlanta, most of its surface water is committed to use for the necessary thermoelectric plants cooling.

- **Florida**: Florida wants enough surface water to reach the Apalachicola Bay to sustain its multi-million-dollar seafood industry, which is under severe ecological stress resulting from lower river flows and saltwater intrusion. Apalachicola Bay is an estuary and lagoon located on the northwest coast of Florida. The Apalachicola Bay system includes St. George Sound, St. Vincent Sound, and East Bay, covering an area of about 208 square miles. The river is part of the Apalachicola-Chattahoochee-Flint (ACF) River basin, a critical component of water supply for Georgia, Alabama, and Florida.

Georgia, Alabama, and Florida have fought over the use of water in the Apalachicola-Chattahoochee-Flint River Basin (ACF), which is managed by the U.S. Army Corps of Engineers' operation of Lake Lanier's Buford Dam and four other similar dams. Lake Lanier is just north of Atlanta. The Corps built Lake Lanier in the 1950s with Congressional authorization for flood control, navigation, and hydropower. Meanwhile, Lake Lanier became the main drinking water supply for metro Atlanta. Alabama and Florida have argued that Georgia withdraws too much water from lake and basin and is not sharing the water fairly. The three states have turned to the courts to try to resolve the conflict.

Box 13.3 Charges and court decisions on the Alabama-Georgia-Florida water cases

In 1990, Alabama began the "Tri-State Water Wars" when it filed suit against the USACE's regulations of water flow for the Chattahoochee River, from Lake Lanier in, Georgia to Alabama and Florid dams in the ACF and ACT water basins. A number of court cases followed.

In 2017, the [USACE] released an updated ACF Water Control Manual to guide operations of the Corps' five reservoirs, dams, and navigation locks in the basin to meet Congressionally authorized purposes of power generation, flood control, navigation, water supply, recreation, and fish and wildlife. In the same year, the Alabama and the National Wildlife Federation et al. filed separate legal challenges against the Corps, alleging that the Corps did not follow National Environmental Policy Act (NEPA) procedures when revising the Water Control Manual.

In January 2021, the state of Georgia and the Corps finalized a $70 million perpetual contact for access to storage space [of surface water] in Lake Lanier...for 254,170 acre feet and up to 222 million gallons of water per day. This contract does not guarantee that the water will be there, just that the state has access to storage space. Georgia will develop individual sub-contracts with the Lake Lanier's five water utilities. It is worth noting that the contract is based on now outdated and inflated need and population projections through 2050.

In April 2021, the Supreme Court unanimously denied Florida's request and dismissed its complaint. The Supreme Court found that Florida had failed to prove that Georgia's water use had caused the decline in the Apalachicola Bay oyster fishery or harmed riverine and plant species in the Florida.

In August 2021, a federal judge rejected the lawsuit lodged by Alabama and the National Wildlife Federation, claiming that the Corps did not follow administrative procedure or fully comply with other environmental laws when preparing the Water Control Manual that governs the operations of Buford Dam/Lake Lanier, West Point Dam/Lake and the three other federal dams in the ACF. In August 2021, the Corps approved a new water supply accounting policy that will allow public water managers to credit "made inflow" into Corps reservoirs from an upstream reservoir and/or wastewater treatment plants. Alabama appealed the court's decision.

Source ACT (2021).

In 2009, a federal district court judge ruled against Georgia, deciding that water supply was not an authorized purpose of Lake Lanier. The judge gave Georgia three years to reach a water-sharing agreement with Alabama and Florida and to get Congressional approval.

In 2011, the Eleventh Circuit Court of Appeals reversed the 2009 district court decision, ruling that water supply is an authorized purpose of Lake Lanier, on par with hydropower, navigation, and flood control. Furthermore, the appeals court gave the Corps one year to determine the extent to which it would operate Lake Lanier to meet water supply needs in addition to the other authorized purposes. Some examples and court decision are included in Box 13.3.

Conclusion

The southeast, normally the wettest region of the coterminous United States, has suffered from many of the same water stress and excessive withdrawals of surface water resources in the western states, including protracted drought and climate change. The southwest is a good example of how, even in generally wet areas, climate change and population growth, taken together, lay the foundation for future water crises. It is not just arid areas or those where industrialization and urban pollution have fouled water resources. The southwest has seen little of these past degradations, or experienced these arid climatic conditions yet the challenges presented by overreliance on groundwater and agricultural practices that fail to replenish supplies, places a future stable supply of water at considerable risk.

Similarly, the lack of investment in aging water systems plague this area. The problems in Jackson, Mississippi mirror those in Flint and Benton Harbor Michigan illustrating a pattern of inadequate funding for infrastructure repairs and replacement particularly in lower income and predominately minority communities. Groundwater from the Mississippi Valley alluvial aquifer system irrigates a large southern agriculture industry, together with meeting the water needs of population growth. The demand for groundwater in the southeastern states has exceeded the aquifer recharging rates of these region's aquifer systems for decades. Excessive agriculture irrigation withdrawals have resulted in declining groundwater levels of more than 50 feet in parts of eastern Arkansas. Groundwater withdrawals for municipal service have resulted in declines of more than 300 feet since the 1940s have occurred in southern Arkansas and northern Louisiana. The Alabama, Georgia, and Florida have long fought over the use of water in the Apalachicola-Chattahoochee-Flint River Basin.

Groundwater supplies the water for 75% of all Mississippi's water needs, with 66% of the total of combined surface and groundwater for irrigated agriculture. The water supply in sections of this state has long been in stress. The water system in Jackson, the state capital, has long been unable to meet the city's water needs as sections of its aged system often fail. These conditions shed light on current problems and become a harbinger of problems ahead for this region.

References

ACT (Atlanta Regional Commission). 2021. Tri-state water wars overview. Available at https://atlantaregional.org/natural-resources/water/tri-state-water-wars-overview/.

AGS (Arkansas Geological Survey). 2020. Water issues. Available at www.geology.arkansas.gov/water/water-issues.html.

APR (American Public Radio). 2021. Water wars continue between Alabama, Georgia, Florida. Available at www.apr.org/news/2021-07-23/water-wars-continue-between-alabama-georgia-florida.

Clark, Brian R., Rheannon M. Hart, and Jason J. Gurdak. 2011. Groundwater availability of the Mississippi Embayment. USGS Professional Paper 1785. https://pubs.usgs.gov/pp/1785/.

Engstrom, Johanna, Sarah Praskievicz, Bennett Beardan and Hamid Moradkhani. 2021. Decreasig water resources in the Southeastern U.S. as observed by the GRACE satellites. *Water Policy*, 33(4): 1017–1029.

EPA (Environmental Protection Administration). 2020. National Pollutant Discharge Elimination System (NPDES): NPDES Stormwater Program. Available at www.epa.gov/npdes/npdes-stormwater-program.

EPA (Environmental Protection Administration). 2022a. History of the Clean Water Act. Available at www.epa.gov/laws-regulations/history-clean-water-act.

EPA (Environmental Protection Administration). 2022b. Basic information about water reuse. Available at www.epa.gov/waterreuse/basic-information-about-water-reuse.

EPA (Environmental Protection Administration). 2022c. What is green infrastructure? Available at www.epa.gov/green-infrastructure/what-green-infrastructure.

EPA (Environmental Protection Administration). 2022d. Harmful algal blooms. Available at www.epa.gov/nutrientpollution/harmful-algal-blooms.

EPA (Environmental Protection Administration). 2022e. What is radon? Available at www.epa.gov/radon/what-radon.

Ericson, James P., Charles J. Vörösmarty, S. Laurence Dingman, Larry G. Ward and Michael Meybeck. 2006. Effective sea-level rise and deltas: Causes of change and human dimension implicdations. *Global and Planetary Change*, 50 (1–2): 63–82.

Kresse, Timothy M., Phillip D. Hays, Katherine R. Merriman, Jonathan A. Gillip, D. Todd Fugitt, Jane L. Spellman, Anna M. Nottmeier, Drew A. Westerman,

Joshua M. Blackstock, and James L. Battreal. 2014. Aquifers of Arkansas—Protection, management, and hydrologic and geochemical characteristics of groundwater resources in Arkansas: U.S. Geological Survey Scientific Investigations Report 2014–5149, 334. Available at https://doi.org/10.3133/sir20145149.

McGraw, George, and Radhike Fox. 2019. Closing the water gap in the United States: A National Action Plan. Available at Dig-Deep_Closing-the-Water-Access-Gap-in-the-United-States_DIGITAL_compressed.pdf.

Torcellini, Paul, Nicholas Long, and Ron Judkoff. 2003. Consumptive water use for U.S. power production. National Renewable Energy Laboratory (NREL TP-550–33905). U.S. Department of Energy.

USGS (United States Geological Survey). n.d. Mississippi Alluvial Plain (MAP): Water use and availability program. Available at www.usgs.gov/centers/lmg-water/science/map?qt-science_center_objects=0#qt-science_center_objects

USGS (United States Geological Survey). 2009a. *Mississippi Embayment Regional Aquifer Study (MERAS)*, ed. Brian R. Clark and Rheannon M. Hart. Available at https://pubs.usgs.gov/sir/2009/5172/pdf/SIR2009-5172.pdf.

USGS (United States Geological Survey). 2009b. Alabama Water use. Available at https://pubs.usgs.gov/fs/2009/3081/pdf/Alabama_2005_water_use_FS_508.pdf.

USGS (United States Geological Survey). 2016. Groundwater and drought. Available at https://water.usgs.gov/ogw/drought/.

USGS (United States Geological Survey). 2018a. Groundwater decline and depletion. Available at www.usgs.gov/special-topic/water-science-school/science/groundwater-decline-and-depletion?qt-science_center_objects=0#qt-science_center_objects.

USGS (United States Geological Survey). 2018b. Wyoming water use in the U.S., 2015. Available at www.usgs.gov/media/images/wyoming-water-use-us-2015.

USGS (United States Geological Survey). 2021a. National water census: Regional groundwater availability studies. Available at www.usgs.gov/mission-areas/water-resources/science/national-water-census-regional-groundwater-availability?qt-science_center_objects=0#qt-science_center_objects.

USGS (United States Geological Survey). 2021b. *Groundwater may decline by a third over next 30 years*. USGS Communications and Publishing, October 28. Available at www.usgs.gov/news/state-news-release/groundwater-flow-colorado-river-may-decline-third-over-next-30-years.

USGS (United States Geological Survey). 2021c. USGS Water Resources. Pennsylvanian and Mississippian Aquifers of the Appalachian Plateaus. Available at www.usgs.gov/mission-areas/water-resources/science/pennsylvanian-and-mississippian-aquifers-appalachian-plateaus#publications.

USGS (United States Geological Survey). 2021d. Mississippi Embayment aquifer system. Available at www.usgs.gov/mission-areas/water-resources/science/mississippi-embayment-aquifer-system.

USGS (United States Geological Survey). 2022. Water use in Alabama, 2005. Report by Lower Mississippi-Gulf Water Services Center, May 19. Available at www.usgs.gov/centers/lower-mississippi-gulf-water-science-center/science/water-use-alabama-2005.

Wikipedia. 2022. The Mississippi River. Available at https://en.wikipedia.org/wiki/Mississippi_River.

CHAPTER 14

Water Crises in the Appalachian Mountain Region

This region has two different sections of a set of water problems in states that follow to a large degree the Appalachian Mountains: an inland section and a coastal section. On the west side of this ancient mountain range is the Appalachian Plateau. The land on the east side of the range is the Atlantic Coastal Plain. For most of the region, portions of the same states are included in both section. The Appalachian Plateau region included in this chapter extends across portions of Alabama, Georgia, Tennessee, Kentucky, Maryland, Ohio, Virginia, a section of Pennsylvania, West Virginia, and New York. Aquifers in this region consist of alternating sequences of fractured sandstone, siltstone, shale, limestone, and coal in Permian-, Pennsylvanian-, and Mississippian-age rock formations (Fig. 14.1).

The Appalachian Plateaus aquifer system extends from New York to Alabama (Fig. 14.1) Pennsylvania, Ohio, Maryland, Virginia, West Virginia, and Kentucky. Small sections of New York and Ohio are also sometimes included but are included in the Mid-Atlantic region here. The region's economy, once highly dependent on coal mining, agriculture, chemical industries, and heavy industry, became more diversified in recent times, and now includes a variety of manufacturing and service industries. The waters used in this section were discharged once use and often polluted, were discharged into the region lowlands, allowed to pass untreated to joint larger rivers and lakes.

D. E. McNabb and C. R. Swenson, *America's Water Crises*,
https://doi.org/10.1007/978-3-031-27380-3_14

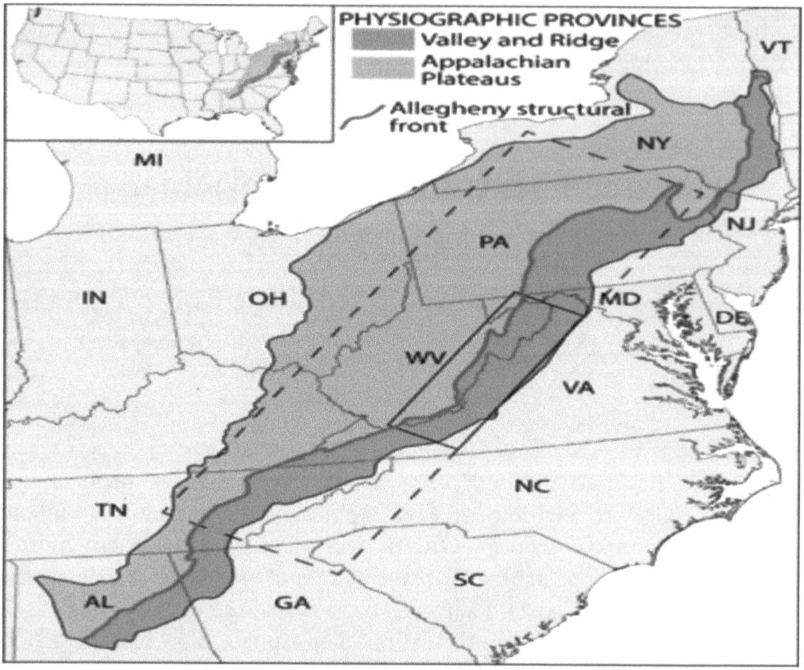

Fig. 14.1 Extent of the Appalachian Plateau aquifer system

Draught Condition in Appalachian Mountain Region

In this region, six states stand out for underlying large shares of the Appalachian aquifer group of systems: Ohio, Pennsylvania, Virginia, West Virginia, North Carolina, and Kentucky. Drought and high temperatures can result in an increase in summer rainstorms throughout this part of the region. An example of the storms that can occur was the July 28, 2022 supreme rainstorm that struck Kentucky. Two days after the storm, more than two dozen residents had been killed, with another 70 or so deaths to follow.

In 2022, 44.5% of Pennsylvania was abnormally dry and 6.9% was under moderate drought conditions—a little more than half of what had been two months earlier. Less than i% of the state (0.2%) was still in severe drought (Table 14.1). In 2016 and 2017, the region was in the midst of a

Table 14.1 August 2022 drought conditions in the Appalachian Mountain Region

Sections of State Included					
State	*D0:D4 Abnormally Dry*	*D1:D4 Moderate Drought*	*D2:D4 Severe Drought*	*D3:D4 Extreme Drought*	*D4 Exceptional Drought*
Pennsylvania	44.5%	6.9%	0.0%	0.0%	0.0%
North Carolina	28.9%	0.0%	0.0%	0.0%	0.0%
Ohio	5.0%	0.0%	0.0%	0.0%	0.0%
Kentucky	4.3%	1.6%	0.2%	0.0%	0.0%
Virginia	4.2%	0.0%	0.0%	0.0%	0.0%
West Virginia	0.0%	0.0%	0.0%	0.0%	0.0%

Source NIDIS Drought.gov

megadrought. In 2016, wildfires were burning more than 100,000 acres across seven states in the southern Appalachian Mountains from October to July. Of the 44 ongoing blazes, at the end of July, the biggest fire struck northern Georgia and western North Carolina, where whole towns have fled as part of a slate of mass evacuations. All of which, weather scientists said, was pretty weird. It rarely occurs for this many fires burning in as much area in the Southeast, according to U.S. Forest Service. Normally, the entire region is subject to a wet climate, and is particularly so in that southern Appalachian region where the fires were occurring. The fires occurred while the Southeast was experiencing a, once in a generation, drought, with the area seeing little to no rainfall at all. Conditions were still exceptionally dry in Pennsylvania at the end of August 2022, with only Pennsylvania and Kentucky reporting any moderate drought and only Kentucky reporting less than two-tenths of one percent in severe drought.

Box 14.1 The effects of mountain-top mines and valley fills on aquatic ecosystems

Mountaintop removal mining removes the soil and rock over a coal seam (i.e., the overburden) to expose the coal. The mountain or ridge top is prepared for mining by building access roads, clearing all trees and stockpiling topsoil for future use in reclamation. Then, explosives are used to blast the entire top of the mountain or ridge to expose and mine one or more coal seams. As much as 300 vertical meters (1,000 ft) of overburden are removed. With it, any springs, ephemeral, intermittent, and small perennial streams on the mountain's surface are also removed.

The overburden removed during mountaintop mining cannot all safely be put back into place because of the overall volume of the material and because the volume increases when the rock is broken up. Some of the overburden is stored on the mined surface in "back-stacks" and used to re-contour the surface. The excess overburden is disposed of in constructed fills in valleys or hollows adjacent to the mined site. These fills bury additional springs and streams.

Both water flow and sediment discharges are altered by the process. The heavy equipment used to mine and move the overburden compacts the bare soils, forming a large, relatively impervious surface on the mined site that increases surface runoff. Surface runoff is diverted into ditches and sediment ponds, replacing natural subsurface flow paths.

Water flows out of the ditches through notches, or is directed toward the valley fill. Depending on the construction and degree of compaction of the valley fill, the water then either percolates through porous fill material or flows through ditches and coarser rock drains within, under, or beside the fill. The effluent that emerges downstream of the ditches and below the downgradient.

Mountaintop mines and valley fills lead directly to 5 principal alterations to stream ecosystems:

1. Springs, intermittent streams and small perennial streams are permanently lost with the removal of the mountaintop the soil buried in the valley fill.
2. Concentrations of major chemical ions are persistently elevated downstream.
3. Degraded water quality reaches levels that are acutely lethal to standard laboratory test organisms.
4. Selenium (Se) concentrations are elevated, reaching concentrations that have caused toxic

5. Effects in fish and birds, macroinvertebrate and fish communities are consistently and significantly degraded.

Source U.S. EPA (2011)

Appalachian Mining and Water Degredation

The two main mining methods in the Appalachian Mountain region are underground mining and surface mining (EPA 2016). There are three underground mining methods: drift mines, slope mines and shaft mines, and four surfacing mining methods: mountaintop mine, contour mines, high wall or auger mines, and surface mines. All can have some impact on surface water and groundwater. Because of its impact on water resources, surface mining is most controversial.

Mountaintop Mining

Mining has long been a source of livelihood in many sections of Appalachia. The peak years of mining occurred during the First World War as demand for coal grew as a fuel to manufacture the industrial military needs of the nation. Large and small mining companies recruited skilled European immigrants to meet the growing demand for coal. Migrants from the Deep South also found work in the Appalachian mines. When coal became the chief manufacturing fuel during the First World War coal boom, there were nearly 12,000 mines operating in the United States employing more than 700,000 men (the moonlit road.com). In addition to underground and pit mines, mountaintop-mining methods was introduced for mining in the steep terrain of central Appalachian coalfields throughout Appalachia. A result is the discharge of overburden and other unusable materials to get at the layers of coal. Mountaintop mining operations became concentrated in eastern Kentucky, southern West Virginia, western Virginia, eastern Ohio, and scattered areas of eastern Tennessee. The method used in these locations takes place in five main steps (EPA 2011):

- The layers of rock and dirt above and between the coal seams are removed down to the surface of the coal seam. The removed rock and soil is moved to an adjacent valley.
- The now-accessible coal seams can be removed; any remaining mixed soil and rocks are removed and deposited in an adjacent valley.
- Removal of rock and soil to the next layer of coal begins; large-scale earth moving equipment excavates and removes all coal from lower levels.
- Re-grading the former hill or mountain begins as coal excavation continues.
- Once all the coal is removed final regrading takes place and the area is revegetated.

To finish the action at the mining site, the mining operator is required to complete the regrading and revegetation to what are considered to what are "approximate original contours" (AOC). However, the recovery requirement may be "waived by the agency that has granted a permit for the operation as specified in the surface Mining Control and Reclamation Act of 1977. AOC may not include recovering any of the small streams and other water resources covered by the removed, deposited in the nearby valley and, thereby, lost.

The EPA considers mountaintop mining and valley fills to be the most harmful to water resources method of mining. Box 14.1 is a brief section of the 2011 EPA report on the industry and some of the harmful environmental products associated with the process (EPA 2011). Citing a 2005 environmental impact study the 2011 report, this mining method and the discharge of overburden into small valleys or narrow gaps between or adjacent to the mining site. The material above the coal seam is the "overburden," a mix of rock and earth. The fill concentrates harmful chemicals that are, on average, about 10 times higher downstream of the surface mine that in streams in unmined watersheds" (EPS 2011, 1–5). Rain moving through the fill carries the pollutants to beyond the mining area. The overfill thus results in the loss of many headwater streams, concentrates of chemicals that kills aquatic life in streams, and pollutes the groundwater in the personal wells of downstream communities and residents with private groundwater wells.

Mining Effects on Water Systems

Underground and mountaintop mining of coal disturbs in much of Appalachia the natural groundwater flow system when the mines are active because artificial drains are constructed to collect the unwanted water and discharge it at various surface collection areas. These drainage systems continue in abandoned underground mines. Mountaintop redirection of surface water systems can create new fractures and thus increase permeability, collecting distributing chemicals. The collecting the water removes it from its normal flow patterns, such that the regional water table is lowered, and groundwater flow directions are changed in some cases. Groundwater tends to flow toward mines, which are usually dewatered by pumping. Abandoned mines often collapse, which causes fracturing of the rocks that overlie the mine and might be accompanied by an appreciable depression on the land surface. These conditions are likely to enhance recharge to the groundwater system and to reduce surface runoff and evapotranspiration. Freshwater associated with coal-mining operations is usually acidic. Sulfur-bearing minerals, such as pyrite, that are present in the coal are exposed to air in mines. The oxidized sulfur combines with water to form sulfuric acid. The acidic water commonly contains large concentrations of iron, manganese, sulfate, and dissolved solids and is highly colored. An exception is in the southern coal fields of West Virginia where the coal is low in sulfur, mine drainage tends to be alkaline, and water from working or abandoned mines is commonly used for public supply (USGS 1997).

Storm Influence on Freshwater

Destructive summer storms are the other side to water shortages: too much water too fast instead of not enough water. For example, the extreme rainstorm that hit eastern Kentucky with historic levels of flooding overnight on Thursday, July 28, 2022, resulting in death and destruction in at least three counties. Climate warming was identified as having contributed to the severity of the storm. According to the National Weather Service, the 12-hour deluge in eastern Kentucky had only a 0.1% or less chance of happening there in any year, meaning that this event had less than a 1-in-1,000 chance of happening in this area in any year. The North Fork Kentucky River at Jackson reached almost 14.5 feet above flood stage, topping its previous record from Feb. 4, 1939. As of July 31, at least 26 persons were known dead and dozens more were missing. With entire homes under water, Kentucky Governor Andy Beshear predicted the number of persons killed from the storm could exceed 75 before the counting was over.

Weather scientists have linked the increases in extreme weather to climate change. They also predict that even more challenges to come. According to a NASA report, severe storms like the July storm form when an air mass becomes unstable. This happens when air in the lowest layers becomes very warm and humid, or air in the upper layers is unusually cold, or if both occur. Rising near-surface air in an unstable air mass expands and cools, making it warmer than its environment, which causes it to rise even farther. Climate change with more increases in temperature is resulting in many such extreme rainstorms. If enough water vapor is present, some of this vapor condenses into a cloud, releasing heat, which makes the air parcel even warmer, forcing it to rise yet again. When the water vapor becomes rain or hail, it fuels the storm, leaving loss of life and physical damage behind. In Kentucky, for example, one river rose from its normal one to two feet depth to about 18 feet in just 10 hours. Along with heavy rainfall, mudslides were also reported in the July storm, along with the much flooding in the Appalachian mountain regions of eastern Kentucky. The flooding was strong enough to destroy three entire towns in three counties were destroyed.

Although the Kentucky storm was not defined as a thunderstorm, according to the U.S. weather service, thunderstorms are a common occurrence in the southern U.S., They are particularly common in Alabama, Mississippi, and Florida any time of the day, month or year, and can also occur in hot weather in other regions, including Kentucky. Less than 1% of the 2,000 or so storms that occur around the world in any one year are classified as "severe." Severe thunderstorms are defined as thunderstorms that produce one inch hail or larger and/or strong wind gusts of 58 mph or greater. All thunderstorms generate potentially deadly lightning. A small percentage of severe thunderstorms produce tornadoes with little or no lightning at all. Across the region, severe thunderstorms typically occur during the late afternoon and evening hours but can develop at any time. While severe thunderstorms can occur any month of the year, the peak Severe Weather Season is during the spring months of March, April, and May. Alabama, Mississippi, and northwest Florida also have a secondary Severe Weather Season in the fall that typically runs from November through December.

Box 14.2 Why the Atlantic coast particularly vulnerable to sea level water crises

One of the reasons why this coast receives so many severe storms is its location. The coast is in the path of many hurricanes. Another reason is the shore along the Atlantic coast is flat and close to sea level. Still another is including thousands of square miles of marshes and other productive wetlands, plus many low-lying cities. In addition, much of the land along the Atlantic coast is sinking, which magnifies the local effect of sea level rise.

The United States land in is still adjusting to the loss of ice after the last ice age, which peaked about 20,000 years ago. During the Ice Age, thick sheets of ice covered areas of what is now Canada and the northern United States. The weight of all that ice depressed the land beneath it. This in turn caused the land farther south (particularly the Mid-Atlantic region from North Carolina to New York) to bulge upward. After the ice melted and the extra weight of the ice removed, northern areas of the U.S. began to rise, and the Mid-Atlantic region started to sink. This very slow process continues today. Every day, it sinks a tiny bit more.

Source EPA (2021)

Water Crises on the Coastal Plain

The states in the Atlantic Coastal Plain share similar problems, among which are extreme rainstorms that damage and pollute surface watercourses and saltwater intrusion of groundwater resources. Virginia, Maryland, West Virginia, North Carolina and South Carolina, Georgia, and Florida. All the states are susceptible to extreme weather events including hurricanes, as well as saltwater intrusion of their shore-based groundwater aquifers. Despite the extreme rainstorms, coastal communities and agriculture are encountering decliners in surface water resources they have long depended upon groundwater and are often find themselves forced to find new resources (Fig. 14.2). An answer why some sections of the coastal plain are sinking and others are rising is included in Box 14.2 (EPA 2021).

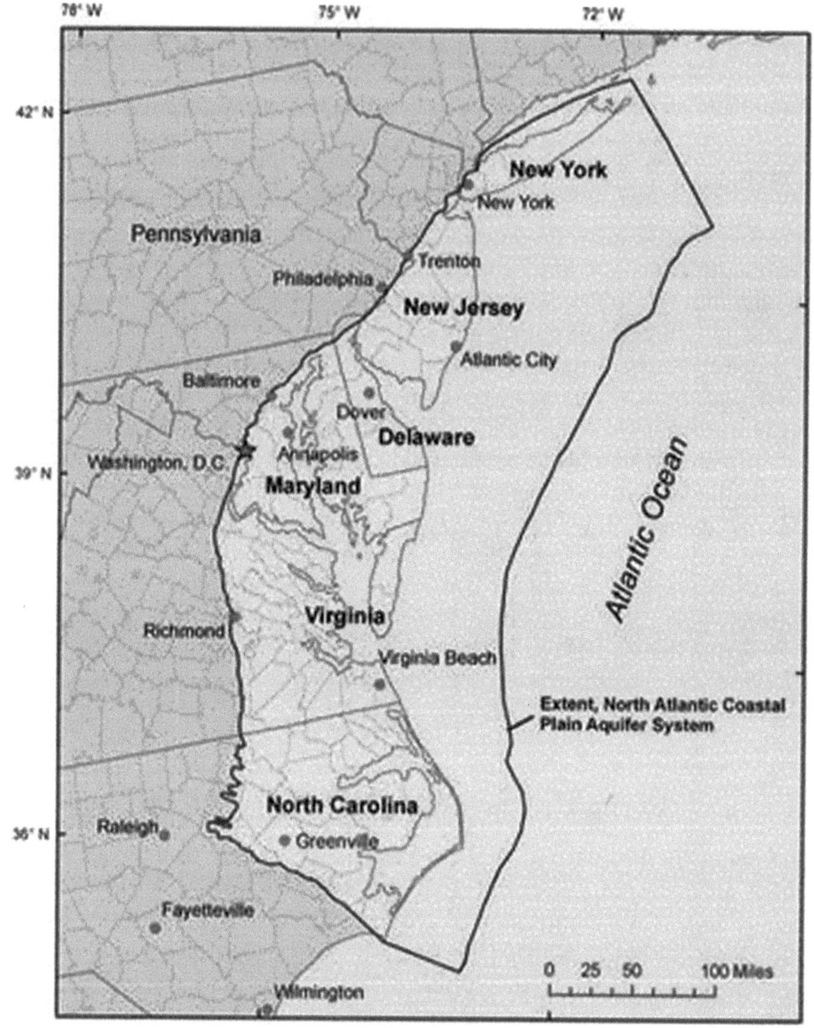

Fig. 14.2 Range of the North Atlantic Coastal Plain aquifer system (*Source* Public Domain map, Area of the Northern Atlantic Coastal Plain Aquifer)

Appalachian Aquifers

The Appalachian Plateau region extends over most of West Virginia, more than one-half of Pennsylvania, and small parts of westernmost Virginia and Maryland (USGS 1997). The province is bounded on the east and southeast by the Valley and Ridge Province and by a narrow strip of the Central Lowland Province in Erie County, Pennsylvania. In most places, the eastern boundary of the Appalachian Plateaus is marked by an escarpment called the Cumberland Escarpment in Virginia and the Allegheny Front in West Virginia, Maryland, and Pennsylvania. The altitude of the Appalachian Plateaus Province is higher than that of the Valley and Ridge Province, as well as the Central Lowland Province.

Fresh groundwater generally circulates only to shallow depths in the Appalachian Plateaus Province. In much of the area, saline water or brine is not far below the land surface with only a thin transition zone between the freshwater and saltwater. In the Pittsburgh, Pennsylvania area, wells drilled deeper than 100 feet below the level of the nearest major stream might yield saline water. A new industry began in southeastern corner of what is now West Virginia when saltwater springs were found along the Kanawha River in 1777.

The Atlantic Coastal Plain aquifer system extends along much of the Eastern portions of New Jersey, Delaware, Pennsylvania, Virginia, and North Carolina. It consists of a sequence of layered sedimentary aquifers, which were deposited starting around 100 million years ago and continuing today. This system is an important water source for domestic/municipal supply and industry in population centers throughout coastal North Carolina, Maryland, Virginia, Delaware, and New Jersey. It is not without significant loss of freshwater, however. Concentrated, localized pumping has led to a reversal of flow direction (toward the wells instead of Eastward) in some of the aquifer units throughout the region. In addition to overarching concerns about the sustainability of withdrawals that exceed recharge rates, the flow reversal has led to local saltwater intrusion, whereby saline ocean water infiltrates the aquifer, rendering it non-potable in some locations.

The Northern section of the Atlantic Coastal Plain (NACP) aquifer system occupies a land area of more than 30,000 square miles and a total area of 50,000 square miles along the eastern seaboard of the United States from Long Island, New York, southward to the northeastern part of North Carolina. In many locations along the Plain, excessive withdrawals of groundwater have resulted in water-level declines from100 to 200 feet levels (DePaul et al. 2008).

Aquifer level declines of more than 100 feet have been measured in New Jersey, Delaware, Maryland, Virginia, and North Carolina. Regional declines in water levels were most widespread in the deeper aquifers. Within the Upper, Middle, and Lower Potomac aquifers, water levels had declined up to 200 feet in southern Virginia and to more than 100 ft in New Jersey, Delaware, Maryland, and North Carolina. Substantial water-level declines were also evident in the regional Lower Chesapeake aquifer in southeastern New Jersey and many others. Since 1980, water levels in many areas have continued to drop. However, in some places conservation measures have served to limit withdrawals in affected areas, moderating or stabilizing water-level decline, and in some cases, resulting in substantial recovery. In other cases, increases in groundwater pumpage have resulted in continued rapid decline in water.

Recharge if the occurs by both natural and managed infiltration on land across much of the coastal plain; groundwater flow in the subsurface is mainly to the East along the sediment layers. One interesting consequence of this flow pattern is that there may be a sizable freshwater resource offshore that could be accessed by drilling in relatively shallow water on the continental shelf. During the last ice age, when conditions were substantially wetter than today and a nearly mile-thick ice sheet covered the northern extent of the aquifer system, recharge was probably even larger—and thus may have forced freshwater several tens of km offshore, where that "fossil" water may remain today.

The Atlantic Coastal Plain aquifer system is an important water source for domestic/municipal supply and industry in population centers throughout coastal North Carolina, Maryland, Virginia, Delaware, and New Jersey. However, concentrated, localized pumping has led to a reversal of flow direction (toward the wells instead of eastward to the sea) in some of the aquifer units throughout the region. In addition to overarching concerns about the sustainability of withdrawals that exceed recharge rates, the flow reversal has led to local saltwater intrusion, whereby saline ocean water infiltrates the aquifer and, in some cases, renders it non-potable.

Saltwater Intrusion

Under natural conditions, the seaward movement of freshwater flowing from the higher elevation of the Appalachian Mountains and Plateau keeps ocean saltwater from encroaching on freshwater coastal aquifers. Where the fresh and saline waters meet actually is referred to as the zone of dispersion or the zone of transition. Groundwater pumping can reduce freshwater flow toward coastal areas and cause saltwater to be drawn toward the freshwater zones of the aquifer. Saltwater intrusion then decreases freshwater storage in the aquifers, and, in extreme cases, can result in the abandonment of wells.

Saltwater intrusion occurs by many ways, including lateral encroachment from coastal waters, on-flow of stormwater during extreme storms, and the removal of the freshwater barrier allowing saltwater to migrate into the discharging wells. The intrusion of saltwater caused by withdrawals of freshwater from the groundwater system can make the resource unsuitable for use. This often occurs in areas of farms that withdraw coastal groundwater for irrigated agriculture. These coastal states are increasingly concerned over the threat of losing access to groundwater due to seawater intrusion. Most of the nation's east Atlantic coast is subject to these crises. The range of the North Atlantic Coastal Plain aquifer system is shown in Fig. 14.2.

Although almost all coastal states are threatened with saltwater intrusion due to excessive groundwater withdrawals, Florida, which is not in the Northern Atlantic Coastal Plain Aquifer System, is one of the few that has had some success in doing something about the intrusion. In Florida, saltwater has intruded into groundwater supplies through different ways. For example, saltwater has encroached into aquifers because fresh groundwater levels have decreased relative to sea level, allowing saltwater to flow toward the freshwater. Florida's saltwater inland canals leave an open pathway to saltwater intrusion. Other methods include leakage between aquifers, and upwelling of saltwater from depth also have impacted freshwater aquifers. Water managers in Florida are using information from local, state, tribal, and federal saltwater-intrusion monitoring networks to prevent and reverse saltwater intrusion (USGS 2019).

Although Florida is not always grouped with the central Atlantic Coast region, it is an example of a state bordering on the Atlantic Ocean and has had saltwater intrusion. Florida has had to deal with saltwater that has encroached into aquifers because of rising ocean levels. Sea levels rose 2.2 centimeters annually from 2011 to 2015 as groundwater levels have decreased from discharges for agriculture. Lower groundwater levels relative to sea level have allowed higher-level ocean water to flow toward the freshwater. Also, saltwater has invaded through inland canals, leakage between aquifers, or even upwelling of saltwater from depth also have impacted freshwater aquifers. Water managers in Florida are using information from local, State, Tribal, and Federal saltwater-intrusion monitoring networks, such as from the USGS, to prevent and reverse saltwater intrusion.

Groundwater Depletion in North Carolina

North Carolina's humid climate and substantial underground water resources helped make the state's reputation as water-rich state. However, as appeared too common in the Atlantic Coastal Plain region, excess withdrawals of groundwater in North Carolina resulted in some water shortages while drought was decreasing available surface or storm water for aquifer recharging. The activity originally considered a positive social condition was becoming a liability: Population growth was forcing increases in withdrawals by the popularity of the North Carolina coast as a desirable living region. Population growth was increasing demand by the public sector (EPA 2020).

Concern over declining groundwater resources in northeast North Carolina section of the Coastal Plain led the State to form the Central Coastal Plain Capacity Use Area (CCPCUA) for management and sustainability of the groundwater supply. Ten years after formation of the CCPCUA, a study on the effectiveness of the group's efforts at groundwater use sector to reducing the total groundwater withdrawals (Manda and Klein 2014). Increased height in water tables in excess of 70% was seen in many of the counties included in the study, with smaller gains were found in most of the other study areas. Table 14.2 lists the withdrawals by five chief sectors.

The largest percent of the groundwater withdrawals were for the public sector and for mine dewatering and supplies are groundwater withdrawals. Mine dewatering can apply to some, but not limited to mines that dewater from pits, including quarries, coal and mineral mines, and mines with stormwater discharges.

A combination of rapid population growth, drought, and surface water storms and uncontrolled water withdrawals for reducing water in the

Table 14.2 Percentages of all groundwater withdrawals by sector in CCPCUA study area

Sector	*North Carolina Total*	*Northeast Study Total*
Public Supply	50.9%	36.5%
Mine Dewatering	37.7%	55.5%
Industrial	7.3%	3.3%
Agriculture	3.7%	4.2%
Recreation	0.4%	0.5%

Source Manda and Klein (2014)

region's mines. Experts predict that if present growth and water use trends continue the freshwater needed may become to satisfy their water needs in the coming decades. North Carolina is one of the fastest-growing states in the nation, experiencing population growth of nearly 17% between 2000 and 2009. In just a little more than a year from April 2020 to July 2021, North Carolina added 112,000 residents. The total population in July 2021 was 10.6 million, making North Carolina the 9th largest state in the nation (Cline 2021). This growth shows no sign of slowing, and research indicates that by 2030 more than 12.2 million people will reside in North Carolina. North Carolina's population boom has not only increased demand for water but will also result in new developments in currently undeveloped forest and wetland areas—environments crucial to ensuring the quantity and quality of the state's water supply.

North Carolina Storm Damage

North Carolina has suffered from its location in the path of a large number of tropical cyclones. Not only have these storms caused damage to surface water resources, their damage to reservoirs and infrastructure have resulted in many water-related crises. North Carolina had suffered from the 52 tropical cyclones known to have affected the state between 2000 and 2019. The most active month during this nearly 20-year period has been September, with 20 total storms, while November is the least active month with two storms. The most intense storm since 2000 was Hurricane Isabel in 2003, which was the most destructive storm at the time, although the storm that caused the most fatalities was Hurricane

Florence in September 2018. Hurricanes with some impact on North Carolina since 2020 and 2021 period were:

- August 3–4, 2020: Hurricane Isaias hit the coast of North Carolina. Four people were killed in the state.
- October 29–30, 2020: Hurricane Zeta damaged hundreds of homes, uprooted trees, and caused flash flooding in certain areas in the state.
- November 13, 2020: Hurricane Eta interacted with a cold front as an extratropical storm and dropped heavy rainfall in the state. The heavy rains caused flash flooding, which is to blame for nine deaths.
- July 8, 2021: Hurricane Elsa caused damage to the central and eastern part of the state. It dropped 5 inches of rain and spawns multiple tornadoes in eastern North Carolina.
- August 16–17, 2021: Tropical Storm Fred dumps up to 12 inches of rain in the mountain region of North Carolina, at least six deaths were caused by flash flooding. Tornadoes developed in the central part of the state.
- August 21, 2021: Hurricane Henri drowned two people due to related rip currents.

Conclusion

Climate change, changing precipitation cycles, and increasing number of severe stormwater water events and other associated water issues in the United States are not only a concern of the west. Water problems in the Appalachian Mountain region and its coastal regions are as much a problem for these states as the problems in the western states are for that section. Appalachia's problems are the same as what exists in more of the mid-west and eastern seaboard. Both the east and the west suffer from droughts and other climate change/global warming conditions. The east coast must deal with algae blooms that kill fish and are harmful to humans that are common in the offshore Northeast. The Appalachian region of the United States has been experiencing severe water contamination due to mining and chemical spills for the past several years. In addition to algae bloom issues throughout the Midwest and increasingly offshore of the northeast lakes and ocean shores, the Appalachian region of the United States continues to experience severe both surface water and groundwater contamination from mining and chemical spills from industry for the past hundred years or longer.

Severe storms are also common in the Appalachian regions. These killer storms occur increasingly in the eastern United States not only kill people and destroy buildings. They include tornados, hurricanes, windstorms, floods, and other destructive weather events. As they pass through a community, they pollute and otherwise destroy freshwater resources. The powerful winds destroy property and play havoc with water resources. The flooding caused by the storms can kill dozens of people and demolished homes and business structures. It also creates a crisis for drinking water supplies. After the storms pass over and through a community, they leave behind ponds full of livestock waste, scores of dead chickens and hogs, raw sewage from wastewater treatment plants and coal ash ponds. The EPA lists at least 23 drinking water systems in the state that had temporarily halted their operations, and another 21 were forced to operate with advisories on boiling their drinking water. Many of the water systems are unable to resume distribution of clean, safe drinking water for months or longer.

The fracking for natural gas in Appalachia and elsewhere produces large amounts of fracking waste. The process produces large amounts of liquid and solid waste, such rock and drilling lubricant left over from the process of drilling a well, wastewater and sand from the fracking and production processes and variety of chemicals, metals, excess salts, and carcinogens like benzene and other harmful waste. Wells can continue to generate wastewater for years and contaminated equipment sent to landfills will leach toxins into the environment.

References

Cline, Michael. 2021. 2021 population estimates show NC growth continues. Release of NCOSBM (North Carolina Office of State Budget and Management). www.osbm.nc.gov/blog/2021/12/22/2021-population-estimates-show-nc-growth-continues.

DePaul, Vincent T., Donald E. Rice, and Otto S. Zapecza. 2008. Water level in aquifers of the Atlantic Coastal Plain, predevelopment to 2000. USGS Document. Available at www.usgs.gov/publications/water-level-changes-aquifers-atlantic-coastal-plain-predevelopment-2000

EPA (Environmental Protection Administration). 2011. The effects of mountaintop mines and valley fills on aquatic ecosystems of the central Appalachian coalfields. Available at https://ofmpub.epa.gov/eims/eimscomm.getfile?p_download_id=501593.

EPA (Environmental Protection Administration). 2016. Saving water in Oregon. Available at www.epa.gov/sites/default/files/2017-02/documents/ws-ourwater-oregon-state-fact-sheet.pdf.

EPA (Environmental Protection Administration). 2020. National Pollutant Discharge Elimination System (NPDES): NPDES Stormwater Program. Available at www.epa.gov/npdes/npdes-stormwater-program.

EPA (Environmental Protection Administration). 2021a. Building sustainable water infrastructure. Available at www.epa.gov/sustainable-water-infrastructure/building-sustainable-water-infrastructure.

EPA (Environmental Protection Administration). 2021b. A closer look: Temperature and drought in the Southwest. Available at www.epa.gov/climate-indicators/southwest.

EPA (Environmental Protection Administration). 2021c. Climate change indicators: Weather and climate. Available at www.epa.gov/climate-indicators/weather-climate.

EPA (Environmental Protection Administration). 2021d. Climate change indicators: Snowpack. Available at www.epa.gov/climate-indicators/climate-change-indicators-snowpack.

EPA (Environmental Protection Administration). 2021e. A closer look: Land loss along the Atlantic Coast. Available at www.epa.gov/climate-indicators/atlantic-coast.

Manda, Ales K., and Wendy A. Klein. 2014. Rescuing degrading aquifers in the Central Coastal Plain of North Carolina (USA): Just process, effective groundwater management policy, and sustainable aquifers. *Water Resource Research* 50: 5662–5677. https://doi.org/10.1002/2013WR015242.

USGS (United States Geological Survey). 1997. Appalachian Plateaus aquifers. Ground Water Atlas of the United States: Delaware, Maryland, New Jersey, North Carolina, Pennsylvania, Virginia, West Virginia. HA 730-L.

USGS (United States Geological Survey). 2019. Saltwater intrusion. Available at www.usgs.gov/mission-areas/water-resources/science/saltwater-intrusion

CHAPTER 15

Water Stress in Mid-Atlantic Region

Seven states are included in the Middle Atlantic region: Pennsylvania, New York, New Jersey, Virginia, West Virginia, Maryland, and Delaware. The estimated population of the seven states in 2022 was 60.6 million. Two of the important factors contributing to water stress in the region are protracted drought and hydraulic fracturing (Fig. 15.1).

For most of the first two decades of the twenty-first century, the states in this region have been relatively free of any prolonged severe drought. However, by the middle of August 2022, only West Virginia was drought-free and without reports of abnormally dry conditions. Four of the seven states reported some moderate drought and only New York and New Jersey had small sections with severe drought. With 32.2% in moderate drought, New Jersey had the largest portion of the region under moderate drought, followed by New York with 20.7% of the state.

Drought in the Mid-Atlantic Region

New Jersey and New York were the two states in this region with any portions of the state under drought conditions greater than moderate drought. With Delaware and Pennsylvania, they were the only four states with any drought beyond moderate drought (Table 15.1). Arid regions and regions under prolonged drought and those with high competition

D. E. McNabb and C. R. Swenson, *America's Water Crises*,
https://doi.org/10.1007/978-3-031-27380-3_15

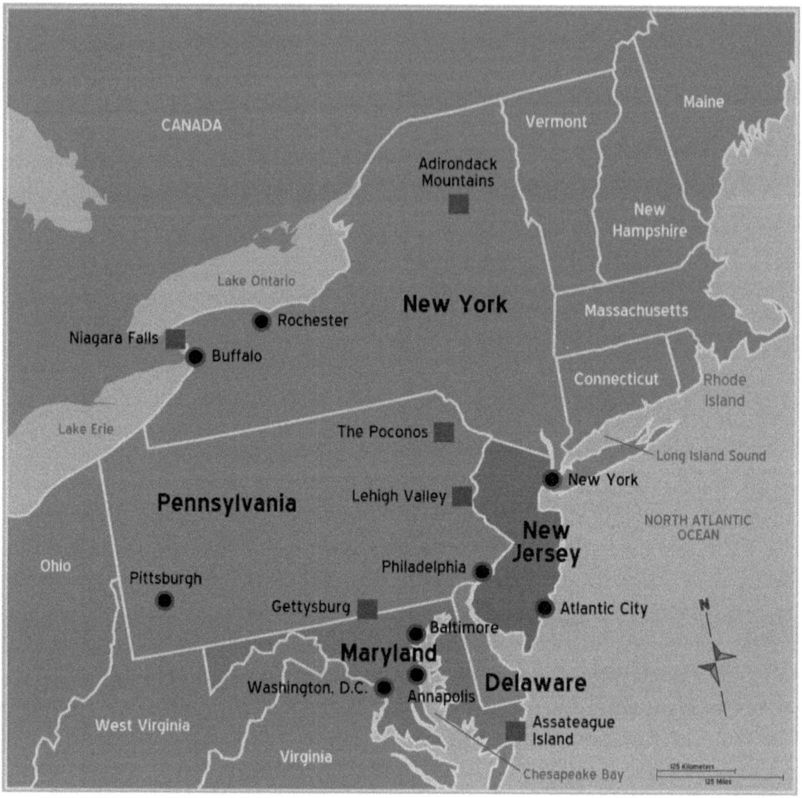

Fig. 15.1 Sections of the Mid-Atlantic and the Northeast regions (*Source* No source. Available at https://mapsof.net/uploads/static-maps/USA_Mid_Atlantic_map.png)

for access to the limited water resources are having to cope with other pressures for the limited supply of freshwater.

The relatively recent practice or hydraulic fracturing for shale oil and natural gas also compete for water. For example, a study of the impact on groundwater supplies associated with more than 25,000 shale oil and shale gas mines from January 2011 to September 2012 found that 65.8 billion gallons of water was used for fracturing processes during the length

Table 15.1 Mid-Atlantic region drought levels, August 2022

Sections of state included

State	*D0:D1 abnormally dry (%)*	*D1:D4 moderate drought (%)*	*D2:D4 severe drought (%)*	*D3:D4 extreme drought (%)*	*D1 exceptional drought (%)*
New Jersey	87.3	32.2	8.1	0.0	0.0
New York	54.7	20.7	3.2	0.0	0.0
Delaware	41.4	7.1	0.0	0.0	0.0
Pennsylvania	40.4	6.9	0.0	0.0	0.0
Maryland	7.0	0.0	0.0	0.0	0.0
Virginia	4.3	0.0	0.0	0.0	0.0
West Virginia	0.0	0.0	0.0	0.0	0.0

Source NIDIS Drought.gov

of the study. Nearly half of the wells were located in water basins with high or extremely high water stress. Wells in Colorado and Texas were in basins with very high water stress (extremely high stress was defined as 80% of the available water in the area is already used for any single or mixed combination of municipal, industrial, or agriculture applications). In Colorado, 92% of the 3,862 wells were in extremely high water stress levels. In Texas, more than half of its 11,634 wells were in high or extremely high water stress areas (Freyman and Salmon 2013).

Drought in New Jersey

For its freshwater, New Jersey depends on water supplies from both ground and surface sources in about equal quantity. In an undated report of the state's water problems, the USGS New Jersey Water Services Center, the state's water supply sources are allocated to such an extent that the state must have near-average rainfall every year to avoid instituting drought-control measures on its water-supply users. In what are considered "normal years," New Jersey receives an average of forty-four inches of precipitation annually, making it one of the nation's wetter states. New Jersey has a large number of known hazardous waste sites per square mile, with 638 multiple sources, multiple media, multiple contaminant sites. The state also has over 1,100 permitted point-source pollution discharges to its waterways, as well as a number of designated flood hazard areas. Again, all of these factors have led to a state which has inherently highly utilized, and at times overly stressed, water resource systems. The water in many of northern New Jersey water systems were reported to have exceeded tough new standards for PFAS, a family of chemicals that have been linked to cancer and other ailments.

In 2022, continued drought saw two of North Jersey's largest water storage facilities, the Wanaque Reservoir and the Oradell Reservoir, were faced with declining water levels continuing the steep decline that appeared in early June. Construction of the Raymond Dam in 1928 formed a lake located within New Jersey along the Wanaque River. In addition to the river, the reservoir receives water from the Pompton Lakes intake, which takes water from the Ramapo River, and the Two Bridges intake, which takes water from the Pompton River. The reservoir, the second largest reservoir in New Jersey by volume and area, is part of the largest reservoir system in the state. In 2022, 173 million gallons per day (MGD) could be taken from the reservoir. However, a proposal to increase withdrawals to 190 MGD is being questioned, as operators of the system believe doing so could result in an increase in pollution in the already drought-impacted reservoir's water supply.

Drought in New York

Although the state of New York is seeing some lessening of the published drought conditions, nearly 4% of the state is still in abnormally dry conditions and 15.6% of the state was under moderate drought. Much of that drought pattern is in New York County and New York City. The City is still facing drought conditions, according to the National Weather Service. Parts of southern Brooklyn and southern Queens, and all of Staten Island, are experiencing severe drought conditions, while the Bronx, Manhattan, and other parts of Queens and Brooklyn are seeing "moderate drought" conditions.

During the 2022 summer, New York City had just 8.82 inches of rainfall. On average, the usual summer amount is 13 inches. During the second week of September 2022, the city's reservoirs were at 74% of capacity, 12% below normal for this time of year. According to the National Drought Mitigation Center (NDMC), this July southern Brooklyn was the 28th driest in the last 128 years, with 1.67 fewer inches of rain than normal.

The last period of drier weather occurred in December 2001, when reservoir water-storage levels were at 44% of storage totals. A month later DEP issued drought warnings and drought emergency warnings throughout the region. Over the next eight months, increased precipitation and reduced water consumption alleviated drought conditions. Normally, over the past more than 100 years, conditions were restored on January 2, 2003, ending a 14-month drought event.

Drought in Pennsylvania

Pennsylvania's average climate has warmed by 1.8° over the last 100 years. The state's climate has become wetter, with an annual average precipitation increase of 10%. Despite these changes in precipitation, climate change is also resulting in periodic droughts for the state.

An additional 8% increase in precipitation is predicted to occur by 2050. With this increase in normal precipitation, the number of extreme storm events are expected. These forecasts suggest more surface water will be available in the future. The state's water infrastructure managers are concerned that the money needed to pay for these present and future water problems may not be available. Securing, collecting, storing, treating, and distributing water to meet the needs of all sectors must be found. Coping with the current water infrastructure problems is already difficult in this area and will only become more so in the next decade.

Drought in the Delaware River Basin

The Delaware River begins in two branches in New York State's Catskill Mountains (Fig. 15.2). The river flows 419 miles (674 km) into Delaware Bay in New Jersey and Delaware. The Basin feeds reservoirs in New York, New Jersey, and Delaware.

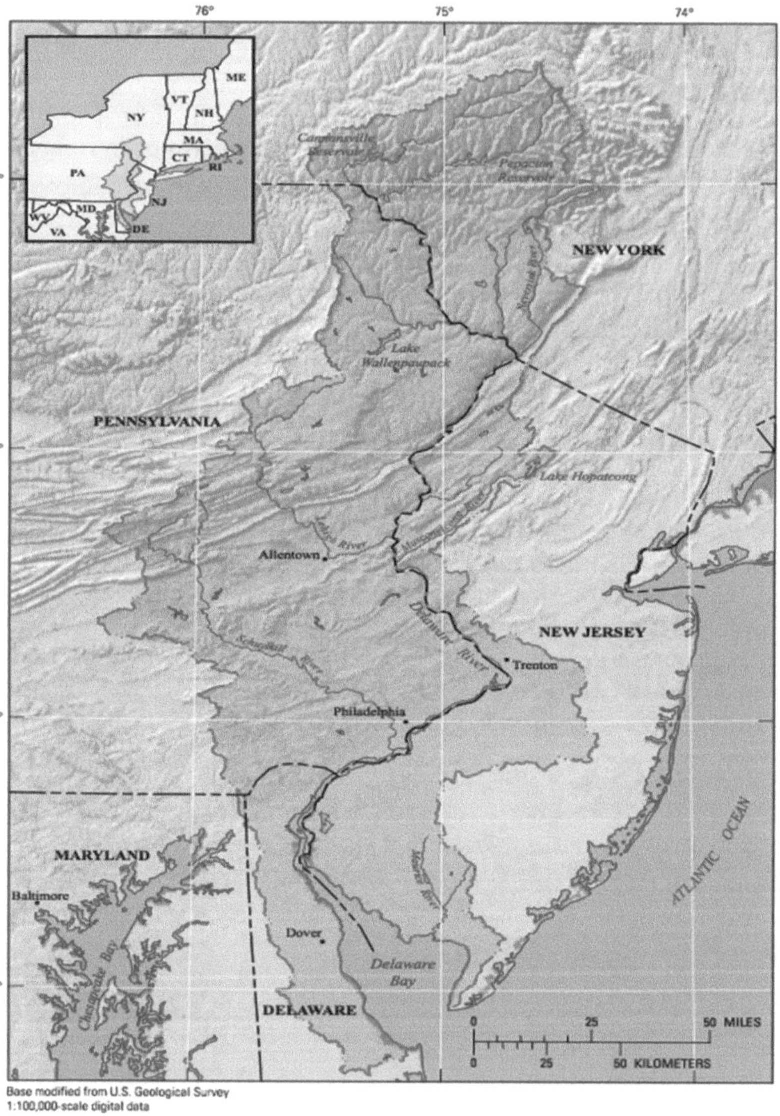

Fig. 15.2 USGS Public domain map of the Delaware River Basin (*Source* USGS 2010)

Box 15.1 Turn of the century drought in the Delaware River Basin

"A drought in the Delaware River Basin during late 1998 and most of 1999 had a major effect on surface and subsurface components of the hydrologic system...From September 1998 to August 1999, the accumulated precipitation deficiency was greater than 12 inches in parts of the basin above Trenton [New Jersey]. Flows in some streams, mainly in the middle and lower parts of the basin, deceased to levels near or less than those measured during the drought of the 1960s, the most severe drought of record in the basin. On several dates in August 1999, combined shortage in three New York City water supply7 reservoirs decreased by 2 billion gallons per day. The drought also affected groundwater levels with below normal recharge and elevated rates of evapotranspiration that produced record low levels in much of the basin. The drought continued to mid-September 1999 when remnants of Tropical Storm Floyd delivered drenching rains throughout the basin."

"The drought affected, in varying degrees, a broad area from Virginia to Maine. In 2000, the Delaware River provided water for municipal, industrial and other uses for more than 7 million residents. Water diversions from the basin provide water to nearly 10 million people outside the basin." New York City withdraws the most water. About one half of the City's total winter supply comes from three large Delaware River basin reservoirs (Pepacton, Cannonsville and Neversink).

Source Paulachok et al. (2000)

Until 1996, the drought that spread across the Basin in the 1960s was the most severe drought of record. The most severe drought in nearly a century occurred during 1995–2002, when the Basin's largest surface water supply, ran dry at its mouth, and declined to the lowest flow on record since 1912. The region has also suffered through severe drought periods in 1998–1999, again in 2002–2003. Box 15.1 includes a brief introduction to a 2000 Interior Department-USGS study of the 1998–1999 drought.

A moderate drought occurred in 2008, a severe drought again in the 2012–2013. Drought returned to the Delaware River in late 2016 and again in 2020. As drought spread across the Basin, it often affected parts of the State of Delaware. For example, in September 2022, moderate drought took hold in Delaware, quickly reaching to 8.24% of the state. By the end of September, lack of rain and abnormally high summer temperatures had resulted in damage to sections of the state's non-irrigated agriculture. Particularly affected were the state's non-irrigated corn and soy bean crops.

The Delaware Aqueduct delivers approximately half of the New York City's drinking water. In 2014, the system was leaking as much as 35 million gallons per day. On June 9, 2022, the re-occurring drought resulted in postponement of the planned aqueduct shutdown for one year to October 2023. Portions of the aqueduct's infrastructure installed more than 100 years ago can need maintenance or repair that must be carried out. The most needed repairs are on the City Water Tunnel No. 1, completed in 1917, and City Water Tunnel No. 2, completed in 1936. Because of concerns that a reduction in water pressure might damage part of the system, neither tunnel had been shut down for repairs (Box 15.2).

Box 15.2 Water shortage by infrastructure failures

New York City's water supply system travels daily through a complex system of reservoirs, aqueducts, tunnels, and water mains to New York City as well as many upstate communities. Due to the system's age, size, and extent, parts of it can be subject to failure. A water shortage can occur from a failure at any point in the water systems infrastructure. Depending on its location and scale, a failure can have varying degrees of impact. For example, a water main break may only affect a specific area of the city, while a dam failure or a collapse of tunnels or aqueducts could affect the distribution of water to all of New York City.

Of particular concern [in 2014 was] the Delaware Aqueduct, which delivers approximately half of the city's drinking water—500 million gallons per day. It is leaking as much as 35 million gallons per day. [On June 9, 2022, it was announced that the planned aqueduct shutdown has been postponed for one year to October 2023.] New technologies have permitted DEP to perform inspections of it, and a major project to repair it is under way. Among other portions of the system that may need maintenance or repair, most notable are City Water Tunnel No. 1, completed in 1917, and City Water Tunnel No. 2, completed in 1936. They have been delivering water to our homes, businesses, and institutions continuously since they were placed in service. Because of concerns that a reduction in water pressure might damage part of the system, neither tunnel has ever been shut down for inspections and repairs.

Water main breaks occur frequently. Some of the causes include the age of the mains and pipe material. For instance, 46% of the City's 6,785 miles of water main were built before 1941. The majority of mains are composed of unlined cast iron or cement-lined cast iron, which is susceptible to internal corrosion and leaks.

Source NYC Hazard Risk Landscape, 2014

Such problems occur in many of the country's water main water delivery systems. Breaks occur frequently, some due to the age of the mains and pipe material. For example, 46% of the 6,785 miles of main system was built before 1941. The majority of mains are composed of unlined cast iron or cement-lined cast iron, which is susceptible to internal corrosion and leaks.

Fracking Barred in the Basin

On February 15, 2021, the Delaware River Basin Commission voted to outlaw fracking in the basin. The Delaware River is the longest free-flowing river east of the Mississippi. Supporters of the ban led by the NRDC worked for close to ten years of efforts to protect the watershed, the Delaware River Basin Commission first proposed today's fracking ban in 2017. Since then, NRDC and its associate groups have waged a comprehensive campaign to support a full ban on fracking. New York State had earlier banned fracking across the state (Guerrero and Ong 2021) as a leader in the move away from fossil fuel infrastructure and toward a cleaner and greener future.

Water Stress and Hydraulic Fracturing

Hydraulic fracturing is the process of altering underground rock formations to increase the flow of oil or natural gas to the well. The process involves fracturing the formation surrounding the wellbore and placing sand or other granular material in those fractures to prop them open. The process has made it possible for the production of oil and natural gas in areas where conventional technologies have proven ineffective. "Recent studies estimate that up to 95% of natural gas wells drilled in the next decade will require hydraulic fracturing" (ALL Consulting, LLC 2012).

This technology combines hydraulic fracturing with horizontal drilling. High-volume Hydraulic Fracturing (HVHF) requires high volume withdrawals of surface water or groundwater over short durations (1 or 2 days); therefore, local stress could occur in streams and groundwater, particularly during periods of drought and high demand. In the humid Marcellus Shale play region (which spans parts of Ohio, West Virginia, Pennsylvania, New York, and Maryland), small streams are a common water source for HVHF activities. Small surface water catchments are prone to water stress because streamflow is low while withdrawal rates remain high. Consequently, small catchments tend to be more at risk to water stress during drought in the absence of regulations that prohibit withdrawals during low-flow periods (Entrekin et al. 2018).

The Clean Water Act

The Clean Water Act requires states to produce an Integrated Water Quality Report (IWQR) for the EPA at least once every three years. Every U.S. state is responsible for performing a water quality study and publishing a result in a freshwater resource report every two years. The report groups water resources by any one of four use categories: (1) aquatic life, (2) recreation, (3) consumption fish, and (4) potable water. The Water Quality Condition report identifies surface and groundwater waterbodies that have attained or were failing to meet standards that the findings must be submitted to the Federal EPA for approval by April 1 of every even-numbered year. A water body (i.e., a river, stream, lake, reservoir, etc.) is considered to be "impaired" if it fails to meet one or more water quality standards.

The number of resources in Pennsylvania's 2022 study, for example, were 3% or 2,418 miles more than the 2020 findings. The findings included 27,886 miles of streams and rivers as impaired for failing to meet their required standard (Scicchitano and Crowley 2022).

The study found an estimated 33% of the total miles of rivers and streams were impaired. The 2022 study covered a total of 85,568 miles of Pennsylvania's streams and 242 major publicly accessible lakes that accounting for 125,119 surface water acres, identifying resources that were below the specific standard set that class. The report also identified status changes achieved since the 2020 study. While nearly 935 miles of streams fell back from attaining status to impaired, 120 miles improved from impaired to sustaining status, and 1,395 stream miles either changed classification or remained within the same designation or saw the source of pollution changed. Since 2004, the first year of the program, approximately 920 miles of streams and 28,000 acres of lakes were moved from impaired to meeting all required standards (DEP 2022).

Pennsylvania Groundwater Protection

Sources of groundwater contamination are included in the bi-annual Integrated Water Quality Report (IWQR). DEP identified 31 contamination sources for major sources for groundwater pollution. Of this total, DEP identified 13 as high priority sources. Sites identified as sources of groundwater pollution included as priority problems include industrial facilities, surface liquid impoundments including centralized impoundments at unconventional well sites, underground storage tanks, hazardous waste sites, landfills, waste piles, aboveground storage tanks, manure/fertilizer applications, chemical facilities, septic systems, acid mine drainage, and abandoned oil and gas wells. In addition to these sources, several studies with total coliform contamination grew from 30 to 90% of private water wells in the state.

Hydraulic Fracturing in Pennsylvania

Since 2011, the hydraulic fracturing process ("fracking") of mining for oil and natural gas in Pennsylvania and other states has increased the domestic production of oil gas. However, in some locations in the state, it has also been associated with some real and some possible adverse health conditions. The first peer-reviewed study of the association between fracturing and the health of infants close to where the fracturing process occurred was published in 2017. The design included study on hydraulic fracturing's health impacts on more than 1.1 million births in Pennsylvania from 2004 to 2013. The researchers compared infants born to mothers living near drilling site to infants living further from the site both before and after the drilling began. The study found a lower levels of health among infants born to mothers living up to 3 kilometers (1.8 miles) from a hydraulic fracturing site than with infants and mothers living farther than 3 kilometers from a well site. The largest impacts were to babies born within 1 kilometer (0.6 miles) of a site. These babies were 25% more likely to be born at a low birth weight. The negative health effects for infants born from 1 to 3 kilometers from the site were about a third to a half of those within 1 kilometer to the site. Infants born to mothers living beyond 3 kilometers saw little to no impact. Negative health impacts included a greater incidence of low-birth weight babies, significant declines is average birth rate, and several other measures of infant health, The researchers concluded that hydraulic fracturing could affect human health through several channels, including water quality and other channels (Currie et al. 2017).

Criticized Report

In 2015, the EPA was criticized for conclusions in a study of the potential impacts of the poetical impact of hydraulic fracturing on drinking water. A new study of the fracking effects in Pennsylvania was carried out in response to a request by the U.S. Congress for the EPA to study the relationship between hydraulic fracturing and drinking water. The study relied on current scientific literature and data in articles published in science and engineering journals, federal and state government reports, non-government organization reports and industry publications. The draft report's section on Major Findings included above and below ground actions ("mechanisms") by which hydraulic fracturing activities have the potential to impact drinking water resources. These included: (1) water withdrawals when, or in areas with, low water availability; (2) spills of hydraulic fracturing fluids and produced water; (3) fracturing directly into underground drinking water resources; (4) below ground migration of liquids and gases; and (5) inadequate treatment and discharge of wastewater. The investigators did not find evidence that any of the activities led to widespread systematic impacts on drinking water resources in the United States (EPA 2015).

The disagreement with the EPA draft report was that the conclusion did not include all the available findings, particularly in Pennsylvania. The reason cited for this discrepancy was mismanaged record-keeping and reporting by the Pennsylvania Department of the Environment (DEP). Based on previously unreported fracking complaints unearthed by the non-profit *Public Herald*, fracking water complaints from 17 of 40 fracking counties became open to the public. However, Pennsylvania's official tally of water contamination remained at only 271 for all 40 counties. Contrary to the EPA fracking study's conclusion, the prevalence of drinking water contamination appears to be much higher than previously reported (Peltier 2016).

A 2014 study researched the connection between unconventional shale gas development (i.e., fracturing) and drinking water supply problems in Pennsylvania. The study examined data from the beginning of fracturing in 2008 to 7,245 wells in place in 2014. During this period, one-fifth of all gas wells drilled in the state received at one non-administrative notice of violation. No cases of subsurface groundwater transport of fracking or flowback fluids into documented water supply reports. Of the 1,000 complaints between 2008 and 2012, most were about impact to water supplies in areas of oil and/or gas production (Brantley et al. 2014).

Chesapeake Bay Watershed Issues

The Chesapeake Bay Watershed (CBW) coincides with most of the Mid-Atlantic region. The Mid-Atlantic Regional Integrated Sciences and Assessments (MARISA) program is a collaboration funded by NOAA through RAND and researchers at Pennsylvania State University, Johns Hopkins University, Cornell University, and the Virginia Institute of Marine Science on climate and weather in the watershed. The series collects information from regional climate centers, news and weather information, and regional-specific climate datasets for the benefit of policymakers, practitioners, residents, and community leaders in the Chesapeake Bay Watershed (Fig. 15.3).

Projections of weather and climate variability and change in the Chesapeake Bay Watershed come from the best available scientific information. The 2019 Autumn edition of MARISIA, the organization's quarterly reported on salt and fresh water and climate change impact conditions in the Chesapeake Bay Watershed up to the Fall of 2019. The following section in this chapter centers on data distributed by MARISIA plus other federal, state, and local government organizations, with some information from private organizations.

The Watershed covers approximately 68,500 square miles and is home to more than 18 million people. Agriculture, the largest land use in the watershed after forestland lands, covers nearly a third of the CBW. More than 83,000 farms and ranches in the CBW contribute more than $10

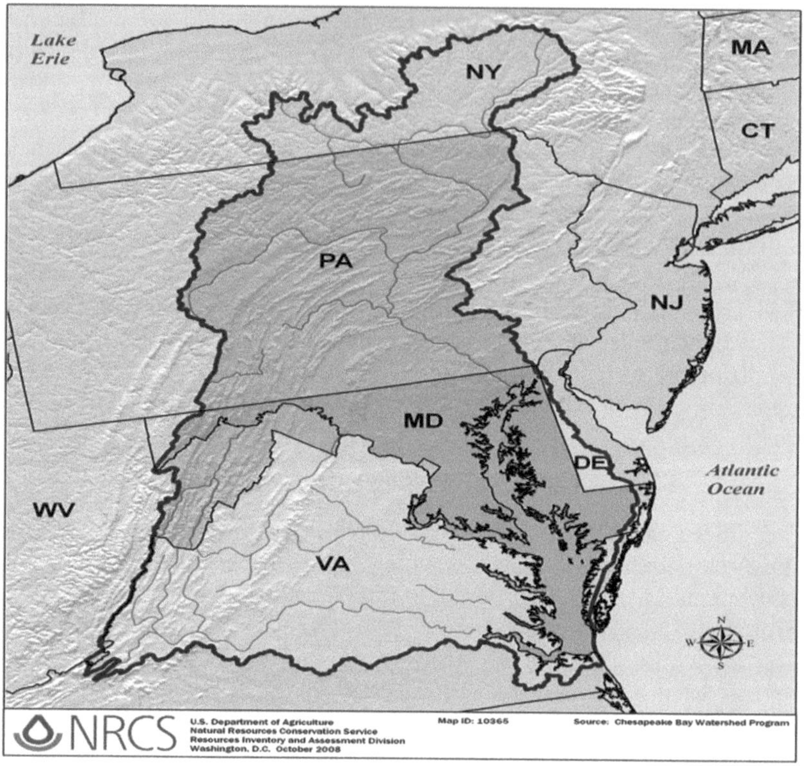

Fig. 15.3 NRCS map of the full scope of the Chesapeake Bay Watershed

billion in annual production to the region's economy. However, agriculture has also been identified as a source of nutrients and sediment entering the Bay. Activities such as over-application of fertilizer and pesticides contribute pollutants to local waterways and ultimately to the Chesapeake Bay (USDA 2019).

The CBW is an important area for agricultural production in the Northeast, and a vibrant and healthy agricultural sector is critical to restoring and maintaining the health of the CBW.

Summer of 2019 in the watershed was turning out to be one of the hottest years on record. September and October of the year were among the ten hottest months on record for several locations including Washington DC and Harrisburg, Pennsylvania. With the heat, September of that year ranked among the top ten all-time driest months for Baltimore, Maryland, and the Capital. Drought levels of moderate and severe appeared in Virginia, Maryland and Delaware, and portions of West Virginia and Pennsylvania.

Chesapeake Bay Watershed Issues

Water supply issues, polluted source discharges into Chesapeake Bay, the future sustainability of the watershed under the impact of climate change in the Mid-Atlantic region are among the issues that lead to concerns about future availability, purity and cost, much like issues of concern in most of the United States. All states in the region either border on the Atlantic Ocean or depend on water resources in the Chesapeake Bay Watershed, or both. The Chesapeake Bay watershed covers almost 65,000 square miles, reaching parts of New York, Pennsylvania, Delaware, Maryland, Virginia, and West Virginia. All the precipitation that falls in this area eventually drains into the Chesapeake Bay, which leads directly into the Atlantic Ocean. With climate change transforming the water cycle, the consequences for this watershed and the many others in the region are profound. Figure 15.2 is a U.S. Department of Agriculture's Natural Resources Conservation Service (NRCS) map of the full scope of the watershed,

A significant portion of land in the watershed is devoted to agriculture. A continuing period with more extreme precipitation events combine to establish a variety of crises for the states in this region. Increasing annual precipitation in parts of the region combined with heavier precipitation events is accompanied by more floods and stormwater runoff, both of which can carry increased sediment loads, pathogens, and nutrients from fertilizer directly into the bay, contributing to a more for harmful algae blooms.

Unlike population increases in some warmer and drier states, the Chesapeake Bay Watershed provides water for three of the top ten states with the greatest population losses. In 2019, ten states saw population loss, according to the U.S. Census Bureau. In states like West Virginia was ranked first in the percentage of lost population, New York was fourth and New Jersey was tenth. Between 2010 and 2018, West Virginia's population fell by about 47,000 people or around 2.8%. This is due to a combination of more deaths than births and net migration.

Conclusion

The Clean Water Act requires states to produce an Integrated Water Quality Report (IWQR) for the EPA at least once every three years. Every U.S. state is responsible for performing a water quality study and publishing a result freshwater resource report every two years. The report groups water resources by any one of four use categories: (1) aquatic life, (2) recreation, (3) consumption fish, and (4) potable water. The Water Quality Condition report must be submitted to the Federal EPA for approval by April 1 of every even-numbered year. A water body (i.e., a river, stream, lake, reservoir, etc.) is considered to be "impaired" if it fails to meet one or more water quality standards in bi-annual freshwater status.

The fracturing ground mining program, fracking, is a growing threat to the availability of clean freshwater in significant parts of this region. Where undertaken, fracking seriously alters the groundwater flow patterns and regime, and has the added negative impact of leaving polluted water behind. While the outcome of studies detailing the negative impacts of fracking has been challenged, the evidence is persuasive that this practice is harmful to the water resource of the region and a potential danger to human health. Moreover, the purpose of fracking is to add to the supply of petroleum and gas for consumption. The consumption and burning of both are known to be major contributors to the climate change that is accelerating our water crisis. Not only is fracking directly impacting water quality in this region; it is furthering the change in climatic conditions that are deteriorating supply and quality of water globally.

References

ALL Consulting, LLC. 2012 The modern practices of hydraulic fracturing: A focus on Canadian resources. Prepared for Petroleum Technology Alliance Canada and Science and Community Environmental Knowledge Fund. Tulsa, OK.

Brantley, Susan L., Dave Yoxtheimer, Sina Arjmand, Paul Grieve, Jon Pollak, Garth T. Llewellyn, Jorge Abad, and Cesar Simon. 2014. Water resource impacts during unconventional shale gas development: The Pennsylvania experience. *International Journal of Coal Geology* 126: 140–156.

Currie, Janet, Michael Greenstone, and Katherine Meckel. 2017. Hydraulic fracturing and infant health: New evidence From Pennsylvania. *Science Advances* 3 (12). Available at https://doi.org/10.1126/sciadv.1603021.

DEP (Pennsylvania Department of Environmental Protection). 2022. 2022 Integrated water quality report. Available at https://storymaps.arcgis.com/stories/b9746eec807f48d99decd3a583eede12.

Entrekin, Sally, Anne Trainor, James Saiers, Lauren Patterson, Kelly Maloney, Joseph Fargione, Joseph Kiesecker, Sharon Baruch-Mordo, Katherine Konschnik, Hannah Wiseman, Jean-Philippe Nicot, and Joseph N. Ryan. 2018. Water stress from high-volume hydraulic fracturing potentially threatens aquatic biodiversity and ecosystem services in Arkansas, United States. *Environmental Science & Technology* 52 (4): 1681–2418.

EPA (Environmental Protection Administration). 2015. Assessment of the potential impacts of hydraulic fracturing for oil and gas on drinking water resources: *Executive Summary*. Working draft edition only, Available at www.epa.gov/sites/default/files/2015-06/documents/hf_es_erd_jun2015.pdf.

Freyman, Monika, and Ryan Salmon. 2013. Hydraulic fracturing and water stress: Growing competitive pressures ofr water. *Ceres*, Mat 1. Available at https://search.issuelab.org/resource/hydraulic-fracturing-water-stress-growing-competitive-pressures-for-water.html.

Guerrero, Marisa, and Kimberly Ong. 2021. Fracking banned in the Delaware River basis. *NRDS*, February 25. Available at https://www.nrdc.org/experts/marisa-guerrero/fracking-banned-delaware-river-basin.

Peltier, Laural. 2016. Pennsylvania fracking water contamination much higher than reported. *EcoWatch*, February 4. Available at www.ecowatch.com/pennsylvania-fracking-water-contamination-much-higher-than-reported-1882166816.html.

Scicchitano, Eric, and Mike Crowley. 2022. Report: 1 in 3 miles of Pennsylvania streams, rivers' impaired.

USDA (United States Department of Agriculture). 2019. Irrigation & water use. USDA Economic Research Service circular. Available at www.ers.usda.gov/topics/farm-practices-management/irrigation-water-use/.

CHAPTER 16

Water Stress in New England

The climate of the New England states varies over the 500 mile range from northern Maine to the southern tip of Connecticut. New England includes Maine, New Hampshire, Vermont, Rhode Island, Massachusetts, and Connecticut. Although they can differ significantly, they are often included among the Nation's "wet" states. They constitute 15 of NOAA'S 344 climate divisions and climatic sub-regions. This monthly precipitation data have been compiled, division-by-division, since 1895 (EPA 2014). The data help predict when drought events and water-related emergencies are likely to occur. The section map in Fig. 16.1 indicates the distribution of the severe weather events in New England regions. The 125-year records of weather changes show that a severe drought occurs in this region about every five to six years. In addition to the 2022 drought, droughts also occurred in 2016 and 2020.

With the exception of these drought years, the New England climate divisions tend to follow precipitation and water resource differences between the NOAA three-state groupings. For example, a lower state group of Connecticut, Rhode Island, and Massachusetts have similar precipitation and temperature pattern that are different than the common patterns of an upper group that includes New Hampshire, Vermont, and Maine. The upper (or northern) three states are usually drier and the lower (or southern) three states are wetter. This chapter follows

D. E. McNabb and C. R. Swenson, *America's Water Crises*,
https://doi.org/10.1007/978-3-031-27380-3_16

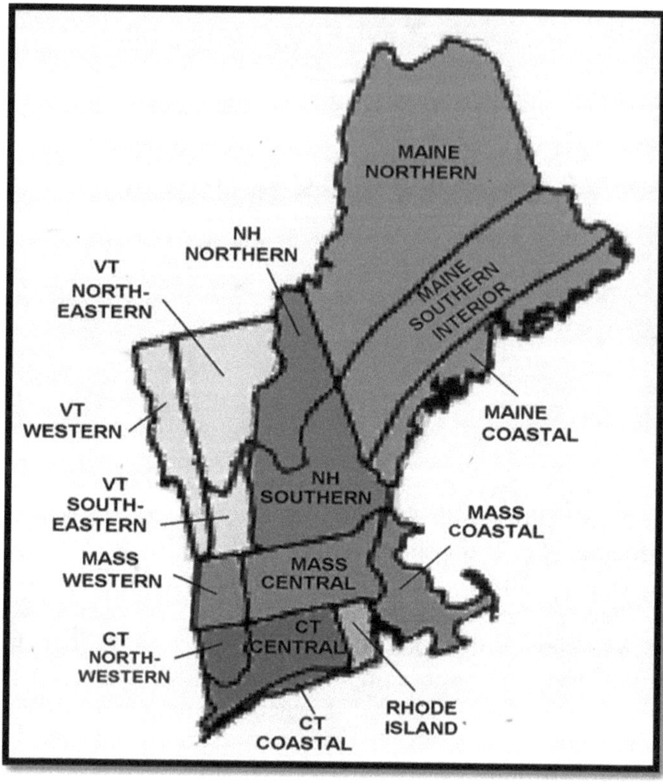

Fig. 16.1 NCDC map of New England, showing the 15 NOAA weather units (*Source* NCEI; Fisk [2019])

this two-section plan in discussions on drought, surface water, and groundwater.

The location of the region in the Northeastern corner of the coterminous United States is shown in Fig. 16.1 (USGS mapWGS84). As of the first week of August 2022 all of the states except Maine reported more than half of their state was abnormally dry with little or no measurable drought, while less than a quarter of Maine was abnormally dry, with 40% of the state in some level of drought conditions with just 8% of the state was in severe drought.

Periodic Drought Records

Periods of drought have been common occurrences in New England, despite the region's reputation as a wet section of the coterminous 48 states. Since the beginning of the current century, periods of droughts have been recorded in 2016, 2020, and in 2022. Table 16.1 lists peak years of drought in New England and the areas most affected, as whether it was in states or a region most affected by the drought. The 2020 drought was one of the most severe. Below average and infrequent rainfall from May through September 2020 led to an extreme drought across much of New England, with some areas experiencing flash drought in northern Maine, southeastern Massachusetts, northern Rhode Island, and northeastern Connecticut. The abrupt change in precipitation from at or above average in early May to as much as four inches below average in September 2020 resulted in declines in surface water streamflows and declines in groundwater level has been characterized as a flash drought at some locations in New England (Lombard et al. 2020).

2020 drought compared in extent with other historic New England droughts, including those in 1995 and 1998–2002. Reductions in average rainfall from 1981 to 2010 over the six-state region ranged from 3 inches to 12.5 inches below average. Much of the rain that did fall during the summer of 2020 occurred during a few heavy storms, with little of the rain able to soak into the dry soil.

Table 16.1 Sixty-plus off-and-on cycles of droughts in New England with section affected

Peak year/s	*Month and states most severely affected*
1961–1969	June 1964 and July (Regional)
1991	June and July (Regional)
1995	June and September (Regional)
1998	May (Maine, New Hampshire, Vermont)
1999	June to August (Regional)
2001	June to August (Regional)
2002	June to August (Regional)
2016	June to August (Massachusetts, Rhode Island, Connecticut)
2020	June and September (Regional)

Source Lombard et al. (2020) and USGS (2020)

Table 16.2 State levels of drought by the category, August 2022

Sections of state included					
State *Inclusions*	*D0:D4 abnormally dry* *D0–D4 (%)*	*D1:D4 moderate drought* *D1–D4 (%)*	*D2:D4 severe drought* *D2–D4 (%)*	*D3:D4 extreme drought* *D3–D4 (%)*	*D4 exceptional drought* *D4 only (%)*
All New England	80.20	58.32	18.17	0.0	0.0
Maine	23.3	39.9	8.1	0.0	0.0
New Hampshire	95.5	85.7	13.3	0.0	0.0
Vermont	57.1	17.5	0.0	0.0	0.0
Rhode Island	100.0	100.0	100.0	0.0	0.0
Massachusetts	100.0	95.2	57.2	0.0	0.0
Connecticut	90.8	66.8	26.2	0.0	0.0

Source NIDIS Drought.gov

By the end of August, all of Massachusetts was in some level of drought for the first time in more than seven years, with nearly 40% in the extreme drought category, 94.5% in the severe drought stage, compared to 57.2% at this level just two weeks earlier. None of the state was in extreme drought at the beginning of the month illustrating the "flash" nature of the drought in the Northeast (a quick onset and relatively brief period affected). Since July, parts of New England, New York, and New Jersey in the mid-Atlantic region were from two to six inches below their normal rainfall, further reducing declining streamflow and reducing groundwater levels (Table 16.2).

State Water Resource Issues

In the six New England states, pumping groundwater for domestic supply and for meeting the demand of other uses has had minimal effect on the water table, the base flow of streams, and in coastal areas has caused saline groundwater to move inland. Many locations on the eastern US Atlantic coast are experiencing similar effects to their coastal aquifers were excessive withdrawals has resulted in saltwater intrusion due to local groundwater depletion. In Maine, New Hampshire and Vermont, for example, all use-sectors depend on groundwater for more than 50%

of their water needs. Surface water is a secondary resource for public drinking water.

Sustainability of the traditional streamflow is dependent upon annual snowmelt for providing the region's spring and summer streamflow and for maintaining the groundwater aquifers for irrigated agriculture. Groundwater fills in for the missing spring and summer rainfall. Without streamflow, the aquifers cannot be recharged. The states that depend more on groundwater are listed in Table 16.3.

Climate change in the Northeast states has resulted in smaller snowpacks than usual—for a combined 2020–2021 and 2021–2022 winters, the snowpack was from four to seven feet below normal for many northern Maine locations, followed by an earlier than normal snowpack melting. Decreases in river ice, and precipitation falling as rain occurring often instead of snow resulted in some declines in streamflow. These changes have resulted in denser and thinner late-winter snowpack. More than half of the northern three states depend upon groundwater for the largest share of their freshwater. Continued drought and reductions in

Table 16.3 Public drinking water systems use of groundwater and surface water, 2014 and 2022 estimates

State	*Year*	*Groundwater population*	*% of total population*	*Surface water population*	*% of total population*	*Total population*
Connecticut	2014	496,847	15%	2,431,829	85%	3,312,314
	2022	541,847		2,815,467		3,612,314
Maine	2014	470,023	51%	453,755	49%	921,613
	2022	698,271		670.888		1,369,159
Massachusetts	2014	1,899,945	27%	5,163,938	73%	7,036,833
	2022	1,899,841		5,136,890		7,126,375
New Hampshire	2014	293,534	55%	1,056,316	45%	1,334,247
	2022	733,386		600,411		1,389,741
Rhode Island	2014	231,735	22%	866,404	78%	1,053,341
	2022	243,395		862,946		1,106,341
Vermont	2014	338,233	54%	274,300	46%	626,358
	2022	297,319		297,426		646,245
Tribal Systems[a]	2014	197	NA	77,000	NA	77,197
	2022	197[a]		77,100		77,197[a]
Region Totals	2014	3,730,514		4,414,256		8,144,770
	2022	4,414,256		10,461,028		15,87,528

Source EPA (2014), Tribal Systems are 2022 data only

snowpack have cut into the annual surface water. Where snow is the primary precipitation for one-third of the year, the annual snowpack is smaller and is melting a week or more earlier than in the past. However, changes in the aquifers in the northern three New England states may not have been as significant as the changes that can occur in the more populated states.

Warmer weather results in snowmelt occurring sooner and changes in precipitation patterns have reduced summer snowmelt. Changes in surface water streamflow in the New England's winter and spring seasons result in reductions in surface water for spring and summer agriculture. Surface water streamflows that had remained average during the winter and spring of 2020, Rainfalls varied thought the summer, resulting some reductions in the length and severity of droughts in much New England. However, beginning in May and staying low for the rest of the summer, lower amounts of drought did become a water supply concern. Although streamflows and groundwater levels began to recover in October, full recovery would not take place until precipitation began in the following fall and winter. Groundwater levels during this period declined, but at a slower and more gradual decline rate than the streamflows. The following examples of groundwater conditions prove a broader picture of the effects in individual states during the 2020 drought.

State Water Issues

Research has shown that each of the New England states had both common and individual water problems. The earliest settlers adopted the water resource that was available at the time the settlers arrived using the available technology of the time. Groundwater wells were dug by hand, so a water table near the surface was necessary. Collecting surface water was easier, but more difficult when the winter ice froze the resource. Today's issues are a reflection of the needs of the local populations and the water needs of the regional society. Brief descriptions of the major water issues and the problems extant in each of the six-state follow, beginning with the most populous of the states, Massachusetts.

Massachusetts

Massachusetts still supports both an agricultural and an industrial economy. The water problems of these industries differ primarily in availability. However, as populations have grown (12.9% between 1990 and 2015), freshwater resources have come under significant stress in many areas, particularly during periods of drought. Continuing drought has left 2022 as one of the driest years of the last 100 years. According to the Drought.Gov system 100% of the state was suffering from moderate drought, although a substantial drop in the August report of 100% of the state under severe drought, by the end of September, it had declined to just 72% of the state.

Small farms in Massachusetts account for 94.2% of all farms in Massachusetts while family or individually owned farms account for 79.7% of Massachusetts farms. According to the 2017 state Agricultural Census, although the number of farms in New England has declined by close to 3%, the dependence upon irrigated agriculture has increased by various amounts. For example, the number of Massachusetts farms that depends on groundwater increased a total of 92.9% between 1974 and 2017. Prolonged and repeated droughts have made irrigation increasingly necessary on farms in Massachusetts. The total number of acres that was irrigated rose between 2012 and 2017 increased by 2.1%. In both the 2012 and 2017 census, slight increases in acreage irrigated had also occurred, despite a decreasing trend the previous two census years.

In 2012, the Massachusetts Water Infrastructure Finance Commission identified a funding gap of at least $21 billion for addressing the Commonwealth's water infrastructure needs. With the upcoming federal stormwater permit requirements, the funding gap will only continue to grow. For small rural water systems, permitting requirements are extremely difficult to meet, groundwater is common in many of the state's rural systems. New EPA treatment methods are adding to their stressed water supply difficulties. In a recent study of groundwater wells in Massachusetts, 21 of 62 wells in Massachusetts had monthly water levels at or below the 25% level. In September, the number had increase to 32 of the 62 wells were below the 25 percentiles. Nearly, 75% of the public or municipal systems rely of surface water.

Townsend, a small city in the Northeast region of Massachusetts, was required to reduce the freshwater it could provide to its nearly 1,130 residents. The Water Department had to shut down its Harbor Trace well in

August 2021 and to install new treatment plant to eliminate the high levels of PFAS (the "forever" plastics pollutant) in the water. The well is the system's largest supplier of water, producing one million gallons per day. The town's new water treatment plant is expected to be available in the Spring 2024. Until then, current wells were being worked to their limit which causes discolored water at times especially from 6 to 9 am and 5 pm to 9 pm where water demand is the greatest. In January 2022, Massachusetts announced it awarded more than $420,000 in grants to help 13 communities with water system interconnection and water mitigation projects. The grants are part of the Water Management Act (WMA) Grant Program (WaterWorld 2022).

New Hampshire

New Hampshire Department of Environmental Services issued the following report: New Hampshire's waterways continued to be adversely impacted by wastewater discharges and non-point sources of pollution such as urban and agricultural runoff, septic systems, stormwater from construction activities and urbanized areas, water withdrawals, and atmospheric deposition. Water experts have learned that the quality, quantity, and ecology of both surface and groundwater are affected by all of the activities occurring within a particular watershed. As a result, there is an increasing need to address water resources on a watershed basis through close collaboration among various state and local organizations.

From 55 to sixty percent of New Hampshire's water users depend on groundwater for their drinking water supplies. The state's Groundwater Protection Act, passed in 1991, authorizes NHDES to regulate large groundwater withdrawals and commercial discharges of wastewater and establishes best management practices that must be employed by activities that are considered potential contamination sources.

Rhode Island

The drought was expanding in both the south and western directions, to include all but one of the six geographic sections of the state and would thereafter include the Connecticut River Valley region. This region jointed the Northeast, the Northeast and Central regions assigned with Level 3 drought conditions. The Cape Cod region was raised to a Level 2 drought, while the Western region and Island regions remained at Level 1 drought. Describes the identified effects of the prolonged drought.

At 78% of the water used in Rhode Island is taken from existing surface water resources, the state is New England's most dependent upon surface water for its freshwater supplies. The traditional dependence on surface water of this section has protected most of the state's groundwater aquifers from excessive withdrawals and the follow up of severe depletion. Although as the climate changes the precipitation patterns are becoming less reliable in New England which has traditionally been one of the Nation's wet states, with many dependable surface water and readily accessible groundwater resources. The current century is already consistently warmer than in the past decades and undergoing spring and summer air temperatures that are occurring more often than in the recent past. Surface water supplies are undergoing continued stress and are expected to be even more difficult during the rest of this century (Dudley and Hodgkins 2002; Dudley et al. 2010, 2017). Among the effects occurring now are these.

Although Rhode Island depends upon groundwater for approximately 22% of its annual freshwater supply, declines in well levels were significant enough to be a concern. Water levels in 20 wells of the 27 wells were at or below the 25% average. In August 2022, levels in 3 wells were below a 25-year average, and in September, 12 wells were below the 25-year average.

Like Massachusetts, in less than a month Rhode Island slipped from a little more than half of the state into all but a sliver of land—nine tenths of 1%—into having all 100% of the state into extreme drought. Providence, Rhode Island's largest city with a population in 2020 of 1.058 million, gets all its drinking water from the state's largest reservoir, the Y-shaped Scituate Reservoir, which is located 15 miles east of the city. The Scituate Reservoir has tributaries that flow in and out of several other reservoirs. Anticipating a continuation of the drought and increased pressure placed on the small state's water supply, A statewide water advisory message was aired in August 2020: While the state's water supply is designed to withstand drought, Rhode Islanders were encouraged to become aware of the current and future climate and water and other resource conditions. As a precaution, the state's residents businesses were urged to adopt water conservation measures. People were urged to follow these voluntary actions:

- Watering lawns no more than is needed. The average lawn needs only one inch of water weekly.
- Avoiding watering during the warmest part of the day—10 a.m. and 2 p.m.—when water is more likely to evaporate.
- Sweeping driveways and sidewalks rather than spraying them with a hose.
- Matching your washing machine and dishwasher settings to the appropriate load size so you do not use more water than required.

Vermont

In the middle of the 2022 summer, Vermont was the least affected by drought of the six New England states, with only 17.5% of the state in moderate drought and only a little over half of the state reported as abnormally dry. The state is almost equally divided between surface and groundwater uses. Vermont has just a few more than 625,000 residents, large numbers of whom receive their water from their own private groundwater wells. The three driving sectors of economy are tourism, manufacturing, and agriculture. Northwest Vermont is home to the state's largest economic center, Burlington, with other cities and towns in the region such as Essex, St. Albans, and Middlebury being home to major employers including universities, hospitals, and technology firms. The agriculture sector has grown and developed beyond dairy farming and maple syrup to include food producers large and small.

In 2016, groundwater was the source for approximately 80% of Vermont's drinking water. Since then, concerns over health and safety issues have forced an increase in the number of water safety actions in response to Clean Water requirements. The Vermont Geological Survey issue a summary in 2016 that included these sections of a description of the development of better groundwater management: "In the late 1970s public water systems began testing new wells for interference with neighboring wells. In the 1980s groundwater quality problems emerged ranging from naturally occurring radionuclides and arsenic to man-made solvents and petroleum products. Clean up efforts began to address groundwater contamination from waste disposal sites, and source water protection area mapping and protection planning began for public water system wells. In the 1980s and 1990s rules were adopted to better protect new wells from septic systems following increased concerns of pathogens. Groundwater issues now include areas of low yielding wells, lowering health standards for naturally occurring elements, and concerns about locating and protecting future municipal sources of drinking water. Today virtually all wells and water systems are permitted under the Vermont Water Supply Rules" (VGS 2016, 8).

Connecticut

Who depends on groundwater in Connecticut varies from time to time and occupation to occupation. There are more than 300,000 wells in this small state. Most of them—from 15 to 23% of the population—are the private residential water wells producing water for home consumption. One estimate is that is that these private wells provide water for a million residents; another says that more than half a million rural residents, 5,500 of whom are the owners and operators of small independent farms, are the major users of groundwater users. Regardless of the total number, most are drilled into the bedrock that lies below the thin layer of glacier-ground soil that underlies the surface of most of New England. According to the EPA, the groundwater from these wells can be contaminated from the manufactured products such as gasoline, oil, road salts, and chemicals get into the groundwater and the PFAS "forever" plastics and other chemicals used in agriculture and industry.

Connecticut's lakes, streams, and reservoirs constitute the nearly 85% of all freshwater use in the state. University research into surface water and climate warming suggests that as the climate continues to get warmer, these natural water resources and reservoirs are "warming relatively quickly over time and that surface, average, and bottom water temperatures may continue to increase by 0.44, 0.30, and 0.16 °C per decade, respectively. Future projections of thermal stratification indicate that significant increases may occur in July–September and that stratification is likely to begin 2–4 weeks earlier and last 2–4 weeks longer by mid-century. High-risk cyanobacteria [often incorrectly referred to as blue-green algae blooms and their toxins] ...occurred in three of the study reservoirs and are correlated with the occurrence of high surface water temperatures, cool bottom water temperatures, and high total Relative Thermal Resistance to Mixing (RTRM). Days per year with extreme high-water temperatures and total RTRM are projected to increase significantly in the future. This shift may favor increased dominance of certain cyanobacteria species that tend to grow best in these extreme conditions, especially in reservoirs where cyanobacteria blooms are already a concern" (Mullin et al. 2020).

Maine

Maine is almost evenly divided between a dependence upon groundwater (49%) and surface water (51%). According to the National Ground Water Association (NGWA), individual households' use of 31.6 million gallons per day (mgd) is 37.3% of the total groundwater, the largest share of the 84.8 mgd daily total for all users. With 27.5 mgd, public supply—water withdrawn by public and private water suppliers that provide water to at least 25 people or have a minimum of 15 connections; public-supply water is used for domestic, commercial, and industrial purposes—is the second largest user of groundwater at 27.4% of the daily total. The use of 5.97 mgd withdrawn for irrigation is a relatively minor portion of the total, and thermoelectric use is less than one percent of all Maine's groundwater withdrawals is the smallest user.

Wherever a significant amount of groundwater is withdrawn for public use, winter icing of roadways and structure deicing chemicals can pollute the region's groundwater. Also, in locations where combined sewer and stormwater runoff, antifreeze, animal waste, and chemicals can pollute untreated groundwater. Combined untreated wastewater and storm water runoff may end up in outdated municipal combined sewer overflow systems that cannot handle heavy rainstorms (MCA 2022). The number of and severity of major storms is increasing with climate change.

In accordance with the EPA's Clean Water regulations, Maine's groundwater is also subject to many previously unmonitored water pollutants and other problems. In published results of a research study in central Maine, for example, 53 out of 1093 (4.8%) private bedrock well water samples tests, uranium was found in naturally occurring locations that were above the EPA's maximum contaminant level (MCL) for drinking water; and 226 (29%) out of 786 samples showed radon in limits above the EPA's alternative maximum level. Radon is described by the USGS as a naturally occurring gas produced by the radioactive decay of radium. Radon is radioactive and decays to form polonium, which is also radioactive and cancer-causing.

Conclusion

The New England region is part of a nationwide arrangement of 344 Climate Divisions organized state-by-state, monthly precipitation statistics have been compiled, division-by-division since. With the exception of severe drought years, which occur about once every five to six years, the New England climate divisions tend to show annual precipitation differences between the two three-state groupings: A southern state group of Connecticut, Rhode Island, and Massachusetts compared with a northern that includes New Hampshire, Vermont, and Maine section.

Trends in New England groundwater levels are available for these four-year periods: 20-year (1991–2010), 30-year (1981–2010), 40-year (1971–2010), and 50-year (1961–2010). In most of the wells that were tested, little change or increases in groundwater levels over the test period was reported. Annual rises in groundwater levels in the 77 wells were common in all of the periods, with the greatest increase occurring in the 20-year period. Changes, if any, tended to follow the last of the summer annual snowpack. The conclusion reached was that in both the 20-year and 30-year records increases in groundwater water table or no substantial change at most wells had occurred in most wells and that increases in groundwater were common in most all periods. Forestation of large regions of the northern states and reductions in withdrawals of groundwater for agriculture likely contributed to the increase (Dudley and Hodgkins 2013).

In the 20 different types of soil, average recharge rates ranged from a low of 12.02 inches per year in Evergreen Forest soils to a high of 23.53 inches per year in scrub/shrub soil (Nielsen and Westenbroek 2020). The data were obtained by calculations of the 25-year mean annual potential recharge, median annual potential recharge, and the minimum and maximum potential recharge for the 25-year simulation. The statewide range in the minimum values was very compressed (about 2–20 in/year)—almost everywhere in the State experienced a drought-year low recharge less.

Surface water resources are the primary water resource in the states in the southern New England region: Massachusetts, Connecticut, and Rhode Island. Surface water resources in these three states are by far the most important water source. Moreover, the states are concerned with the impact of population increases than are the northern three states. The 2022-estimated total population of the southern three states was 11.7 million. Under most climate conditions, the region has benefited from a sufficient supply of freshwater from a mix of surface water, with supplemental groundwater resource to meet the needs of all water use sectors. However, natural and social conditions are beginning to place much of the region under periodic conditions of water stress. Climate change has gendered several periods of moderate to severe drought, while population increase is moving some of the water resources in the region from stress toward an over withdrawal water crisis.

In the states that depend on surface water for more than as much as two-thirds of their freshwater supply, groundwater replaces the stressed surface water. In Rhode Island, surface water accounts for 78% of freshwater withdrawn; in Connecticut it is 85%; and in Massachusetts, surface water is nearly 75% of the total. In August of 2022, all three of these states were suffering from a prolonged drought. In August 9, 2022, the state was beginning the sixth month of less than normal rainfall, together with higher than traditional temperatures.

As with all industrialized areas and land where past and present agricultural activity has relied on chemical fertilizers, pollution contributes to surface and groundwater degradation in New England. The impacts of climate change are also adding stress to the long-standing water procurement practices here, particularly in those areas where there is a reliance on small distribution systems, be they surface water catchments or groundwater wells.

References

Dudley, Robert W., and Glenn A. Hodgkins. 2002. Trends in streamflow, river ice, and snowpack for coastal river basins in Maine during the 20th century. Water-Resources Investigations Report 2002-4245. https://doi.org/10.3133/wri20024245.

Dudley, Robert W., Glenn A. Hodgkins, James B. Shanley and Thomas J. Mack. 2010. Quantifying effects of climate change on the snowmelt-dominated groundwater resources of Northern New England. U.S. Geological Survey Fact Sheet 2010–3104. Available at https://pubs.usgs.gov/fs/2010/3104.

Dudley, Robert W. and Glen A. Hodgkins, 2013. Historical groundwater trends in Northern New England and relations with streamflow and climatic conditions. *Journal of the American Water Resources Association* (JAWRA) (2013) 1–15. https://doi.org/10.1111/jawr.12080

Dudley, Robert W., Glenn A. Hodgins, M.R. McHale, M.J. Kolian, and Benjamin Renard. 2017. Trends in snowmelt-related streamflow timing in the coterminous United States. *Journal of Hydrology* 547: 208–221.

EPA (Environmental Protection Administration). 2014. EPA New England regional climate adaptation plan. EPA publication number 100K14001H.

EPA (Environmental Protection Administration). 2022a. History of the Clean Water Act. Available at www.epa.gov/laws-regulations/history-clean-water-act

EPA (Environmental Protection Administration). 2022b. Basic information about water reuse. Available at www.epa.gov/waterreuse/basic-information-about-water-reuse.

EPA (Environmental Protection Administration). 2022c. What is green infrastructure? Available at www.epa.gov/green-infrastructure/what-green-infrastru cture

EPA (Environmental Protection Administration). 2022d. Harmful algal blooms. Available at www.epa.gov/nutrientpollution/harmful-algal-blooms

EPA (Environmental Protection Administration). 2022e. What is radon? Available at www.epa.gov/radon/what-radon

Lombard, Pamela J., Jane R. Barclay, and Dee-Ann E. McCarthy. 2020. Drought in New England. USGS Open-File Report 2020-1148. https://doi.org/10.3133/ofr20201148.

MCA (Maine Conservation Alliance). 2022. Clean water. Available at https://protectmaine.org/priorities/clean-water#:~:text=Maine%20waters%20are%20too%20often,cannot%20handle%20heavy%20rain%20storms.

Mullin, Cristina A., Christine J. Kirchhoff, Guiling Wang, and Penny Vlahos. 2020. Future projections of water temperature and thermal stratification in Connecticut reservoirs and possible implications for Cyanobacteria. *Water Resources Research* 56: 11. Online. https://doi.org/10.1029/2020WR027185

Nielsen, Martha G., and Stephen M. Westenbroek. 2020. Groundwater recharge estimates for Maine using a Soil-Water-Balance model—25-year average, range, and uncertainty, 1991 to 2015. USGS Scientific Investigations Report 2019-5125

USGS (United States Geological Survey). 2020. Below and infrequest rinfall from May through September led to an esxtreme hydrologicl drought across through much of New England, with some qareas experienceing a flash (quick-start) drought. Available at https://www.usgs.gov/centers/new-england-water-science-center/science/new-england-drought-2020#:~:text=Below%20average%20and%20infrequent%20rainfall,station%2001109070%2C%20Dighton%2C%20Massachusetts.

VGS (Vermont Geological Survey). 2016. Vermont water use data and research. VGS Department of Environmental Conservation. Available at https://Vermont-WUDR-Workplan.pdf.

WaterWorld. 2022. Massachusetts awards $420K for water infrastructure. Available at www.waterworld.com/drinking-water/infrastructure-funding/press-release/14223613/massachusetts-awards-420k-for-water-infrastructure.

Bibliography

Abdalla, Charles W., Joy R. Drohan, and John C. Becker. 2010. *River basin approaches to water management in the Mid-Atlantic states*. University Park, PA: Pennsylvania State University.

ACT (Atlanta Regional Commission). 2021. Tri-state water wars overview. Available at https://atlantaregional.org/natural-resources/water/tri-state-water-wars-overview/.

ADECA (Alabama Dept. of Economic and Community Affairs). 2016. Water withdrawals in Alabama, 1960 to 2015. Available at https://adeca.alabama.gov/wp-content/uploads/2015-Water-Use-Report-Fact-Sheet.pdf.

AGS (Arkansas Geological Survey). 2020. Water issues. Available at www.geology.arkansas.gov/water/water-issues.html.

AGU (American Geophysical Union). 2019. US infrastructure unprepared for increasing frequency of extreme storms. *ScienceDaily*, August 1. Available at www.sciencedaily.com/releases/2019/08/190801120209.htm.

Ajaz, Ali, Sumon Datta, and Scott Stoodley. 2020. High plains aquifer—State of affairs of irrigated agriculture and role of irrigation in the sustainability paradigm. *Sustainability* 12 (9): 3714.

Alcamo, Joseph, Martina Flörke, and Michael Maerker. 2007. Future long-term changes in global water resources driven by socio-economic and climate changes. *Hydrological Sciences Journal* 52 (2): 247–275.

ALL Consulting, LLC. 2012. The modern practices of hydraulic fracturing: A focus on Canadian resources. Prepared for Petroleum Technology Alliance Canada and Science and Community Environmental Knowledge Fund. Tulsa, OK.

D. E. McNabb and C. R. Swenson, *America's Water Crises*, https://doi.org/10.1007/978-3-031-27380-3

Allen, John T. 2018, Climate change and severe thunderstorms. Oxford Research: *Climate Science*, online. Available at https://doi.org/10.1093/acrefore/9780190228620.013.62.

Alley, William M., Thomas E. Reilly, and O. Lehn Franke. 1999 (and 2013 update). Sustainability of groundwater resources. USGS Survey Circular 1186. Available at https://pubs.usgs.gov/circ/circ1186/pdf/p1-5.pdf.

American Rivers. 2021. 69. Dams removed in 2020. Available at www.americanrivers.org/2021/02/69-dams-removed-in-2020/.

Ao, Yufei Z., Nathan P. Hendricks, and Landon T. Marston. 2021. Growing farms and groundwater depletion in the Kansas High Plains. *Environmental Research* 16 (8). Available at https://doi.org/10.1088/1748-9326/ac1816.

APR (American Public Radio). 2021. Water wars continue between Alabama, Georgia, Florida Available at www.apr.org/news/2021-07-23/water-wars-continue-between-alabama-georgia-florida.

Apurv, Tushar and Ximing Cai. 2020. Impact of droughts on water supply in U.S. watersheds: The role of renewable surface and groundwater resources. *Earth's Future* 8 (10). Available at https://doi.org/10.1029/2020EF001648.

Arata. Akira. 2020. Water crisis in India—issues, causes, and consequences. USPC: Union Public Service Commission reprint for the Civil Services of the Government of India. Available at https://upscfreematerials.org/water-crisis-issues-causes-and-consequences/.

Arthur, Michael and Demian Saffer. n.d. Atlantic *Coastal Plain aquifer system.* InTeGrate, PennState University. Available at www.e-education.psu.edu/earth111/node/923.

ASCE (American Society of Engineers). 2021. Report card for America's infrastructure: Our nation's drinking water infrastructure. Available at https://infrastructurereportcard.org/wp-content/uploads/2020/12/National_IRC_2021-report.pdf.

ASDSO (Association of State Dam Safety Officials). 2019. Dam failure and incident investigation guide. Available at https://damsafety-prod.s3.amazonaws.com/s3fs-public/files/Dam%20Failure%20and%20Incident%20Investigation%20Guide_V%202.0_FINALdocx.pdf.

Averyt, Kristen, James Meldrum, Peter V. Caldwell, Ge Sun, Steven G. McNulty, and Annette Huber-Lee. 2013. Sectoral contributions to surface water stress in the coterminous United States. *Environmental Research Letters* 8 (2–013). Available at https://doi.org/10.1088/1748-9326/8/3/035046/pdf.

Azuara, Josué M., Jay R. Lund, Mustafa S. Dogan, Ian Buck. 2019. Statewide effects of ending long-term groundwater overdraft in California. *Journal of Water Resources Planning and Management* 145 (9). https://doi.org/10.1061/(ASCE)WR.1943-5452.0001096.

Ball, Vanessa. 2017. The Colorado River ecosystem: Conflict and conservation. *Outdoor Project*. Available at www.outdoorproject.com/articles/colorado-river-ecosystem-conflict-and-conservation.

Bauer, Carl L. 2020. The long view of the water/energy nexus: Hydropower's first century in the U.S.A. *Natural Resources Journal* 60 (2): 173–206.

Bessire, Lucas. 2021. How did Kansas become ground zero for the imminent water crisis? *Literary Hub*. Available at https://lithub.com/how-did-kansas-become-ground-zero-for-the-imminent-water-crisis/.

Bicak, Charles J. 2002. The rigors of existence on the Great Plains: The role of water. In *Water on the Great Plains*, ed. Peter J. Longo, and David W. Yoskowitz, 3–15. Lubbock, TX: Texas Tech University.

Bischoff-Mattson, Zachary, Gillian Maree, Coleen Vogel, Amanda Lynch, David Oliver, and Deon Terblanche. 2020. Shape of a water crisis: Practitioner perspectives on urban water scarcity and 'Day Zero' in South Africa. *Water Policy* 22: 193–210.

Boin, Arjen. 2017. Crisis management. *Encyclopedia Britannica*. Available at www.britannica.com/topic/crisis-management-government.

Bormido, Hanna. 2019. These countries are the most at risk from a water crisis. Available at www.bloomberg.com/graphics/2019-countries-facing-water-crisis/.

Boulé, David. 2021. *A brief history of the navel orange in California*. Oroville, CA: Merlo Farming Group (from original in the *Sacrament Bee*). Available at www.merlofarminggroup.com/brief-history-navel-orange-california-%E2%80%93-sacramento-bee.

Bradford, Nick. 2021. *The increasing demand and decreasing supply of water*. Washington, DC: National Environmental Education Foundation. Available at www.neefusa.org/nature/water/increasing-demand-and-decreasing-supply-water.

Brantley, Susan L., Dave Yoxtheimer, Sina Arjmand, Paul Grieve, Jon Pollak, Garth T. Llewellyn, Jorge Abad, and Cesar Simon. 2014. Water resource impacts during unconventional shale gas development: The Pennsylvania experience. *International Journal of Coal Geology* 126: 140–156.

Breeze, Paul. 2014. *Power generation technologies*, 2nd ed. Amsterdam, Netherlands: Elsevier.

Brikowski, Y.H. 2008. Doomed reservoirs in Kansas, USA? Climate change and groundwater mining on the Great Plaines lead to unsustainable surface water storage. *Journal of Hydrology* 354 (1–4): 90–101.

Brooks, H.E. 2013. Severe thunderstorms and climate change. *Atmospheric Research* 123: 129–138.

Burns, Erick R., David S. Morgan, Rachael S. Peavler, and Sue C. Kahle. 2010. Three-dimensional model of the geologic framework for the Columbia Plateau Regional Aquifer System, Idaho, Oregon, and Washington. USGS Scientific

Investigations Report 2010–5246. Available at https://pubs.usgs.gov/sir/2010/5246/pdf/sir20105246.pdf.

Butts, Ed. 2021. Water system risk & resilience. *Water Well Journal*, July 15. Available at https://waterwelljournal.com/water-system-risk-and-resilience/.

Carle, David. 2016. *Water in California*. Oakland, CA: University of California Press.

CAST (Council for Agricultural Science and Technology). 2019. Aquifer depletion and potential impacts on long-term irrigated agricultural productivity. Issue Paper IP63 (February 2019). Available at www.cast-science.org/publication/aquifer-depletion-and-potential-impacts-on-long-term-irrigated-agricultural-productivity/.

Castellano, Isaac M. 2020. *Unauthorized water use and the new future of water*. Cham, Switzerland: Palgrave Macmillan.

CDC (Center for Disease Control). 2016. Types of agriculture and water use. Available at https://www.cdc.gov/healthywater/other/agricultural/types.html.

CDWR (California Department of Water Resources). 2021. California's groundwater update. Available at file:///C:/Users/David/Downloads/calgw2020_highlights.pdf.

CGS (Colorado Geological Survey). 2022. Colorado groundwater atlas. Available at https://coloradogeologicalsurvey.org/water/colorado-groundwater-atlas/#gwa00.

Chambers, Jeanne C. 2008. Climate change and the Great Basin. Reno, NV: USDA Forest Service, Rocky Mountain Research Station, Reno, NV.

Chang, Fi-John., Chien-Wei. Huang, Su-Ting. Cheng, and Li-Chiu. Chang. 2017. Conservation of groundwater from over-exploitations: Scientific analyses for groundwater resources management. *Science of the Total Environment* 598: 828–838.

Christensen, Niklas S., Andrew Wood, Nathalie Voisin, Dennis P. Lettenmaier, and Richard N. Palmer. 2014. The effects of climate change on the hydrology and water resources of the Colorado River basin. *Climatic Change* 62: 337–363. https://doi.org/10.1023/B:CLIM.0000013684.13621.1f.

CIA (Central Intelligence Agency). 2022. Filed listing: Climate. Available at www.cia.gov/the-world-factbook/field/climate/.

Cilluffo, Anthony, and Neil G. Ruiz. 2019. World's population is projected to nearly stop growing by the end of the century. *Pew Research Center*. Available at www.pewresearch.org/fact-tank/2019/06/17/worlds-population-is-projected-to-nearly-stop-growing-by-the-end-of-the-century/.

City of New Orleans. 2021. *Infrastructure failure: Water systems*. Hazard Mitigation Plan. Available at https://ready.nola.gov/hazard-mitigation/hazards/infrastructure-failure-water-systems/.

Clark, Brian R., Rheannon M. Hart, and Jason J. Gurdak. 2011. Groundwater availability of the Mississippi embayment. USGS Professional Paper 1785. https://pubs.usgs.gov/pp/1785/.

Climate Impact Group. 2020. *How will climate change effect water in Washington? Section 6*. Seattle, WA: University of Washington, College of the Environment. Available at https://cig.uw.edu/wp-content/uploads/sites/2/2020/12/snoveretalsok2013sec6.pdf.

Climate.gov. 2021. Climate change to increase water stress in many parts of U.S. Available at https://www.climate.gov/news-features/featured-images/climate-change-increase-water-stress-many-parts-us.

Cline, Michael. 2021. 2021 population estimates show NC growth continues. Release of NCOSBM (North Carolina Office of State Budget and Management). www.osbm.nc.gov/blog/2021/12/22/2021-population-estimates-show-nc-growth-continues.

Cline, Sara and Rebecca Boone. 2021. Seattle hits record high amid Pacific Northwest heat wave. AP (Associated Press Online), June 27. www.aol.com/news/historic-heat-wave-blasts-pacific-193908457-164303551.html.

Clow, David M. 2011. Changes in the timing of snowmelt and streamflow in Colorado: A response to recent warming. *Journal of Climate* 23 (9): 2293–2306.

Cross, Wyatt F., John LaFave, Alex Leone, Whitney Lonsdale, Alisa Royem, Tom Patton, and Stephanie McGinnis. 2017. Water and climate change in Montana. Chapter 3: 2017. *Montana Climate Assessment*. Available at https://leg.mt.gov/content/Committees/Interim/2019-2020/Water-Policy/Meetings/Sept-2019/MontanaClimateAssessment-Water.pdf.

CRT (Climate Resilience Toolkit). 2021. The northern Great Plains. Available at https://toolkit.climate.gov/regions/northern-great-plains.

Currie, Janet, Michael Greenstone, and Katherine Meckel. 2017. Hydraulic fracturing and infant health: New evidence from Pennsylvania. *Science Advances*, 3 (12). Available at https://doi.org/10.1126/sciadv.1603021.

Curtis, Jennifer A., Erick R. Burns, and Roy Sando. 2020. Regional patterns in hydrologic response, a new three-component metric for hydrologic analysis and implications for ecohydrology, Northwest Volcanic Aquifer Study Area, USA. *Journal of Hydrology: Regional Studies*. Available at https://doi.org/10.1016/j.ejrh.2020.100698.

Dai, Aiguo and National Center for Atmospheric Research Staff (eds.). 2020 (Last modified 12 December 2019). The climate data guide: Palmer Drought Severity Index (PDSI). Available at https://climatedataguide.ucar.edu/climate-data/palmer-drought-severity-index-pdsi.

Data.gov. 2020. Dams (national). U.S. Department of Transportation Data. Available at https://catalog.data.gov/dataset/dams-national.

Deeds, Neil E., and Scott Hamline. 2015. Final conceptual model report for the high plains aquifer system groundwater availability model. TWDB Report. Available at www.twdb.texas.gov/groundwater/models/gam/hpas/HPAS_GAM_Conceptual_Report.pdf.

Degnan, James, and MaryLynn Musgrove. 2021. Groundwater quality in the Colorado Plateaus Aquifers, Western United States. U.S. Geological Survey (USGS) National Water-Quality Assessment. USGS Fact Sheet 2021-3012. Available at https://pubs.usgs.gov/fs/2021/3012/fs20213012.pdf.

De Gisi, Sabino, Patrizia Casella, Carmela Maria Cellamare, Marco Ferraris, Luigi Petta, and Michele Notarnicola. 2017. Encyclopedia of sustainable technologies, 53–68.

DEP (Pennsylvania Department of Environmental Protection). 2018. Sources of groundwater contamination. 2022 Water Quality Report. Available at 2018_PA_Integrated_WQ Report_GW_contam_table.pdf.

DEP (Pennsylvania Department of Environmental Protection). 2022. 2022 integrated water quality report. Available at https://storymaps.arcgis.com/stories/b9746eec807f48d99decd3a583eede12.

DePaul, Vincent T., Donald E. Rice, and Otto S. Zapecza. 2008. Water level in aquifers of the Atlantic Coastal Plain, predevelopment to 2000. USGS Document. Available at www.usgs.gov/publications/water-level-changes-aquifers-atlantic-coastal-plain-predevelopment-2000.

Dettinger, Michael D., and Michael L. Anderson. 2015. Storage in California's reservoirs and snowpacks in this time of drought. *San Francisco Estuary and Watershed Science* 13 (2). Available at https://doi.org/10.15447/shews.2015v13issrart1.

Devineni, Naresh, Upmanu Lall, Elius Etienne, Daniel Shi, and Chen Xi. 2015. America's water risk: Current demand and climate variability. *Geophysical Research Letters* 42 (7): 2285–2293. https://doi.org/10.1002/2015GL063487.

De Villiers, Marq. 2000. *Water: The fate of our most precious resource*. Boston: Houghton Mifflin.

Dieter, Cheryl A., Molly A. Maupin, Rodney R. Caldwell, Melissa A. Harris, Tamara I. Ivahnenko, John K. Lovelace, Nancy L. Barber, and Kristin S. Linsey. 2018. Estimated use of water in the United States in 2015. USGS Circular 1441. Available at https://pubs.er.usgs.gov/publication/cir1441.

Dimick, Dennis. 2014. If you think the water crisis can't get any worse, wait until the aquifers are drained. Available at https://news.nationalgeographic.com/news/2014/08/140819-groundwater-california-drought-aquifers-hidden-crisis/.

Ding, Grace K.C., and Sumita Ghosh. 2017. Sustainable water management—A strategy for maintaining future water. *Encyclopedia of Sustainable Technologies* 4. Available at https://doi.org/10.1016/8975-0-12-409548-991201741-X.

Dogan, Mustafa S., Ian Buck, Josue Medellin-Azuara, and Jay R. Lund. 2019. Statewide effects of ending long-term groundwater overdraft in California. *Journal of Water Resources Planning and Management* 145 (9). https://doi.org/10.1061/(ASCE)WR.1943 5452.0001096.

Doršner, Kamals (Modified by Matthew R. Fisher). 2018. Water supply problems and solutions, Chapter 7. *Essentials of Environmental Science*. Available at https://openoregon.pressbooks.pub/envirobiology/chapter/7-2-water-supply-problems-and-solutions/.

Drought.ca.gov. 2022a. California drought upgrade. February 22, 2022. Available at https://drought.ca.gov/media/2022a/02/CA-Drought-Update-2-22-22.pdf.

Drought.ca.gov. 2022b. California drought upgrade, April 11. Available at https://drought.ca.gov/media/2022b/02/CA-Drought-Update-4-1-22.pdf.

Dudley, Robert W., and Glenn A. Hodgkins. 2002. Trends in streamflow, river ice, and snowpack for coastal river basins in Maine during the 20th century. Water-Resources Investigations Report 2002-4245. https://doi.org/10.3133/wri20024245.

Dudley, Robert W., Glenn A. Hodgkins, James B. Shanley, and Thomas J. Mack. 2010. Quantifying effects of climate change on the snowmelt-dominated groundwater resources of Northern New England. U.S. Geological Survey Fact Sheet 2010–3104. Available at https://pubs.usgs.gov/fs/2010/3104.

Dudley, Robert W., Glenn A. Hodgins, M.R. McHale, M.J. Kolian, and Benjamin Renard. 2017. Trends in snowmelt-related streamflow timing in the coterminous United States. *Journal of Hydrology* 547: 208–221.

EGLE (Michigan Department of Environment, Great Lakes and Energy). 2020. *Preliminary report on the Edenville Dam Failure and response efforts*. Lansing, Michigan: Environmental Assistance Center. Available at www.michigan.gov/documents/egle/egle-EdenvilleDamPreliminaryReport_700997_7.pdf.

EIA (U.S. Energy Information Administration). 2021. Hydropower explained. Available at www.eia.gov/energyexplained/hydropower/where-hydropower-is-generated.php.

EIC (Illinois Environmental Council). 2021. Protecting clean water in Illinois is essential to community, economic and ecosystem health. Available at https://ilenviro.org/clean-water/.

England, Thomas. n.d. *History of Ogallala*. Ogallala, NB: Keith County News. Available at www.keithcountyne.gov/business/keith_county_historical_photos_pre_1950.php.

Environment Go! 2021. 15 Main causes of pollution. Available at https://environmentgo.com/main-causes-of-water-pollution/.

EPA (Environmental Protection Administration). n.d. Climate impact in the Midwest. Available at https://climatechange.chicago.gov/climate-impacts/climate-impacts-midwest.

EPA (Environmental Protection Administration). n.d. Mew England's drinking water. Available at www3.epa.gov/region1/eco/drinkwater/ne_drinkwater.html#main-content.

EPA (Environmental Protection Administration). 2011. The effects of mountaintop mines and valley fills on aquatic ecosystems of the central Appalachian coalfields. Available at https://ofmpub.epa.gov/eims/eimscomm.getfile?p_download_id=501593.

EPA (Environmental Protection Administration). 2013. Saving water in New Mexico. Available at www.epa.gov/sites/default/files/2017-02/documents/ws-ourwater-new-mexico-state-fact-sheet.pdf.

EPA (Environmental Protection Administration). 2014. EPA New England regional climate adaptation plan. EPA publication number 100K14001H.

EPA (Environmental Protection Administration). 2015. Assessment of the potential impacts of hydraulic fracturing for oil and gas on drinking water resources: *Executive Summary*. Working draft edition only, Available at www.epa.gov/sites/default/files/2015-06/documents/hf_es_erd_jun2015.pdf.

EPA (Environmental Protection Administration). 2016. Saving water in Oregon. Available at www.epa.gov/sites/default/files/2017-02/documents/ws-ourwater-oregon-state-fact-sheet.pdf.

EPA (Environmental Protection Administration). 2017. What climate change means for Montana. Available at https://19january2017snapshot.epa.gov/sites/production/files/2016-9/documents/climate-change-mt.pdf.

EPA (Environmental Protection Administration). 2017a. Climate impacts in the Northwest. Available at https://19january2017snapshot.epa.gov/climate-impacts/climate-impacts-northwest_.html.

EPA (Environmental Protection Administration). 2017b. State water facts. Available at https://19january2017snapshot.epa.gov/www3/watersense/our_water/state_facts.html#content.

EPA (Environmental Protection Administration). 2017c. Saving water in Washington. Available at www.epa.gov/sites/default/files/2017c-02/documents/ws-ourwater-washington-state-fact-sheet.pdf.

EPA (Environmental Protection Administration). 2018a. Agricultural water use. *EPA EnviroAtlas*. Available at https://enviroatlas.epa.gov/enviroatlas/DataFactSheets/pdf/ESN/Agriculturalwateruse.pdf.

EPA (Environmental Protection Administration). 2018b. What are the trends in the quality of drinking water and their effects on human health? Available at www.epa.gov/report-environment/drinking-water.

EPA (Environmental Protection Administration). 2020. National Pollutant Discharge Elimination System (NPDES): NPDES Stormwater Program. Available at www.epa.gov/npdes/npdes-stormwater-program.

EPA (Environmental Protection Administration). 2021a. Water reuse research Available at www.epa.gov/water-research/water-reuse-research.

EPA (Environmental Protection Administration). 2021b. Building sustainable water infrastructure. Available at www.epa.gov/sustainable-water-infrastructure/building-sustainable-water-infrastructure.

EPA (Environmental Protection Administration). 2021c. A closer look: Temperature and drought in the Southwest. Available at www.epa.gov/climate-indicators/southwest.

EPA (Environmental Protection Administration). 2021d. Climate change indicators: Weather and climate. Available at www.epa.gov/climate-indicators/weather-climate.

EPA (Environmental Protection Administration). 2021e. Climate change indicators: Snowpack. Available at www.epa.gov/climate-indicators/climate-change-indicators-snowpack.

EPA (Environmental Protection Administration). 2021f. A closer look: Land loss along the Atlantic Coast. Available at www.epa.gov/climate-indicators/atlantic-coast.

EPA (Environmental Protection Administration). 2022a. History of the Clean Water Act. Available at www.epa.gov/laws-regulations/history-clean-water-act.

EPA (Environmental Protection Administration). 2022a. Basic information about water reuse. Available at www.epa.gov/waterreuse/basic-information-about-water-reuse.

EPA (Environmental Protection Administration). 2022b. What is green infrastructure? Available at www.epa.gov/green-infrastructure/what-green-infrastructure.

EPA (Environmental Protection Administration). 2022c. Harmful algal blooms. Available at www.epa.gov/nutrientpollution/harmful-algal-blooms.

EPA (Environmental Protection Administration). 2022d. What is radon? Available at www.epa.gov/radon/what-radon.

Esterhay, Carl A. 1981. Restoring the water quality of the Great Lakes: The Joint Commitment of Canada and the United States, 4 Can.-U.S. L.J. 208. Available at: https://scholarlycommons.law.case.edu/cuslj/vol4/iss/22.

Fahey, Jim. 2022. Wyoming water supply outlook—May 1, 2022: Hydrologic forecast discussion and additional graphics. Available at https://storymaps.arcgis.com/stories/ece3959a21844692816bbe8da25c64a8.

Famiglietti, James S. 2014. The global groundwater crisis: Groundwater depletion the world over poses a far greater threat to global water security than is currently acknowledged. *Natural Climate Change* 4 (2114): 945–948.

Fang, Wendi, Ying Peng, Derek Muir, Jun Lin, and Xiaowei Zhang. 2019. A critical review of synthetic chemicals in surface waters of the US, the EU and China. *Environment International* 131: 104994.

FEMA (Federal Emergency Management Administration). 2020. National Inventory of Dams. Available at www.fema.gov/emergency-managers/risk-management/dam-safety/national-inventory-dams.

FEMA (Federal Emergency Management Administration). 2021. National Inventory of Dams. Available at www.fema.gov/emergency-managers/risk-management/dam-safety/national-inventory-damst.

Firley, Stephanie Worley. 2012. *The water web: More about WaSSi*. USDA; CompassLine. Available at www.srs.fs.usda.gov/compass/2012/02/02/the-water-web-more-about-wassi/

Fisk, Charles. 2020. K-means cluster analysis identification of idealized relative anomaly patterns in annual total precipitation across New England states' NCDC climate divisions encompassing the 1895–2018 period of record. November 2019 26th conference on Probability and Statistics, for the 100th American Meteorological Society's January 20, 2020 Annual Meeting at Boston, MA.

Foti, Romano, Jorge A. Ramirez, and Thomas C. Brown. 2014. A probabilistic framework for assessing vulnerability to climate variability and change: The case of the US water supply system. *Climatic Change* 125: 413–427. https://doi.org/10.1007/s10584-014-1111-6.

Frankel, Jeremy. 2018. Crisis on the high plains: The loss of America's largest aquifer—The Ogallala. *University of Denver Water Law Review at the Sturm College of Law*. Available at http://duwaterlawreview.com/crisis-on-the-high-plains-the-loss-of-americas-largest-aquifer-the-ogallala/.

Freyman, Monika. 2014. *Hydraulic fracturing & water stress: Water demand by the numbers*. Boston, MA: Ceres.

Gallagher, Katherine. 2021. States with the biggest agriculture industry. *Stacker*, April 3. Available at https://stacker.com/stories/3183/states-biggest-agriculture-industry.

GAO (U.S. Government Accountability Office). 2014. Freshwater: Supply concerns continue, and uncertainties complicate planning. Available at www.gao.gov/assets/gao-14-430.pdf.

GAO (U.S. Government Accountability Office). 2016. Municipal water scarcity: Using technology to improve distribution system efficiency and tap nontraditional water sources. Available at www.gao.gov/assets/gao-16-474.pdf.

Gauron, Trevor, and Rebecca Coulter. 2021. Self-audits an efficient way to maintain stormwater management programs and bring communities to a place of preparedness. *Stormwater* 22 (5): 22–25.

George, Peter G., Robert E. Mace and Rima Petrossian. 2011. Aquifers of Texas. TWDB Report 389. Available at file:///C:/Users/David/Downloads/R380_AquifersofTexas%20(1).pdf.

Gleick, Peter H. 1996. Basic water requirements for human activities: Meeting basic needs. *Water International* 21: 83–92.

Global Water Forum. 2015. Water for thermal power plants: Understanding a piece of the water energy nexus. https://globalwaterforum.org/2015/06/22/water-for-thermal-power-plants-understanding-a-piece-of-the-water-energy-nexus/.

Gollehon, Noel, and Bernadette Winston, 2013. *Groundwater irrigation and water withdrawals: The Ogallala Aquifer Initiative.* Washington, DC: US Department of Agriculture. Available at www.nrcs.usda.gov/Internet/FSE_DOCUMENTS/stelprdb1186440.pdf.

Green, Donald E. 1992. A history of irrigation technology used to exploit the Ogallala aquifer. In *Groundwater exploitation in the high plains*, ed. David E. Kromm and Stephen E. White, Chapter 2: 28–43. Lawrence, Kansas: University Press of Kansas.

Haacker, Erin M. K., Kayla A. Cotterman, Samuel J. Smidt, Anthony D. Kendall, and David W. Hyndman. 2016. Effects of management areas, drought, and commodity prices on groundwater decline patterns across the high plains aquifer. *Agricultural Water Management* 218: 259–273.

Hall, Dorothy K., Christopher J. Crawford, Nicolo E. DeGiolamo, George A. Riggss, and James L. Foster. 2015. Detection of earlier snowmelt in the Wind River Range, Wyoming using Landsat imagery, 1972–2013. *Remote Sensing of Environment* 162: 45–54.

Hanasaki, Naota, S. Kanae, T. Oki, K. Masuda, K. Motoya, N. Shirakawa, Y. Shen, and K. Tanaka. 2008. An integrated model for the assessment of global water resources—Part 1: Model description and input meteorological forcing. *Hydrology and Earth System Sciences* 12 (2): 1007–1025.

Harrington, Lisa M. B., Max Lu and Donald E. Kromm. 2010. Milking the plains: Movement of large dairy operations into southwestern Kansas. *The Geographical Review* 100 (4): 1–19.

Haseeb, Jamal. 2017. What is consumptive use of water and factors affecting it. Available at www.aboutcivil.org/consumptive-use-of-water.html.

Hatchett, Benjamin J. 2021. Seasonal and ephemeral snowpacks of the conterminous United States. *Hydrology* 8 (1): 32. https://doi.org/10.3390/hydrology8010032.

Hearney, James P., David J. Sample, and Leonard Wright. 2002. Costs of urban stormwater control. U.S. EPA, National Risk Management Research Laboratory. Available at https://play.google.com/store/books/details/Costs_of_Urban_Stormwater_Control?id=q3EFWwJJGrMC&gl=US.

Heatherly, Larry G. 2021. *Irrigation and aquifer depletion*. Starkville, MS: Mississippi Soybean Promotion Board (MSPB). Available at www.mssoy.org/article/irrigation-and-aquifer-depletion.

Hedberg, Jonas, Eva Blomberg, and Inger O. Wallinder. 2019. In the search for nanospecific effects of dissolution of metallic nanoparticles at freshwater-like conditions: A critical review. *Environmental Science and Technology* 53 (8): 4030–4044. https://doi.org/10.1021/acs.est.8b05012.

Heggie, Jon. 2020. Why is America running out of water? Available at www.nationalgeographic.com/science/article/partner-content-americas-looming-water-crisis.

Hernandez, Jarro H., Prasanna H. Marek, Terry A. Howell, and Wonsook Ha. 2013. Groundwater levels in Northern Texas High Plains: Baseline for existing agricultural management practices. *Texas Water Journal* 4 (1): 22–34.

Herring, David. 2021. Climate change: Global temperature projections. NOAA: Climate.gov. Available at www.climate.gov/news-features/understanding-climate/climate-change-global-temperature-projections.

Hinds, Julian. 1932. The Colorado River Aqueduct. *The Military Engineer* 24 (134): 15–19.

Ho, Michelle, Upmanu Lall, Maura Allaire, Naresh Devineni, Hyun H. Kwon, Indrani Pal, David Raff, and Dave Wagner. 2017. The role of dams in the United States of Americas. *Water Resources Research*.

Hodgkins, Glenn A., Robert W. Dudley, and Thomas G. Huntington. 2005. Changes in the number and timing of days of ice-affected flow on Northern New England Rivers, 1930–2000. *Climate Change* 71: 319–340.

Holden, Emily. 2020. Michigan: Threat of toxic contamination looms after dam failures trigger flooding. *The Guardian* (US Edition), May 20. Available at www.theguardian.com/us-news/2020/may/20/michigan-flooding-dam-toxic-contamination-threat.

Hoover, Joseph H. 2011. The vital role of electrical energy for Arizona ester services. In *The water-energy nexus in the American West*, ed. Douglas S. Kenney and Robert Wilkinson, 165–176. Cheltenham, UK: Edward Elgar.

Hrozencik, R. Aaron, and Marcel Aillery. 2021. Trends in U.S. irrigated agriculture: Increasing resilience under water supply scarcity. U. S. Department of Agriculture Economic research Service Bulletin Number 229.

IDNR (Indiana Department of Natural Resources). 2021. Potentometric surface mapping. Available at www.in.gov/dnr/water/ground-water-wells/assessment-maps-and-publications/potentiometric-surface-mapping-148000/potentiometric-surface-mapping-148000-overview/

Iowa DNR (Iowa Department of Natural Resources). 2022. Water summary update: A Snapshot of water resource trends for August 2022. Issue 135. Available at /www.iowadnr.gov/Portals/idnr/uploads/Water%20Summary%20Update/20220908wsu.pdf.

Ivahnenko, John K. Lovelace, Nancy L. Barber, and Kristin S. Linsey, Ferlemann, Kyle G. 2019. *Disaster response Smartbook 3: Disaster preparedness.* Lakeland, FL: The Lighting Press.

Kahle, Sue C., and John J. Vaccaro. 2015, Groundwater resources of the Columbia Plateau Regional Aquifer System: U.S. Geological Survey Fact Sheet 2015-3063. Available at https://doi.org/10.3133/fs20153063.

KDOA (Kansas Department of Agriculture). 2016. Water use reporting. Available at https://agriculture.ks.gov/divisions-programs/dwr/water-appropriation/water-use-reporting.

KDOA (Kansas Department of Agriculture). 2017. Irrigation water use in Kansas. Available at https://agriculture.ks.gov/docs/default-source/dwr-water-appropriation-documents/ks_irr_2017.pdf?sfvrsn=6b8d88c1_0.

Kottasová, Ivana, Henrik Pettersson and Krystina Shveda. 2020. The world's largest wetlands are on fire. That's a disaster for all of us. *CNN*, November 13. Available at www.cnn.com/2020/11/13/americas/pantanal-fires-climate-change-intl/index.html.

Hutson, Susan S., Nancy L. Barber, Joan F. Kenny, Kristin S. Linsey, Deborah S. Lumia, and molly A. Maupin. 2004. Estimated use of water in the United States in 2000. U.S. Geological Survey Circular 1268.

James, Ian and Steve Reilly. 2015 (updated 2016) Pumped beyond limits, many U.S. aquifers in decline. *Desert Sun*, December 10 (online). www.desertsun.com/story/news/environment/2015/12/10/pumped-beyond-limits-many-us-aquifers-decline/76570380/.

Januta, Andrea, and Daniel Trotta. 2021. Hoover Dam reservoir hits record low, in sign of extreme western U.S. drought. *U.S. News*, June 10. Available at www.reuters.com/world/us/hoover-dam-reservoir-hits-record-low-sign-extreme-western-us-drought-2021-06-10/.

Johannis, Mary, Lorraine E. Flint, Michael D. Dettinger, Alan L. Flint, and Regina Ochoa. 2016. The role of snowpack, rainfall, and reservoirs in buffering California against drought effects. *USGS: Environmental Science.* Available at https://doi.org/10.3133/fs20163062.

Katz, David, Amir Grinstein, Ann Kronrod, and Udi Nisan. 2016. Evaluating the effectiveness of a water conservation campaign: Combining experimental and field methods. *Journal of Environmental Management* 180: 335–347.

Kelly, Morgan. 2020. *Expansion, environmental impacts of irrigation y 23050 greatly underestimated.* Princeton Environmental Institute. Available at www.princeton.edu/news/2020/05/05/expansion-environmental-impacts-irrigation-2050-greatly-underestimated.

Kenny, Douglas, and Robert Wilkinson, eds. 2011. *The water-energy nexus in the American West.* Cheltenham, UK: Edward Elgar.

KGS (Kansas Geological Survey). 2021. Groundwater levels fall in central and western Kansas. Available at www.kgs.ku.edu/General/News/2021/groundwater_levels_fall.html.

Kingsbury, James A., Laura M. Bexfield, Terri Arnold, MaryLynn Musgrove, Melinda L. Erickson, James R. Degnan, Anthony J. Tesoriero, Bruce D. Lindsey, and Kenneth Belitz. 2021. Groundwater-quality and select quality-control data from the National Water-Quality Assessment Project, January 2017 through December 2019. USGS Data Series 1136. Available at https://pubs.usgs.gov/ds/1136/ds1136.pdf.

Konikow, L.F. 2013. Groundwater depletion in the United States (1900−2008): U.S. Geological Survey Scientific Investigations Report 2013−5079. Available at http://pubs.usgs.gov/sir/2013/5079.

Korzendorfer, B.A., and M.A. Horn. 1990. Estimated use of water in the New England states. USGS Water-Resources Investigations Report 94-4252.

Kresse, Timothy M., Phillip D. Hays, Katherine R. Merriman, Jonathan A. Gillip, D. Todd Fugitt, Jane L. Spellman, Anna M. Nottmeier, Drew A. Westerman, Joshua M. Blackstock, and James L. Battreal. 2014. Aquifers of Arkansas—Protection, management, and hydrologic and geochemical characteristics of groundwater resources in Arkansas: U.S. Geological Survey Scientific Investigations Report 2014–5149, 334 . Available at https://doi.org/10.3133/sir20145149.

Kromm, David E. and Stephen E. White. 1992a. The high plains Ogallala region. In *Groundwater exploitation in the high plains*, ed. David E. Kromm and Stephen E. White, Chapter 1: 1–27. Lawrence, Kansas: University Press of Kansas.

Kromm, David E., and Stephen E. White. 1992b. Groundwater Problems. In *Groundwater exploitation in the high plains*, ed. David E. Kromm and Stephen E. White, Chapter 3: 44–63. Lawrence, Kansas: University Press of Kansas.

Lall, Upmanu, Tanya Heikkila, Casey Brown, and Tobias Siegfried. 2008. Water in the 21st century: Defining the elements of global crises and potential solutions. *Journal of International Affairs* 61 (2): 1–17.

Le Brocque, Andrew F., Jarrod Kath and Katheryn Reardon-Smith. 2018. Chronic groundwater decline: A multidecadal analysis of groundwater trends under extreme climate cycles. *Journal of Hydrology* 561: 976–986. Available at https://doi.org/10.1016/j.jhydrol.2018.04.05.

Lee, Uisung, Jeongwoo Han, Amgad Elgowainy, and Michael Wang. 2018. Regional water consumption for hydro and thermal electricity generation in the United States. *Applied Energy* 210: 661–672.

Leltier, Laurel. 2016. Pennsylvania fracking water contamination much higher than reported. *EcoWatch*, February 4. Available at https: www.ecowatch.com/pennsylvania-fracking-water-contamination-much-higher-than-reported-1882166816.html#:~:text=Officially%2C%20Pennsylvania%20reports%20271%20confirmed,within%20six%20months%20of%20drilling.

Liddel, Michael, and Melissa Yencho (eds.). 2021. *Fisheries of the United States*. Silver Springs, MD: National Marine Fisheries Service, NOAA (National Oceanic and Atmospheric Administration). Series No. 2019. Available at https://media.fisheries.noaa.gov/2021-05/FUS2019-FINAL-webready-2.3.pdf?null.

Linquip. 2021. 10 types of water pollution in 2021. Available at www.linquip.com/blog/types-of-water-pollution/.

Lombard, Pamela J., Jane R. Barclay and Dee-Ann E. McCarthy. 2020. Drought in New England. USGS Open-File Report 2020-1148. https://doi.org/10.3133/ofr20201148.

Longo, Peter J., and David W. Yoskowitz (eds.). *Water on the Great Plains*. Lubbock, TX: Texas Tech University.

Lovelace, John K., Martha G. Nielsen, Amy L. Read, Chid J. Murphy and Molly A. Maupin. 2020. Estimated groundwater withdrawals from principal aquifers in the United States, 2015. USGS Circular 1464. Available at https://pubs.er.usgs.gov/publication/cir1.

LSRD (Lower Snake River Dams Options). 2022. Lower Snake River Dams: Benefit replacement draft report. Available at www.lsrdoptions.org/wp-content/uploads/2022/06/LSRD-Benefit-Replacement-Draft-Report_2022-0609.pdf.

Luce, Charles H., Viviana Lopez-Burgos, and Zachary Holden. 2014. Sensitivity of snowpack storage to precipitation and temperature using spatial and temporal analog models. *Water Resources Research* 5 (12): 9447–9462. https://doi.org/10.1002/2013WR014844.

Mack, Elizabeth Q., and Sarah Wrase. 2017a. A burgeoning crisis? A nationwide assessment of the geography of water affordability in the United States. *PLoS ONE* 12 (1). Available at https://doi.org/10.1371/journal.pone.0169488.

Mack, Elizabeth A., and Sarah Wrase. 2017b. Correction: A burgeoning crisis? A nationwide assessment of the geography of water affordability in the United States. *PLoS ONE* 12 (4, April): 21, e0176645. Gale Academic OneFile. Accessed 1 August 2022.

Maddocks, Andrew, and Paul Reig. 2014. *World's 18 most water-stressed rivers*. Washington, DC: World Resources Institute, May 20. Available at www.wri.org/insights/worlds-18-most-water-stressed-rivers.

Malmqvist, Björn, and Simon Rundle. 2002. Threats to the running water ecosystems of the world. *Environmental Conservation* 29 (2): 134–153. Available at https://doi.org/10.1017/S0376892902000097.

Manda, Ales K., and Wendy A. Klein. 2014. Rescuing degrading aquifers in the Central Coastal Plain of North Carolina (USA): Just process, effective groundwater management policy, and sustainable aquifers. *Water Resource Research* 50: 5662–5677. https://doi.org/10.1002/2013WR015242.

Mandel, S., and Z. Shiftan. 1921. *Groundwater resources: Investigation and development*. New York: Academic Press.

MARISA (Mid-Atlantic RISA Team). 2019. Chesapeake Bay watershed climate impacts summary and outlook: Fall 2019. Available at www.midatlanticrisa.org/climate-summaries/2019/12.html.

Martin, Anna Delgado. 2015. Water for thermal power plants: Understanding a piece of the water energy nexus. Available at https://globalwaterforum.org/2015/06/22/water-for-thermal-power-plants-understanding-a-piece-of-the-water-energy-nexus/.

Mass.gov. 2022. Drought status. Available at www.mass.gov/info-details/drought-status#current-drought-status.

Maupin, Molly A., and Nancy L. Barber. 2005. Estimated withdrawals from principal aquifers in the United States. USGS Circular 1279. Available at https://pubs.usgs.gov/circ/2005/1279/ (online).

MCA (Maine Conservation Alliance). 2022. Clean water. Available at https://protectmaine.org/priorities/clean-water#:~:text=Maine%20waters%20are%20too%20often,cannot%20handle%20heavy%20rain%20storms.

McGraw, George, and Radhike Fox. 2019. Closing the water gap in the United States: A National Action Plan. Available at Dig-Deep_Closing-the-Water-Access-Gap-in-the-United-States_DIGITAL_compressed.pdf.

McElvaine, Robert S. 1984. *The Great Depression: America, 1929–1941*. New York: Times Books.

McGuire, Virginia L. 2017. Water-level and recoverable water in storage changes, high plains aquifer, predevelopment to 2015 and 2013–15: U.S. Geological Survey Scientific Investigations Report 2017–5040. Available at https://doi.org/10.3133/sir20175040.

McKee, Governor Dan. 2020. State-wide drought advisory. Available at https://governor.ri.gov/press-releases/governor-mckee-issues-statewide-drought-advisory.

McNabb, David E. 2019. *Global pathways to water sustainability*. London: Palgrave-Springer.

McNabb, David E. 2017. *Water resource management: sustainability in an era of climate change*. London: Palgrave-Springer.

MCTA (Minnesota Center for Environmental Advocacy). 2022. Clean water. Available at https://www.mncenter.org/clean-water.

Melosi, Martin V. 2011. *Precious commodity: Providing water for America's cities*. Pittsburg, PA: University of Pittsburgh Press.

Mickley, Michael. 2018. Updated and extended survey of U.S. municipal desalination plants. U.S. Department of the Interior Desalination and Water Purification Research and Development Program Report No. 207. Denver, Colorado: U.S. Department of the Interior Bureau of Reclamation technical Services Center.

Miller, Dan, and Ken Semmens. 2002. *Waste management in aquiculture*. Morgantown, WV: West Virginia University. Available at https://freshwater-aquaculture.extension.org/wp-content/uploads/2019/08/WasteManagemetninAquaculture.pdf.

Mote, Philip W., Alan F. Hamlet, Martyn P. Clark, and Dennis P. Lettenmaier. 2005. Declining mountain snowpack in western North America. *Bulletin of the American Meteorological Society* 86 (1): 39–40.

Mote, Philip W., David E. Rupp, Sihan Li, Darrin J. Sharp, Friederike Otto, Peter F. Uhe, Mu Xiao, Dennis P. Lettenmaier, Heidi Cullen, and Myles R. Allen. 2016. Perspectives on the causes of exceptionally low 2015 snowpack in the western United States. AGU Publications. *Geophysical Research Letter* 43: 10980–10988. https://doi.org/10.1002/2016GL069965.

Mullin, Cristina A., Christine J. Kirchhoff, Guiling Wang, and Penny Vlahos. 2020. Future projections of water temperature and thermal stratification in Connecticut reservoirs and possible implications for Cyanobacteria. *Water Resources Research* 56: 11. Online. https://doi.org/10.1029/2020WR027185.

Multipure.com. n.d. Top 10 states with the worst public tap water in the United States. Available at www.multipure.com/purely-social/science/top-10-states-worst-public-water-ratings-united-states/.

NASA (National Aeronautics and Space Administration). 2000. Drought: The creeping disaster. https://earthobservatory.nasa.gov/features/DroughtFacts.

NASA (National Aeronautics and Space Administration). n.d. Ocean worlds: Water in the solar system and beyond. Available at https://www.nasa.gov/specials/ocean-worlds/#banner.

NASA. (National Aeronautics and Space Administration). 2010. *The water planet*. NASA Earth Observatory. Available at https://earthobservatory.nasa.gov/images/46209/the-water-planet.

NCA (National Climate Assessment). Chapter 3: Water. Available at https://nca2018.globalchange.gov/chapter/3/.

NCEI (National Centers for Environmental Information). 2022. U.S. Climate regions NOAA release. Available at www.ncei.noaa.gov/monitoring-references/maps/us-climate-regions.

NCCOS (National Center for Coastal Ocean Science). 2017. Phytoplankton Monitoring Network (PMN). Available at https://coastalscience.noaa.gov/research/stressor-impacts-mitigation/pmn/.

NDEA (Nebraska Department of Environment and Energy). 2021. Irrigated acres in Nebraska 1964–2020. Available at https://neo.ne.gov/programs/stats/inf/73b.html.

NDMC (National Drought Mitigation Center). 2022. Measuring drought. Available at https://drought.unl.edu/ranchplan/DroughtBasics/WeatherandDrought/MeasuringDrought.aspx.

Newcomb, Tim. 2020. The 8 most awe-inspiring dams in America. *Popular Mechanics* (online), October 9. Available at www.popularmechanics.com/technology/infrastructure/g2837/7-most-serious-dams-us/.

Nielsen, Martha G., and Stephen M. Westenbroek. 2020. Groundwater recharge estimates for Maine using a Soil-Water-Balance model—25-year average, range, and uncertainty, 1991 to 2015. USGS Scientific Investigations Report 2019–5125.

NGWA (National Ground Water Association). 2020a. Groundwater use in California. Available at https://www.ngwa.org/docs/default-source/default-document-library/states/cal.pdf.

NGWA (National Ground Water Association). 2020b. Groundwater use in Kansas. Available at https://www.ngwa.org/docs/default-source/default-document-library/states/ks.pdf

NIDIS (National Integrated Drought Information System). 2022a. Water supply. Available at www.drought.gov/topics/water-supply#definition.

NIDIS (National Integrated Drought Information System). 2022b. Drought status update for the northeast. Available at www.drought.gov/drought-status-updates/drought-status-update-northeast-1.

NIDIS (National Integrated Drought Information System). 2022c. Research spotlight: Climate-driven megadrought. Available at www.drought.gov/research-spotlight-climate-driven-megadrought.

NIDIS (National Integrated Drought Information System). 2022d. Missouri River Basin. Available at NIDIS www.drought.gov/dews/missouri-river-basin.

Nie, Wanshu, Benjamin F. Zaitchik, Matthew Rodel, Sujay V. Kumar, Mathew C. Anderson, and Christopher Hain. 2018. Groundwater withdrawals: Under drought, reconciling GRACE and land surface models in the United States High Plains aquifer. *Water Resources Research* 54 (8): 5282–5299.

NOAA (National Oceanic and Atmospheric Administration). 2019. Drought: August 2019. National Centers for Environmental Information, September 11. Available at www.ncdc.noaa.gov/sotc/drought/201908.

NOAA (National Oceanic and Atmospheric Administration). 2020. The hydroclimatic cycle. Available at www.weather.gov/media/climateservices/Hydroclimate.pdf.

NOAA (National Oceanic and Atmospheric Administration). 2021. National Weather Service: Climate Prediction Center. U.S. Monthly Drought Outlook

Climate Prediction Center Drought Tendency. Available at www.cpc.ncep.noaa.gov/products/Drought/.

NOAA (National Oceanic and Atmospheric Administration). 2022. State of the climate: Drought for annual 2021 (published online January 2022). Available at https://data.nodc.noaa.gov/sotc/drought/202113.

NOAA (National Oceanic and Atmospheric Administration). 2022. The new U.S. climate normals are here. What do they tell us about climate change? Available at www.noaa.gov/news/new-us-climate-normals-are-here-what-do-they-tell-us-about-climate-change.

NOAA Fisheries (National Oceanic and Atmospheric Administration). n.d. How dams affect water and habitat on the West Coast. Available at https://www.fisheries.noaa.gov/west-coast/endangered-species-conservation/how-dams-affect-water-and-habitat-west-coast.

Noack, L., D. Höning, A. Rivoldini, C. Heistracher, N. Zimov, B. Journaux, H. Lammer, T. Van Hoolst, and J. H. Bredehöft. 2016. Water-rich planets: How habitable is a water layer deeper than on Earth? *Icarus* 277: 215–236. Available at https://doi.org/10.1016/j.icarus.2016.05.009.

NPS (National Park Service). 2020. The South Fork Dam. Available at www.nps.gov/jofl/learn/historyculture/south-fork-dam.htm.

NRCS (Natural Resources Conservation Service). 2022. April 11, 2022 Water Supply Outlook Report. Available at www.nrcs.usda.gov/wps/portal/nrcs/detail/id/newsroom/releases/?cid=NRCSEPRD1909224.

Nunez, Christina. 2020. Water pollution is a rising global crisis. Here's what you need to know. *National Geographic*, 16 March 2010. Available at www.nationalgeographic.com/environment/article/freshwater-pollution.

NWS (National Weather Service). 2022. Oregon water supply summary as of February 8, 2022. Available at www.weather.gov/media/pqr/WaterSupplyOutlook.pdf.

NYC (New York City). 2014. NYC'S risk landscape: Guide to Hazard mitigation. Available at www1.nyc.gov/assets/em/downloads/pdf/hazard_mitigation/nycs_risk_landscape_a_guide_to_hazard_mitigation_final.pdf.

Oboh, Ijeoma P., and Nkonylasua K. Egun. 2017. Groundwater suitability for agriculture: A case study of Aghor, Delta State Nigeria. *Ethiopian Journal of Environmental Studies & Management* 10 (9): 1199–2017.

Omerod, Kerri J., Samantha Redman and Loretta Singletary. 2020. Reclaimed water: Uses and definitions. University of Nevada-Reno College of Agriculture. Available at https://naes.agnt.unr.edu/PMS/Pubs/2020-3813.pdf.

Ong, Kim. New Mexico ground-water quality. USGSS Open File Report 87-0741. Available at https://pubs.usgs.gov/of/1987/0741/report.pdf.

OSM (Open Source Maps). USA OSM Topo Routable (Open Source). Available at www.gmaptool.eu/en/content/usa-osm-topo-routable.

Ouyang, Ying, Yongshan Wan, Wei Jin, Theodor D. Leininger, Gary Feng, and Yuguo Han. 2021a. Impact of climate change on groundwater resource in a region with a fast depletion rate: The Mississippi Embayment. *Journal of Water and Climate Change*. Available at https://doi.org/10.2166/wcc.2021.326

Ouyang, Ying, Wei Jin, Theodor D. Leininger, Gary Feng and Jia Yang. 2021b. Impacts of afforestation on groundwater resource: A case study for Upper Yazoo River watershed, Mississippi, USA. *Hydrological Sciences Journal*. Available at https://doi.org/10.1080/02626667.2021.1876235.

Oyler, Lindi Diane. 2021. *Spatiotemporal observations of water stress in Kansas winter wheat and corn from remotely sensed evapotranspiration and NDWI*. Missouri University of Science and Technology Masters Theses. https://scholarsmine.mst.edu/masters_theses/7979.

Peltier, Laural. 2016. Pennsylvania fracking water contamination much higher than reported. *EcoWatch*, February 4. Available at www.ecowatch.com/pennsylvania-fracking-water-contamination-much-higher-than-reported-1882166816.html.

Pera, Matt. 2021. These are the California crops that use the most water. *The Press Democrat*, June 24. Available at www.pressdemocrat.com/article/specialsections/these-are-the-california-crops-that-use-the-most-water/.

Peronne, Debora, and Scott Jasechko. 2021. *Water wells are at risk of going dry in the US and worldwide*. Waltham, MA: The Conversation Inc. Available at 160147#:~:text=Groundwater%20flows%20through%20tiny%20spaces,into%20lakes%2C%20rivers%20and%20streams.

Philpot, Simone, Keith Hipel, and Peter Johnson. 2016. Strategic analysis of a water rights conflict in the south western United States. *Journal of Environmental Management* 180: 247–256.

Poděbradská, Markéta, Bruce K. Wylie, Michael J. Hayes, Brian D. Wardlow, Deborah J. Bathke, Norman B. Bliss, and Devendra Dahal. 2019. Monitoring drought impact on annual forage production in semi-arid grasslands: A case study of Nebraska Sandhills. *Remote Sensing* 11 (18): 1–25.

Poeter, Eileen, Ying Fan, John Cherry, Warren Wood, and Douglas Mackay. 2020. *Groundwater in our water cycle: Getting to know earth's most important fresh water source*. Guelf, Ontario, Canada: The Groundwater Project.

Potempa, Hanalei. 2022. The climate crisis in Idaho: Experts share environmental and health concerns. *The Arbiter*, February 18.

Rawling, Geoffrey. 2018. Mapping the lifetime of the Ogallala aquifer in east-central New Mexico. *New Mexico Earth Matters*, Summer 2018. Available at https://geoinfo.nmt.edu/publications/periodicals/earthmatters/18/n2/em_v18_n2.pdf.

Reilly, Thomas E., Kevin F. Dennehy, William M. Alley, and William L. Cunningham. 2008. Ground-water availability in the United States. Circular

1323, U.S. Geological Survey. Available at https://pubs.usgs.gov/circ/1323/pdf/Circular1323_book_508.pdf.

Reisner, Marc. 1993. *Cadillac desert: The American West and its disappearing water*. New York: Penguin Books.

Riganti, Curtis. 2021. Contiguous U.S. (drought map). U.S. Drought Monitor, Thursday 24, 2021. Available at https://droughtmonitor.unl.edu/CurrentMap/StateDroughtMonitor.aspx?conus.

Robbins, Tim. 2019. As water scarcity increases, desalination plants are on the rise. *Yale Environment 360*. Available at https://e360.yale.edu/features/as-water-scarcity-increases-desalination-plants-are-on-the-rise.

Robson, Stanley G., and Edward R. Banta. 1985. Ground-water Atlas of the United States: Segment 2, New Mexico. USGS. Available at https://doi.org/10.3133/ha730C.

Rogers, Peter, and Susan Leal. 2010. *Running out of water: The looming crisis and solutions to conserve our most precious resource*. New York: Palgrave Macmillan.

Rosegrant, Mark W., Ximing Cai, and Sarah A. Cline. 2007. Water. global water outlook to 2025: Averting an impending crisis. *New England Journal of Public Policy* 21 (11): Article 11.

Safeopedia.com. 2017. Water stress. Available at www.safeopedia.com/definition/3092/water-stress.

Sakas, M. Elizabeth. 2022. *Colorado's snowpack is starting to melt. Warmer temperatures and drought likely mean another year of struggling water supplies*. Colorado Public Radio. Available at www.cpr.org/2022/04/14/colorado-water-supply-snowpack-melting-drought-rivers-reservoirs/.

Sarni, Will. 2021. Drought, mega drought. *Water and Wastewater Digest*, August 18. Available at www.wwdmag.com/one-water/drought-mega-drought.

Scanlon, Bridget R., Claudia C. Faunt, Laurent Longuevergne, Robert C. Reedy, William M. Alley, Virginia L. McGuire, and Peter B. McMahone. 2012. Groundwater depletion and sustainability of irrigation in the US High Plains and Central Valley. *Proceedings of the National Academy of Sciences* 109 (24): 9320–9325.

Schreiner-McGraw, Adam P., and Hoori Ajami. 2021. Delayed response of groundwater to multi-year meteorological droughts in the absence of anthropogenic management. *Journal of Hydrology* 603 (Part B): 126917. https://doi.org/10.1016/j.jhydrol.2021.126917.

Scott, Michon, and Rebecca Lindsey. 2022. Large declines in the snowpack across the U.S. West. April 7, 2022. NOAA: Climate.gov.

Schulte, Peter, and Jason Harmon. 2014. *Driving harmonization of water-related terminology*. The CEO Water Mandate. Available at https://ceowatermandate.org/wp-content/uploads/2019/11/terminology.pdf.

Seni, Steven J. 1980. Sand-body geometry and depositional systems, Ogallala Formation, Texas: The University of Texas at Austin, Bureau of Economic Geology Report of Investigations No. 105, 36 p. TWDB.

SETP. (Society of Experimental Test Pilots). n.d. Map of SETP U.S. Sections. Available at www.setp.org/setp-sections/map-of-setp-u-s-sections.html.

Shahpurwala, Aiman. 2021. *Washington's freshwater under stress from drought*. Olympia, WA: Washington Water Trust, August 11. Available at www.washingtonwatertrust.org/2021/08/11/washingtons-freshwater-under-stress-from-drought/.

Sharp, Jay. 2022. Water resources in the Southwest. *DesertUSA*. Available at www.desertusa.com/desert-water/water-in-the-southwest.html.

Shaw, Sameer, and Vishal Narin. 2019. Re-framing India's "water crisis": An institutions and entitlements perspective. *Geoforum* 101: 76–79.

Silver, Jonathan. 2021. Decaying infrastructure in the post-industrial city: An urban political ecology of the US pipeline crisis. *Nature and Space* 4 (3): 756–777.

Singurindy, Olga, Brian Berkowitz, and Robert P. Lowell. 2004. Carbonate dissolution and precipitation in coastal environments: Laboratory analysis and theoretical consideration. *Water Resources Research* 40: W04401. https://doi.org/10.1029/2003WR002651.

Scicchitano, Eric and Mike Crowley. 2022. Report: 1 in 3 miles of Pennsylvania streams, rivers' impaired.

SNWA (Southern Nevada Water Authority). 2011. Southern Nevada Water authority Clark, Lincoln and White Pine Counties Groundwater Development Project: Conceptual Plan of Development. Available at http://water.nv.gov/hearings/past/Spring%20-%20Cave%20-%20Dry%20Lake%20and%20Delamar%20Valleys%202011/Exhibits/SNWA%20Exhibits/SNWA_Exh_191_2011%20Plan%20of%20Development.pdf.

Snyder, Daniel T., and Jonathan V. Haynes. 2010. Groundwater conditions during 2009 and changes in groundwater levels from 1984 to 2009, Columbia Plateau Regional Aquifer System, Washington, Oregon, and Idaho. USGS Scientific investigations Report 2010-5040. Available at http://pubs.usgs.gov/sir/2010/5040/index.html.

Sood, Aditya, Prathapar Sanmugam, and Vladimir Smakhtin. 2014. Green and blue water. In *Key concepts in water resource*, Chapter 7: 91–102. Available at www.researchgate.net/publication/271387282_Green_and_Blue_Water/citation/download.

SPE (Society of Petroleum Engineers). 2022. Challenges in reusing produced water. Available at www.spe.org/en/industry/challenges-in-reusing-produced-water/.

Srinivasan, V., E.F. Lambin, S.M. Gorelick, B.H. Thompson, and S. Rozelle. 2012. The nature and causes of the global water crisis: Syndromes from a meta-analysis of coupled human-water studies. *Water Resources Research* 48 (10). https://doi.org/10.1029/2011WR011087.

Stevens, Alison. 2021. Which mountain snowpacks are most vulnerable to global warming? NOAA Climate.gov. Available at www.climate.gov/news-features/featured-images/which-mountain-snowpacks-are-most-vulnerable-global-warming.

Steward, David R., and Andrew J. Allen. 2016. Peak groundwater depletion in the high plains aquifer, projections from 1930 to 2010. *Agricultural Water Management* 170: 36–48.

Stewart, Iris T., Darren L. Ficklin, Carlos A. Carrillo, and Russell McIntosh. 2015. 21st century increases in the likelihood of extreme hydrologic conditions for the mountainous basin of the Southwest United States. *Journal of Hydrology* 529: 340–353.

SWDOE (State of Washington Department of Ecology). n.d. Columbia River water supply projects. Available at https://ecology.wa.gov/Water-Shorelines/Water-supply/Water-supply-projects-EW/Columbia-River-Basin-projects.

Tavernia, Brian G., Mark D. Nelson, Peter Caldwell, and Ge Sun. 2013. Water stress projections for the northeastern and midwestern United States in 2060: Anthropogenic and ecological consequences. *Journal of the American Water Resources Association* 49 (4): 938–952.

TDEC (Tennessee Department of Environment and Conservation). 2020. Tennessee groundwater monitoring and management: Groundwater 305(b) Report. 2020 305(b) addendum. *Status of Groundwater Quality in Tennessee*. Available at www.tn.gov/content/dam/tn/environment/water/drinking-water-unit/wr_wq_report-groundwater-305b-2020.pdf.

Teiken, Katherine. 2012. *Water scarcity in Minnesota a policy analysis and recommendation*. University thesis, Twin Cities: University of Minnesota.

TMR (The Moonlit Road). n.d. Coal mining in Appalachia. *Travelling the Strange Sought Through Story Telling*. Available at www.themoonlitroad.com/coal-mining-appalachia/.

Torcellini, Paul, Nicholas Long, and Ron Judkoff. 2003. Consumptive water use for U.S. power production. National Renewable Energy Laboratory (NREL TP-550-33905). U.S. Department of Energy.

Touton, Camille C. 2022. U.S. Bureau of Reclamation Commissioner Touton's report to the Senate Committee on Energy and Natural Resources, June 14, 2022. Available at www.energy.senate.gov/services/files/6CB52BDD-57B8-4358-BF6B-72E40F86F510.

Tracy, John, Jennifer Johnson, Leonard Konikow, Gretchen Miller, Dana O. Porter, Zhuping Sheng, and Steve Sibray. 2019.Aquifer depletion and potential impacts on long-term irrigated agricultural productivity. Issue Paper 63, CAST. Ames, Iowa.

Trainor, Sally, James Saiers, Lauren Patterson, Kelly Maloney, Joseph Fargione, Joseph Kiesecker, Sharon Baruch-Mordo, Katherine Konschnik, Hannah Wiseman, Jean-Philippe Nicot, and Joseph N. Ryan. 2017. Water stress from high-volume hydraulic fracturing potentially threatens aquatic biodiversity and ecosystem services in Arkansas, United States.

Trapp, Henry Jr., and Harold Meisler. 1992. *The regional aquifer system underlying the Northern Atlantic Coastal Plain in parts of North Carolina, Virginia, Maryland, Delaware, New Jersey and New York.* Washington, DC: USGS, U.S. Department of the Interior.

Trias, Angelo. 2020. Which is which? Differences between crisis, emergencies and disasters. Available at https://portal.edukasyon.ph/blog/which-is-which-differences-between-a-crisis-emergency-and-disaster.

Trimarchi, Maria. 2022. What caused the Dust Bowl? *How Stuff Works Handout* 14.

TWDB (Texas Water Development Board). 2016. Texas aquifers study groundwater quantity: Quality, flow, and contributions to surface water. Available at www.twdb.texas.gov/groundwater/docs/studies/TexasAquifer sStudy_2016.pdf#page125.

UCS (Union of Concerned Scientists). 2011. Climate hot map: Global effects around the world. https://climatehotmap.org/global-warming-effects/water-supply.html.

UDWR (Utah Division of Water Resources). 2012. As winter ends and runoff begins, Utah remains in drought. Available at https://water.utah.gov/water-data/drought/.

UGS (Utah Geological Survey). 2021. Groundwater and wetlands: Major aquifers of Utah. Available at https://geology.utah.gov/water/groundwater/groundwater-aquifers/major-aquifers-of-utah/.

UNESCO (United Nations Educational, Scientific and Cultural Organization). 2014. Water related disasters. Available at https://en.unesco.org/themes/water-security/hydrology/water-related-disasters.

UNW (United Nations-Water). 2010. Cities coping with water uncertainties. UN media brief. Available at www.un.org/waterforlifedecade/swm_cities_zar agoza_2010/pdf/02_water_uncertainties.pdf.

UNW (United Nations-Water). Climate change adaptation: The pivotal role of water. Available at file:///C:/Users/David/Downloads/unw_ccpol_web.pdf.

USA.com. 2022. U.S. average precipitation state rank. Available at www.usa.com/rank/us--average-precipitation--state-rank.htm.

USACE (U.S. Army Corps of Engineers). 2010. Lower Snake River Dams: A value to the nation. Available at www.nww.usace.army.mil/Missions/Lower-Snake-River-Dams/

USACE (U.S. Army Corps of Engineers). 2018. National Inventory of Dams (NID). Available at https://nid.sec.usace.army.mil/ords/f?p=105%3A1%3A%3A%3A%3A%3A%3A%3A.

USBR (U. S. Bureau of Reclamation). 2022a. Colorado River 5-year probabilistic projections. Available at www.usbr.gov/lc/region/g4000/riverops/crss-5year-projections.html.

USBR (U. S. Bureau of Reclamation). 2022b. Bureau of reclamation: About us. Available at www.usbr.gov/main/about/fact.html.

USBR (U. S. Bureau of Reclamation). 2022c. Bureau of reclamation release of 2022 August operating guidelines for Tier 1. Available at www.usbr.gov/lc/region/g4000/24mo.pdf.

USDA (United States Department of Agriculture). 2013. Groundwater irrigation & water withdrawals: The Ogallala aquifer initiative. USDA Economic Series Number 1. Available at www.nrcs.usda.gov/Internet/FSE_DOCUMENTS/stelprdb1186440.pdf.

USDA (United States Department of Agriculture). 2019. Irrigation & water use. USDA Economic Research Service Circular. Available at www.ers.usda.gov/topics/farm-practices-management/irrigation-water-use/.

USDA/NASS (United States Department of Agriculture/National Agricultural Statistics Service). 2021. 2020 state agriculture overview—California. Available at www.nass.usda.gov/Quick_Stats/Ag_Overview/stateOverview.php?state=CALIFORNIA.

USCRT (U.S. Climate Reliance Toolkit). 2019. Available at https://toolkit.climate.gov/regions/great-lakes.

USDM (United States Drought Monitor). 2021. This week's drought summary (Map released: June 24, 2021). Available at https://droughtmonitor.unl.edu/.

USDM (United States Drought Monitor). 2022a. Measuring drought. Available at https://droughtmonitor.unl.edu/About/WhatistheUSDM.aspx.

USDM (United States Drought Monitor). 2022b. National drought summary for May 3, 2022. Available at https://droughtmonitor.unl.edu/Summary.aspx.

USDS (United States Department of Agriculture). 2019. Chesapeake Bay Action Plan 2018–2019 Progress Report Natural Resources Conservation Service Map 10365. Available at www.nrcs.usda.gov/wps/portal/nrcs/detailfull/national/programs/initiatives/?cid=stelprdb1047323.

USFWS (United States Fish and Wildlife Service). n.d. Dam removal: an opportunity for our rivers. Available at www.fws.gov/southeast/pdf/fact-sheet/dam-removal.pdf.

USFWS (United States Fish and Wildlife Service). 2020. Coal mining. Available at www.fws.gov/ecological-services/energy-development/coal.html.

USGCRP (U.S. Global Change Research Program). 2020. *Our changing planet: The U.S. global change research program for fiscal year 2020*. Washington, DC, USA. https://doi.org/10.7930/ocpfy2020.

USGCRP (U.S. Global Change Research Program). 2021. *Impacts, risks and adaptation in the United States*. Volume II. Fourth National Climate Assessment: Water. Washington, DC.

USGS (United States Geological Survey). n.d. California drought. Available at https://ca.water.usgs.gov/california-drought/what-is-drought.html.

USGS (United States Geological Survey). n.d. Hydroelectric power water use. Available at www.usgs.gov/special-topic/water-science-school/science/hydroelectric-power-water-use?qt-science_center_objects=0#qt-science_center_objects.

USGS (United States Geological Survey). n.d. Groundwater decline and depletion. Available at www.usgs.gov/special-topic/water-science-school/science/groundwater-decline-and-depletion?qt-science_center_objects=0#qt-science_center_objects.

USGS (United States Geological Survey). n.d. Desalination. Available at www.usgs.gov/special-topic/water-science-school/science/desalination?qt-science_center_objects=0#qt-science_center_objects.

USGS (United States Geological Survey). n.d. Principal aquifers in the United States. Available at www.usgs.gov/mission-areas/water-resources/science/principal-aquifers-united-states?qt-science_center_objects=0#qt-science_center_objects.

USGS (United States Geological Survey). n.d. How much water is there on Earth? Available at www.usgs.gov/special-topic/water-science-school/science/how-much-water-there-earth?qt-science_center_objects=0#qt-science_center_objects.

USGS (United States Geological Survey). n.d. Mississippi Alluvial Plain (MAP): Water use and availability program. Available at www.usgs.gov/centers/lmg-water/science/map?qt-science_center_objects=0#qt-science_center_objects.

USGS (United States Geological Survey). n.d. Critical water issues of New Jersey. Available at www.usgs.gov/centers/new-jersey-water-science-center/critical-water-issues-new-jersey.

USGS (United States Geological Survey). 1965. Eutrophication of the St. Laurence Great Lakes. Available at www.usgs.gov/publications/eutrophication-st-lawrence-great-lackes.

USGS (United States Geological Survey). 1997. Appalachian Plateaus aquifers. Ground Water Atlas of the United States: Delaware, Maryland, New Jersey, North Carolina, Pennsylvania, Virginia, West Virginia. HA 730-L.

USGS (United States Geological Survey). 2000. Ground Water Atlas of the United States: California, Nevada (HA 730-B). James A. Miller, ed. Available at https://pubs.usgs.gov/ha/730/report.pdf.

USGS (United States Geological Survey). 2009a. *Mississippi Embayment Regional Aquifer Study (MERAS)*, ed. Brian R. Clark and Rheannon M. Hart. Available at https://pubs.usgs.gov/sir/2009a/5172/pdf/SIR2009a-5172.pdf.

USGS (United States Geological Survey). 2009b. Alabama water use. Available at https://pubs.usgs.gov/fs/2009b/3081/pdf/Alabama_2005_water_use_FS_508.pdf.

USGS (United States Geological Survey). 2014. *High plains aquifer resources and information water-level changes in the high plains aquifer, 1980 to 1996*. Nebraska Water Science Center. Available at https://ne.water.usgs.gov/ogw/hpwlms/fs96/hp96fig3.html.

USGS (United States Geological Survey). 2015. Update for the South Carolina Atlantic Coastal Plain Groundwater Availability Model. Available at www.usgs.gov/centers/sa-water/science/update-south-carolina-atlantic-coastal-plain-groundwater-availability-0?qt-science_center_objects=0#qt-science_center_objects.

USGS (United States Geological Survey). 2016a. Groundwater and drought. Available at https://water.usgs.gov/ogw/drought/.

USGS (United States Geological Survey). 2016b. The role of snowpack, rainfall, and reservoirs in buffering California against drought impacts. Available at www.usgs.gov/news/role-snowpack-rainfall-and-reservoirs-buffering-california-against-drought-impacts.

USGS (United States Geological Survey). 2016c. The importance of ground water in the Great Lakes Region. Water Resources Investigations Report 00-4008. Available at https://water.usgs.gov/ogw/pubs/WRI004008/conditions.htm.

USGS (United States Geological Survey). 2016d. *Water use trends in Alabama*. Lower Mississippi-Gulf Water Science Center. Available at www.usgs.gov/centers/lower-mississippi-gulf-water-science-center/science/water-use-trends-alabama.

USGS (United States Geological Survey). 2017. USGS: High plains aquifer groundwater levels continue to decline. Available at www.usgs.gov/news/usgs-high-plains-aquifer-groundwater-levels-continue-decline.

USGS (United States Geological Survey). 2018a. Groundwater decline and depletion. Available at www.usgs.gov/special-topic/water-science-school/science/groundwater-decline-and-depletion?qt-science_center_objects=0#qt-science_center_objects.

USGS (United States Geological Survey). 2018b. Wyoming water use in the U.S., 2015. Available at www.usgs.gov/media/images/wyoming-water-use-us-2015.

USGS (United States Geological Survey). 2019a. Saltwater intrusion. Available at www.usgs.gov/mission-areas/water-resources/science/saltwater-intrusion.

USGS (United States Geological Survey). 2021a. National water census: Regional groundwater availability studies. Available at www.usgs.gov/mission-areas/water-resources/science/national-water-census-regional-groundwater-availability?qt-science_center_objects=0#qt-science_center_objects.

USGS (United States Geological Survey). 2019b. Saltwater intrusion. Available at https://www.usgs.gov/mission-areas/water-resources/science/saltwater-intrusion.

USGS (United States Geological Survey). 2021a. *Groundwater may decline by a third over next 30 years*. USGS Communications and Publishing, October 28. Available at www.usgs.gov/news/state-news-release/groundwater-flow-colorado-river-may-decline-third-over-next-30-years.

USGS (United States Geological Survey). 2021b. USGS water resources. Pennsylvanian and Mississippian Aquifers of the Appalachian Plateaus. Available at www.usgs.gov/mission-areas/water-resources/science/pennsylvanian-and-mississippian-aquifers-appalachian-plateaus#publications.

USGS (United States Geological Survey). 2021c. Mississippi embayment aquifer system. Available at www.usgs.gov/mission-areas/water-resources/science/mississippi-embayment-aquifer-system.

USGS (United States Geological Survey). 2022. Water use in Alabama, 2005. Report by Lower Mississippi-Gulf Water Services Center, May 19, 2022. Available at www.usgs.gov/centers/lower-mississippi-gulf-water-science-center/science/water-use-alabama-2005.

U.S. News. 2021. Worst day: Water crisis deepens on California-Oregon line. Available at www.nbnews.com/news/us-news/worst-day-water-crisis-deepens-california-oregon-line-n267479.

USWA (United States Wheat Associates). 2022 Report on replacing the benefits of the Lower Snake River Dams generates concern. Available at www.uswheat.org/wheatletter/report-on-replacing-lower-snake-river-dams-benefits-generates-concern/.

Van Loon, Anne F. 2015. Hydrological drought explained. WiresWater, April 15. Available at https://doi.org/10.1002/wat2.1085.

Viñas, Maria-Jose 2015. Historical satellite images reveal snow is melting earlier in Wyoming. NASA's Earth Science News Team, National Aeronautics and Space Administration.

Vittek, Shelby. 2021. Western drought forces farmers to make tough decisions. *Modern Farmer*, June 7. Available at https://modernfarmer.com/2021/06/western-drought-forces-farmers-to-make-tough-decisions/.

VGS (Vermont Geological Survey). 2016. Vermont water use data and research. VGS Department of Environmental Conservation. Available at https://Vermont-WUDR-Workplan.pdf.

WAECY (Washington State Department of Ecology). 2016. Washington groundwater level status and trends—Spring 2016. Available at https://Vermont-WUDR-Workplan.pdf.

Walkenhorst, Emily. 2017. Report warns of Arkansas water depletion. *Arkansas Democrat Gazette*, May 22. Available at www.arkansasonline.com/news/2017/may/22/report-warns-of-water-depletion-2017052-1/.

Walls, Margret and Leonard Shabman. 2020. *Federal funding for Dam removal in the United States*. Washington, DC: Resources for the Future. WiresWater. Available at: https://media.rff.org/documents/Federal_Funding_for_Dam_Removal_in_the_United_States.pdf.

Walton, Brett, 2015. Groundwater depletion stresses majority of world's largest aquifers. *Circle of Blue*, June 16. Available at www.circleofblue.org/2015/world/groundwater-depletion-stresses-majority-of-worlds-largest-aquifers/.

WaterWorld. 2022. Massachusetts awards $420K for water infrastructure. Available at www.waterworld.com/drinking-water/infrastructure-funding/press-release/14223613/massachusetts-awards-420k-for-water-infrastructure.

World Atlas. 2022. The eight US states located in the Great Lakes region. Available at www.worldatlas.com/articles/the-eight-us-states-located-in-the-great-lakes-region.html.

World Population Review. 2022. Snowiest states. Available at https://worldpopulationreview.com/state-rankings/snowiest-states.

WRDS (Water Resources Data System). 2021. Wyoming drought. Available at www.wrds.uwyo.edu/drought/drought.html.

Weeks, John B., and Ren Jen Sun. 1986. High plains regional aquifer study. In *Regional aquifer system analysis program of the U. S. geological survey summary of projects, 1978–84*, ed. Ren Jen sun. USGS Circular 1002.

Wen, Fujiang and Xuunhong Chen. 2006. Evaluation of the impact of groundwater irrigation on streamflow in Nebraska. *Journal of Hydrology* 327: 603–617.

Whittemore, Donald O., James J. Butler, Jr., and B. Brownie Wilson. 2018. Status of the high plains aquifer in Kansas. Kansas Geological Survey Technical Series 22. Available at https://www.kgs.ku.edu/Publications/Bulletins/TS22/HPA_tech_series22.pdf.

Wilkinson, Robert. 2011. The water-energy nexus: Methodologies, challenges and opportunities. In *The water-energy nexus in the American West*, ed. Douglas S. Kenney and Robert Wilkinson, 3–17. Cheltenham, UK: Edward Elgar.

Wikipedia. 2022. The Mississippi River. Available at https://en.wikipedia.org/wiki/Mississippi_River.

Willhite, Donald A., and Michael H. Glantz. 1985. Understanding the drought phenomenon: The role of definitions. *Drought Mitigation Center Faculty Publications* 20. http://digitalcommons.unl.edu/driughtfacpubs/20.

WorldAtlas.com. 2021. What percentage of the world's freshwater is frozen? Available at https://www.worldatlas.com/articles/what-percentage-of-earth-s-water-is-frozen.html.

World Nuclear Association, 2021. Nuclear share figures, 2009–2019. Available at www.world-nuclear.org/information-library/facts-and-figures/nuclear-generation-by-country.aspx.

World Water Council. 2021. Water crisis. Available at www.worldwatercouncil.org/en/water-crisis.

Wright, Will. 2019. 'Tip of the spear.' As customers beg for clean water, is a crisis looming in Appalachia? *Herald Leader*, May 5 online. Available at www.kentucky.com/news/local/watchdog/article221701865.html.

WSGS (Wyoming State Geological Survey). 2022. Water resources in Wyoming. Available at www.wsgs.wyo.gov/water/water.

WWSC (Washington Water Science Center). 2017. Glaciers and snowpack. USGS report Available at www.usgs.gov/centers/washington-water-science-center/science/glaciers-and-snowpack

Xiao, Mu, Bradley Udall and Dennis Lettenmaier. 2018. On the causes of declining Colorado River streamflows. *Water Resources Research* 54, https://doi.org/10.1029/2018WR023153.

Yonts, Dean C., Amir Haghverdi, David L. Reichert, and Suat Irmak. 2018. Deficit irrigation and surface residue cover effects on dry bean yield, in-season soil water content and irrigation water use efficiency in western Nebraska high plains. *Agricultural Water Management* 199: 138–147.

Index

0–9
21st Century drought, 80
30 year Climate "Normals", 5
2022 National Weather Service's Water Supply Outlook, 186

A
Adaptation to climate change, 68, 69
Agricultural pollutants, 103
Agricultural water needs by activity, 103, 104
Algae blooms, 132
Appalachian drought, 280
Aquafer depletion and overuse, 114, 116
Aquafer stress, 111, 116, 211
Aquitards, 110
Arizona municipal water use, 208
Arizona water management, 32
Artesian aquifers, 110

B
Barriers to acquiring water, 37
Bellevue Washington dam proposal example, 57, 58
Bureau of Reclamation, 112, 140, 216

C
California aquifer system, 119
California emergency, 191
California Land Subsidence, 191
California Megadrought, 196
California Natural Resources Agency, 183
California reservoirs, 129
California's Colorado Aqueduct, 193
Central Arizona Groundwater Replenishment District (CAGRD), 208, 209
Central Arizona Project (CAP), 208, 209
Central Coastal Plain aquafer, 292
Central Coastal Plain Capacity Use Area ground water withdrawals, 292
Central Plains groundwater depletion, 234

D. E. McNabb and C. R. Swenson, *America's Water Crises*,
https://doi.org/10.1007/978-3-031-27380-3

Central/Southern High Plains drought levels by state, 231
Central Valley Aquafer system, 194
Central Valley Project (CVP), 195
Chesapeake Bay Watershed (CBW), 309–311
City of Chicago water, 31
Classes of re-used water, 165
Clean Water Act (CWA), 10, 24, 25, 97, 306, 312
Clean Water State Revolving Fund (CWSRF), 97
Climate change, 166
Climate disasters, 17, 43, 44
Colorado aquifer system, 119, 123
Colorado River water use patterns, 94
Colorado water crisis, 218
Columbia Basin aquafer, 112
Coming water stress, 16
Competing demand for water, 70
Competing interests for water use, 57, 58, 70
Connecticut groundwater, 325
  300,000 wells, 325
Connecticut surface water quality, 327
Crisis of groundwater depletion, 112
Cycles of drought in New England, 317, 320

D
Dam Failure examples, 143
Dam failures and water loss, 140
Dam Oversight Complexity, 143
Dam removal, 132, 153
Dam Safety Act of 2006, 141
Defining "normal", 79
Delaware Aqueduct, 304
Delaware River Basin, 301–303
Development impact on wetlands, 293
Draught condition in Appalachian Mountain region, 280, 281
Drinking water pollution, 100
Drinking Water State Revolving Fund (DWSRF), 98
Drought, 75–88
  "Creeping Disaster", 75
Drought and water scarcity, 17
Drought conditions by state in the south, 259, 260
Drought definitions, 76, 80
Drought in Arizona, 204
Drought in Pennsylvania, 301
Drought in the southwest, 204
Drought in the west, 182
Drought severity, 77, 78, 80, 84, 85, 87
Drought severity scoring, 78
Dust Bowl, 75

E
Economic development and water availability, 85
Eutrophication, 134, 135
Evaporation and stored water loss, 131

F
Factors shaping the crisis, 56
Failing infrastructure in the Great Lakes region, 247
Farm Relief Act and the Farm Mortgage Act, 11
Financial interests, 86
Flint and Benton Harbor Michigan water crises, 252
Flint Michigan example, 30
Flint Michigan water crisis, 59
Florida water crisis, 271
Forces shaping water demand, 60
Fracking ban, 305
Fracking in Pennsylvania, 307, 308

Fracking water quality impacts, 295, 308

G
General water uses in the United States, 7, 14, 17, 127, 169
Global change research program, 60, 64
Global temperature change, 67
Global water research, 60
Great Lakes region drought levels, 243
Great Lakes-St. Lawrence River basin, 249
Great Lakes water levels, 255
Groundwater aquafers listed, 112
Groundwater depletion, 116
Groundwater depletion in North Carolina, 292
Groundwater in Louisiana and Arkansas, 257, 258, 260
Groundwater in the Great Lakes region, 248, 249
Groundwater loss, 36, 44
Groundwater stress in Texas, Oklahoma and Kansas, 225
Groundwater supplies in Arizona, 209
Groundwater use across the world, 35

H
High Plains aquifer, 225–228, 232, 233, 235, 236, 238
High Plains drought, 229
Hurricane Katrina, 259
Hydro-electric reservoirs, 50
Hydrologic cycles, 6, 19–21
Hydrological structure failures: *Geophysical Research Letters* 2019, 172
Hydrologic Fracturing and water stress, 303
Hydro power in Alabama, 273, 274

I
Indiana drought, 242
Infrastructure failure, 158–161, 173
Infrastructure Report Card 2021, 159
Infrastructure Report Card by the American Society of Civil Engineers, 141
Integrated water portfolios, 206
Irrigated agriculture, 231
Irrigated agriculture by state, 114
Irrigation typology, 116
Irrigation use by state, 18

J
Jackson Mississippi water crisis, 30
Jet Propulsion Laboratory (JPL), 63

K
Karst water problems in Tennessee, 270
Klamath River Reclamation Project, 186

L
Lake evaporation, 207
Lakes Powell, Mead and Pleasant, 205–207, 217, 222
Levels of drought by categories in New England, 317, 324
Lower Mississippi Water Science Center (LMWSC), 268

M
Maine water quality, 326
Massachusetts Water Infrastructure Finance Commission, 321

Massachusetts water use, 321, 328
Measuring water stress, 33
Mid Atlantic drought areas by state, 287, 297, 299
Mid-Atlantic drought impacts, 297
Middle East, 116
Mining impacts on water quality, 285
Minnesota water infrastructure investment needs, 247
Minority communities impacted by infrastructure failures, 252
Mississippi Embayment aquifer, 257, 258, 264
Mississippi River Valley, 257, 258, 262, 264
Missouri River Basin, 244, 245
Montana water use, 182
Mountain top mining and aquatic ecosystems, 281
Municipal water treatment, 173

N
National Centers for Environmental Information (NCEI), 5, 17
National Climate Assessment (NCA), 62
National Climate Data Center, 4
National Dam Inspection Act of 1972, 141
National Drought Mitigation Center (NDMC), 76, 78
National Phytoplankton Monitoring Network (PMN), 132
National Pollution Discharge Elimination System (NPDES), 100, 101
National Renewable Energy Laboratory (NREL), 104
Natural aquafer recharge, 228, 290
Nebraska groundwater stress, 232, 233
Nebraska irrigation, 233
New England periodic drought, 317
New Hampshire, Rhode Island, and Vermont water use, 315
New Mexico's Middle Rio Grande Project, 209
New Mexico groundwater withdrawals, 209, 210, 212
New Orleans example, 258
New York and New Jersey, 297, 301
New York City water use, 300, 303
NOAA, 4, 5, 11
Non-point source pollution, 96, 97
North Dakota wildfires, 227
Northern High Plaines drought levels by state, 229, 232
Number of water and wastewater system in the United States, 160

O
Ogallala Aquifer, 228, 232–235, 238

P
Pennsylvania ground water use, 307
Peoria Arizona example, 206
Pollution, 3, 7, 10, 19, 23–26
Population growth and climate change, 47
Population pressure, 55
Proportions of Southwest states in drought 2022, 204
Public drinking water systems in New England, 319
Public health impacts of water system failures, 161

R
Recent dam failures by year and location, 145
Reclaimed water examples, 165, 169
Recycled water, 166
Recycled water limits, 168

Repeated non-compliance with Safe Drinking Water Act, 265
Re-used, reclaimed and recycled water, 164, 173
Rio Grande Basin, 209, 210
Rio Grande Rift, 126
Rising temperatures, 5, 139

S
Safe drinking water, right or privilege, 252
Saltwater intrusion, 111, 260, 262, 273
Saltwater intrusion in Atlantic region, 291
San Juan-Chama trans-mountain diversion project, 209
Scituate Reservoir, Rhode Island, 323
Sea level rise, 287
Snake River and Washington State examples, 145
Snake River in Idaho, 215
Snowpack loss, 213
Snowpack monitoring, 136, 137
Snowpack water storage, 135
Social political divides, 58
Source sustainability, 23
South America, 116
Southern Nevada Water Authority (SNWA), 216
Southwestern Desert Aquafer, 112
Stored water and fisheries, 133
Stored water quality, 133
Storm impacts on fresh water, 285
Stormwater management challenges, 99
Stormwater management practices, 99
Stormwater pollution, 98, 171
Surface water crisis, 91, 92
Surface water pollution typology, 92, 96, 98

T
Temperature change, 40, 48
Ten most populace nations 2050 estimates, 13, 14
Tennessee water use by category, 269, 270
Texas, 116, 232
Texas Rio Grande Project, 209
The importance of public process, 58
The importance of water assumptions in state decision making, 56
The Palmer Index, 78
Thermal power and water use, 46
The world's common water problems, 9
Tittabawassee and Tobacco rivers, 133, 150, 152
Townsend Massachusetts water department, 322
Tri-State Water Wars: Alabama-Georgia-Florida water disputes, 272, 273
Types of irrigation, 115

U
Upper Colorado Basin, 210, 216
Urbanization, 116
U.S. and Mexico water agreements, 93
U.S. Army Corps of Engineers (USACE), 140–143, 153, 154
US climate trends, 4, 75
U.S. Department of Agriculture's Climate Resilience Tool Kit (CRTK), 229, 230
Utah snowpack, 213

W
Washington Department of Ecology, 184
Washington state groundwater, 124

Washington state water stress, 179, 189
Water as a public good, 58
Water crisis concepts, 8
Water crisis concepts and problems, 8
Water disaster, 31
Water Infrastructure defined, 158
Water pollution typology, 95, 96, 106
Water Resources Development Act of 1986, 141
Water Risk and Resilience Assessment, 32, 33
Watershed management, 92
Water stress terms defined, emergency, crisis, disaster, 6
Water systems failures and funding, 98, 161
Water utility funding, 162
Water withdrawals by category, 15
Water withdrawals by category in Alabama, 267, 269
Weather change effects on water supply, 39
West Coast drought data, 182
Western state's snowfall, 135
Wetland loss, 91, 92, 95, 96
Wildfire impacts to water supply, 228
Willamette, and Snake Rivers, 186
World Water Council (WWC), 30, 37
Wyoming snowpack, 188

FSC
www.fsc.org
MIX
Papier | Fördert
gute Waldnutzung
FSC® C083411

Kolibri 360 GmbH
Ferdinand-Jühlke-Straße 7
99095 Erfurt, Deutschland
produktsicherheit@kolibri360.de